Contents

Part I - Centrifugal Pumps - Basic Principles

Contents

Contents

Part II - Hermetic centrifugal pumps - Basic Principles

Contents

Contents

Contents

Contents

Part III - Rotary displacement pumps - Basic Principles

Contents

Contents

Contents

Part IV - Hermetic rotary displacement pumps

Contents

Foreword

The development of hermetic pumps is inseparably linked to the design of the canned motor. This development is in turn closely associated with the names of the founder of "Hermetic" GmbH, in Freiburg-Gundelfingen, Hermann Krämer (1902 to 1993) and his staff. The present state of development of hermetic centrifugal pumps and rotary displacement pumps and the widespread use they have achieved is due mainly to them.

At the time that industrial production of centrifugal pumps was started—at the turn of the century—the company which Wilhelm Lederle had founded in 1866 was one of the first German pump manufacturers, along with Klein-Schanzlin & Becker, and Weise & Monski, to manufacture this type of pump.

As early as the start of the 1950's, the processing and chemical industries demanded pumps capable of leak-free delivery of toxic, explosive and combustible liquids.

As the planning of nuclear power stations began in the fifties, this also led to the need for hermetically-sealed circuits to meet the increased safety requirements of radioactive substances, thus making shaft seals no longer acceptable for pumps in primary and secondary circuits to meet the increased safety requirements of radioactive, thus making shaft seals no longer acceptable for pumps in primary and secondary circuits.

To meet these demands, the industry examined various possibilities and concluded that the hermetic type of pump with a canned motor offered the best solution even for high system pressures up to 120 MPa and large motor power outputs up to 280 kW. Hermetic pumps meet the requirements of today's environmental technology, the need for environmental protection, and the demands of bio-technology and nuclear energy engineering.

To meet these requirements and the increased safety needs of the chemical industry, of technical process plants and of nuclear engineering, the Lederle company began the development of canned motor pumps based on the canned motor which had already been patented in 1913. The design work was completed in the sixties and these pumps were ready for series production. The "Hermetic" company was founded and, together with other manufacturers, was devoted specifically to manufacturing these pumps as centrifugal and rotary displacement pumps. Further development with regard to monitoring, safety and automation of hermetic pumps and pump systems has continued unabated up to the present time.

The development of the canned motor and permanent magnet coupling also influenced centrifugal pumps and rotary displacement pumps, leading to important new developments in design. In particular, experience with hermetic pumps showed the close relationship and need for harmonization between machinery construction and the construction of electric motors. It is not unusual today to find pump manufacturers with their own highly modern electric machine construction facilities producing all components for canned motor pumps.

In this publication, the author not only presents the development of the features, design characteristics and variety of types of hermetic pumps, but also deals with specific and application-oriented problems of centrifugal pumps and rotary displacement pumps in a simple but comprehensive and graphical manner, which will be valued by every specialist. The author does not restrict himself merely to an illustration of principles but also provides an ample number of examples of designs and applications which enables the details to be understood. This gives this book a welcome distinction from many other specialist works. This book is not a textbook in the usual sense and makes no claim to be, but should really be seen as a treatise on hermetic pumps.

It is to be warmly commended both to students at technical universities and at specialized institutions of higher learning, particularly those students specializing in design as well as engineers working in the field of design, project planning and operation.

It is by no means an everyday event for the knowledge and experience of a life in pump construction, gathered and elucidated from many publications, lectures and presentations both at home and abroad, to be presented to the professional world. In this book that knowledge and experience is devoted to hermetic pumps.

The book is arranged in four sections and can be studied without any particular requirements, but it does need a basic knowledge of flow machines, thermodynamics and electric machines, together with good understanding of design. The text is well amplified by a variety of illustrations. The book is made easier to understand by the fact that all equations are set out as magnitude equations and that the employed symbols are explained at the relevant point together with the corresponding SI units.

I am convinced that this book will reward the careful reading that it deserves.

Prof. Dr.-Ing. Habil. D. Surek
Halle, March 1994

Part I

Centrifugal Pumps

Basic Principles

1. Introduction

Pumps are amongst the oldest working machines used by man. From earliest times up to the middle of the eighteenth century, mainly hydrostatic pumping systems such as land reclamation schemes, piston pumps and rotary piston pumps were used. The centrifugal pump (Fig. 1.2) already described 1689 in the "Acta Eruditorium" by the Marburg professor Denis Papin (Fig. 1.1) met with only limited use by the main customers at that time such as public water supply utilities, the Massachusetts pump (Fig. 13) which came into use about 1818 (Fig. 1.3) being one of the relatively few exceptions [1-1]. The invention of the internal combustion engine and electric motor was a turning point in the use of pumps. Powerful driving machines with high rotational speeds and low power/weight ratio which could be directly coupled were now available, thus enabling the use of hydrodynamic centrifugal pumps with the known advantages of continuous delivery and enabled a flow characteristic line which could be throttled.

Fig 1.1 Denis Papin (1647-1712)

The mathematical, physical basis was provided by the two friends Professors Daniel Bernoulli (1700-1782) and Leonhard Euler (1707-1783), who lectured in Basle and in St. Petersburg (Fig. 1.4 and 1.5). Bernoulli formulated the following valid definition for obtaining the energy of a friction-free flowing liquid:

"This states that in the stationary flow of an ideal fluid in a homogenous gravitational field, the sum of all energies of one and the same thread of flow is always constant with regard to a fluid level of random characteristics."

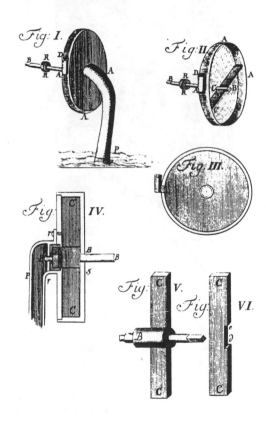

Fig 1.2 The first centrifugal pump presented by Denis Papin in 1689

1

Leonard Euler created the mathematical basis for energy transfer in the rotor of a centrifugal pump by establishing Euler's equation for water turbines in 1756 in his work "Théorie plus complète des machines qui sont mises en mouvement par la réaction d'eau" .

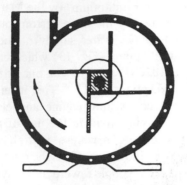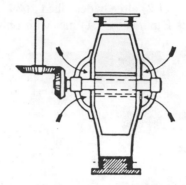

Fig 1.3 The Massachusetts pump, 1818

Fig 1.4 Daniel Bernoulli
(1700 - 1782)

Fig 1.5 Leonhard Euler
(1707 - 1783)

Centrifugal pumps were, however, not manufactured until towards the end of the eighteenth century (Fig 1.6) [1-2].

Industrialization led to a rapid increase in the demand for pumps, particularly centrifugal pumps. Particularly large users up to the present time are municipal utilities, the chemicals, petrochemical-chemicals, and refrigeration industries and refineries. Their requirement for pumps increases steadily and this is accompanied by greater demands on availability, operating safety, freedom from emissions due to leakage losses and low noise. The trend towards more environmentally-friendly pumps has been particularly pronounced in the last four decades, resulting in harmonization of the business interests with those of the legislator and the general public.

Ecological pumps not only have a high degree of availability but are also safe in operation, largely maintenance-free, and therefore also economical.

Fig 1.6 First standard production centrifugal pump from the W. Lederle Engine and Pump Factory from 1898.

This book should encourage engineers involved in planning or in the operation of technical plants to use hermetic pumps. In the relatively short time since such pumps were introduced, it has been shown that they completely meet the stringent requirements of a modern industrial society and are particularly suitable for adaptation to difficult service conditions. The hermetic power transfer from the electric motor to the pump has enabled operation in applications with high system pressures, and high and low temperatures with complete freedom from leaks and therefore enabled processes to be performed which were not previously feasible or could only be achieved at a high equipment cost. Hermetic pumps mean that today's chemical engineer has a new generation of machines which contribute considerably to the safety and environmental-friendliness of his operation.

Development is still ongoing and the steadily increasing number of different pumping tasks opens up a wide field of constructive activity in the development of leak-free pumps.

Bibliography for Chapter 1

[1-1]   Berdelle-Hilge, Ph.   Die Geschichte der Pumpen
                             [The History of Pumps]
                             published by Philip Hilge GmbH, Bodenheim/Rhein 1992

[1-2]   W. Lederle   Engine and Pump Factory,
                             Freiburg, Company Catalogue 1900

2. Centrifugal Pumps

Centrifugal pumps are machines in which fluids are conveyed in rotors (impellers) fitted with one or more blades by a moment of momentum so that pressure is gained in a continuous flow. The rotating impeller transmits the mechanical work from the driving machine to the fluid in the impeller channels. Centrifugal forces are generated by the rotation which force the pumped fluid out of the impeller channels, thus creating a negative pressure zone at the start of the blade into which fluid continuously flows due to the steady atmospheric or tank pressure on the surface of the inlet fluid.

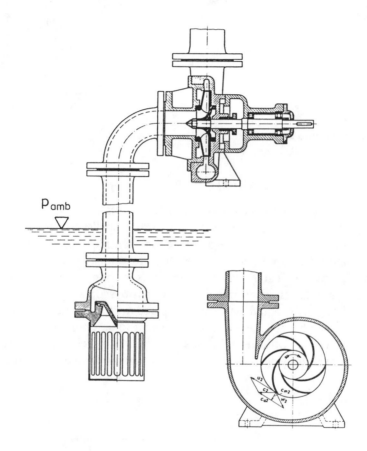

Fig 2.1   Centrifugal pump with suction pipeline and foot valve and a liquid level at atmospheric pressure on the inlet side (principle of operation)

Therefore the volume of the flow is the same as that displaced (Fig. 2.1). The energy transfer (Fig. 2.2) begins at the inlet to the cascade (A_1) and ends on departure from the blade channels (A_2). The pressure and velocity of the pumped liquid is increased on this path. The pressure increase is due to the centrifugal forces and deceleration in the relative velocity w in the impeller channels between the channel inlet and outlet. The strong increase in the absolute velocity c of the fluid during the flow through the impeller channels is partly converted to pressure energy in a diffuser, volute guide or stator after leaving the impeller.

5

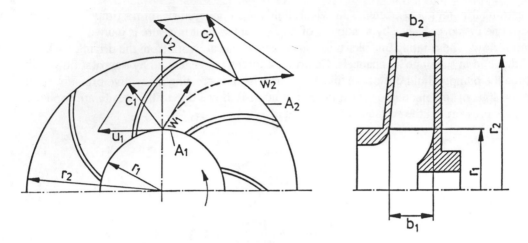

Fig. 2.2   Flow velocities at impeller

2.1 Euler's Equation (from the velocities)

The flow processes in the impeller which lead to formation of the H (Q) line can be mathematically determined by the theoretical assumption that the impeller has an infinite number of infinitely thin blades. A distinction is to be made between movement processes of the absolute velocity c and relative velocity w. The absolute velocity c is that which liquid particles exhibit compared with a static environment. This is communicated to an observer looking through the glass cover of an impeller using stroboscopic lighting. The relative velocity w is the velocity of a liquid particle, compared with the rotating blades, when flowing through the blade channel. This would be evident to an observer looking through the glass cover of the impeller with the aid of a strobe light. The relative velocity w is the velocity of a liquid particle flowing through the impeller channel compared with the rotating blade.

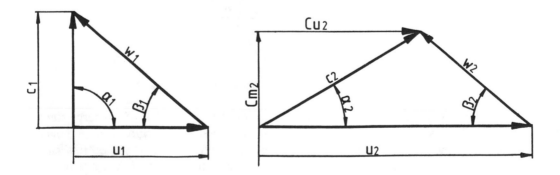

Fig. 2.3 Inlet and outlet triangles on radial impeller

This would be evident to an observer who rotated with the impeller and looked through the glass cover into the blade channels. The peripheral velocity u of the rotating blades at the particular distance from the axis of rotation is also important. The pressure path in the impeller is parabolic, corresponding to the laws of dynamics.

The peripheral velocity u, absolute velocity c and relative velocity w can be illustrated for each point in the impeller channel by means of a velocity triangle (Fig. 2.2 and 2.3). To determine the flow processes mathematically, however, all that is required is the recording of the velocities at the blade inlet and outlet (Fig. 2.3).

Pressure energy is produced on one hand by the torque transmitted from the impeller to the fluid by the change of moment of momentum and on the other hand, in the case of radial impellers, by deceleration of relative velocity w in the blade channel and conversion of absolute velocity c in the diffuser device downstream of the impeller.

The theoretical head H_{th} achieved during the energy transmission therefore consists of the following partial pressures [2-1]:

H_1 = content of centrifugal forces due to change in the peripheral velocity from u_1 to u_2

$$H_1 = \frac{u_2^2 - u_1^2}{2 \cdot g} \qquad (2 - 1)$$

H_2 = content of pressure increase due to deceleration of the relative velocity in the blade channel from w_1 to w_2*)

$$H_2 = \frac{w_1^2 - w_2^2}{2 \cdot g} \qquad (2 - 2)$$

H_3 = content due to increase in the absolute velocity from c_1 to c_2 *)

$$H_3 = \frac{c_2^2 - c_1^2}{2 \cdot g} \qquad (2 - 3)$$

This gives the specific total energy supply from the pressure and velocity energy as a theoretical head which could be achieved with loss-free flow

$$H_{th} = \frac{u_2^2 - u_1^2}{2 \cdot g} + \frac{w_1^2 - w_2^2}{2 \cdot g} + \frac{c_2^2 - c_1^2}{2 \cdot g} \qquad (2 - 4)$$

This equation can be simplified on the basis of the trigonometric relationships of the individual velocities in the velocity triangle, with the aid of the cos principle.

---

*) These are the averaged velocities at points (1) and (2) of the impeller.

$$H_{th} = \frac{1}{g}(u_2 \cdot c_2 \cdot \cos\alpha_2 - u_1 \cdot c_1 \cdot \cos\alpha_1) \qquad (2-5)$$

where g = gravity acceleration [m/s²]

whereby α_1 ; α_2 are the angles between the absolute and peripheral velocity at the blade inlet and outlet (Fig. 2.3 and 2.4 Derivation of formula for H_{th}).
The vortex torque component

$$c_u = c \cdot \cos\alpha \qquad (2-5a)$$

can be introduced here and thus we can also use :

$$H_{th} = \frac{1}{g}(u_2 \cdot c_{u2} - u_1 \cdot c_{u1}) \qquad (2-6)$$

This formula is Euler's equation for centrifugal pumps. It represents a theoretical relationship between the individual values where there is an ideal uniform distribution of all liquid particles in the particular flow cross-sections. However, it also states:

" The theoretical head of a centrifugal pump is independent of the density and
the physical properties of the fluid flowing through it."

This also applies as a good approximation for the head where the flow is subject to losses, provided the physical properties of the fluid do not substantially affect the pump efficiency. Where a centrifugal pump of a normal type is operated at the design point, it can be assumed with reasonable accuracy that the fluid entering the blades is vortex free so that $c_{u1} = 0$. In this case we get a simple form of Euler's equation for equilibrium conditions

$$H_{th} = \frac{u_2 \cdot c_{u2}}{g} \qquad (2-7)$$

whereby c_{u2} is the vortex component of the absolute flow at the impeller outlet.

The energy transmitted to the fluid comprises:

$$\frac{u_2^2 - u_1^2}{2\cdot g} + \frac{w_1^2 - w_2^2}{2\cdot g}$$

the potential pressure head

and

the velocity head

$$\frac{c_2^2 - c_1^2}{2\cdot g} \quad \text{thus:}$$

We get: $\qquad a^2 = b^2 + c^2 - 2\cdot b\cdot c\cdot\cos\alpha$

$$w^2 = c^2 + u^2 - 2\cdot c\cdot u\cdot\cos\alpha$$

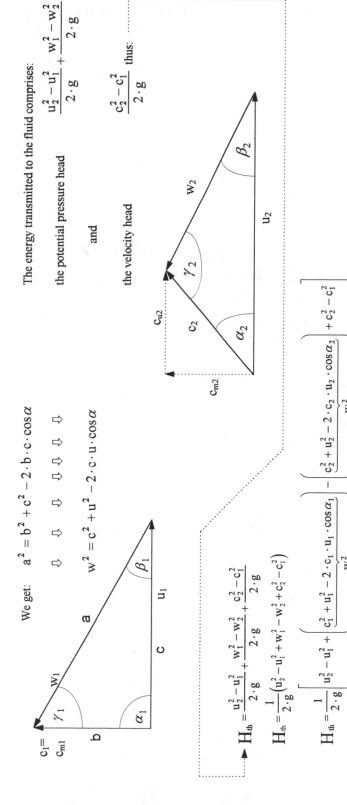

$$H_{th} = \frac{u_2^2 - u_1^2}{2\cdot g} + \frac{w_1^2 - w_2^2}{2\cdot g} + \frac{c_2^2 - c_1^2}{2\cdot g}$$

$$H_{th} = \frac{1}{2\cdot g}\left(u_2^2 - u_1^2 + w_1^2 - w_2^2 + c_2^2 - c_1^2\right)$$

$$H_{th} = \frac{1}{2\cdot g}\left[u_2^2 - u_1^2 + \underbrace{c_1^2 + u_1^2 - 2\cdot c_1\cdot u_1\cdot\cos\alpha_1}_{w_1^2} - \left(\underbrace{c_2^2 + u_2^2 - 2\cdot c_2\cdot u_2\cdot\cos\alpha_2}_{w_2^2}\right) + c_2^2 - c_1^2\right]$$

$$H_{th} = \frac{1}{2\cdot g}\left[u_2^2 - u_1^2 + c_1^2 + u_1^2 - 2\cdot c_1\cdot u_1\cdot\cos\alpha_1 - c_2^2 - u_2^2 + 2\cdot c_2\cdot u_2\cdot\cos\alpha_2 + c_2^2 - c_1^2\right]$$

$$H_{th} = \frac{1}{2\cdot g}\left(-2\cdot c_1\cdot u_1\cdot\cos\alpha_1 + 2\cdot c_2\cdot u_2\cdot\cos\alpha_2\right)$$

$$H_{th} = \frac{1}{g}\left(c_2\cdot u_2\cdot\cos\alpha_2 - c_1\cdot u_1\cdot\cos\alpha_1\right) \text{ where } \alpha_1 = 90° \text{ becomes } \cos\alpha_1 = 0 \text{ thus}$$

$$H_{th} = \frac{1}{g}\cdot u_2\cdot c_2\cdot\cos\alpha_2 \text{ and as } u_2\cdot\cos\alpha_2 = c_{u2} \text{ the final formula for } H_{th} \text{ becomes}$$

$$\boxed{H_{th} = \frac{u_2\cdot c_{u2}}{g}}$$

Fig. 2.4 Derivation of the theoretical head H_{th} from the flow velocities using the cos principle

9

2.2 Euler's equation (according to the moment of momentum principle)

The same result of the energy conversion can also be arrived at with the aid of the moment of momentum principle [2 - 2]. From the angular momentum

$$M_1 = \dot{m} \cdot r \cdot c_u \qquad (2 - 8)$$

we get the moment of the impeller blades as the difference between the angular momentum at the blade inlet (position 1) and at the outlet (position 2). Therefore the drive of the impeller blades requires a drive moment

hence:

$$M_{Sch} = M_{I2} - M_{I1} = \dot{m} \cdot (r_2 \cdot c_{u_2} - r_1 \cdot c_{u_1}) \qquad (2 - 9)$$

Where $\quad \dot{m}$ is the volume rate of flow (e.g. in kg/s)

r_1 and r_2 are the radii at the inlet (1) and outlet (2) of the impeller cascade and

c_{u1} and c_{u2} are the vortex components of the absolute flow at the inlet and the outlet

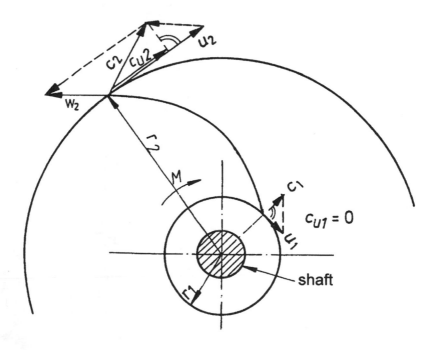

Fig. 2.5 Transmission of torque from the impeller to the flow [2 - 3]

10

At an angular velocity ω the motor power output for the blading, the so-called blade power, is:

$$P_{Sch} = M_{Sch} \cdot \omega = \dot{m} \cdot (u_2 \cdot c_{u2} - u_1 \cdot c_{u1}) \qquad (2 - 10)$$

with the peripheral velocity $u = r \cdot \omega$ and the data at the inlet (1) and (2) of the blading. If the blade power is related to the volume rate of flow, we get the specific blade power

$$Y_{Sch} = P_{Sch}/\dot{m} = u_2 \cdot c_{u2} - u_1 \cdot c_{u1} \qquad (2 - 11)$$

which also corresponds to the specific blade energy (e.g. with the unit Nm/kg), i.e. blade energy per mass unit. If blade power is divided by the weight rate of flow

$$\dot{G} = \dot{m} \cdot g \qquad (2 - 12)$$

we get the specific blade power relative to the weight rate of flow

$$H_{Sch} = P_{Sch}/\dot{G} = \frac{u_2 \cdot c_{u2} - u_1 \cdot c_{u1}}{g} \; \hat{=} \; H_{th} \qquad (2 - 13)$$

which in turn is equal to a specific blade energy (e.g. with a unit m) and thus also represents a theoretical head H_{th}, which would be achieved with a loss-free flow through the blading. Where the flow is subject to losses with a blade resistance η_{Sch}, the centrifugal pump thus achieves only the lower head

$$H = \eta_{Sch} \cdot H_{th} \qquad (2 - 14)$$

For most vortex-free blade suction flow, where $c_{u1} = 0$ and thus $\alpha_1 = 90°$, the relationship for the theoretical head is simplified as follows:

$$H_{th} = \frac{u_2 \cdot c_{u2}}{g} \qquad (2 - 15)$$

2.3 The H (Q) curve (Fig. 2.6)

The associated capacity $Q_{th\infty}$ corresponds to the head $H_{th\infty}$. It results from the meridional velocity c_{m2} and the area of outlet flow A_2 from the impeller

$$A_2 = D_2 \cdot \pi \cdot b_2 \qquad (2-16)$$

this is the outlet area of the impeller with infinitely thin blades

$$Q = c_{m2} \cdot A_2 \qquad (2-17)$$

$$Q = c_{m2} \cdot D_2 \cdot \pi \cdot b_2 \qquad (2-18)$$

We can also use the following for c_{u2}

$$c_{u2} = u_2 - \frac{c_{m2}}{\tan \beta_2} \qquad (2-19)$$

or

$$c_{u2} = u_2 - \frac{Q}{D_2 \cdot \pi \cdot b_2 \cdot \tan \beta_2} \qquad (2-20)$$

thus we get

$$H_{th\infty} = \frac{1}{g} \cdot u_2^2 - \frac{u_2 \cdot Q}{D_2 \cdot \pi \cdot b_2 \cdot \tan \beta_2} \qquad (2-21)$$

This relationship for a theoretical curve applies to a loss-free, ideal flow without pre-rotation in the suction flow and without reverse flows with exchange processes at the impeller outlet. Where the rate of flow is greatly reduced, considerable reverse flows occur at the impeller outlet (Fig. 2.65 and 2.66) so that the values for the design model no longer fully apply to this area.

If, however, there is a restriction to a fictitious blading with an infinite number of blades whose thickness proceeds towards zero, the flow angle at exit β'_2 corresponds to the blade angle β_{2s} over the complete curve range and we get the equation of the Euler curve

$$H_{th\infty} = \frac{u_2^2}{g} - \frac{u_2}{\pi \cdot D_2 \cdot b_2 \cdot \tan \beta_{2s}} \cdot Q \qquad (2-21a)$$

This equation shows that the theoretical head $H_{th\infty}$ is linearly dependent on the capacity. For $H_{0th\infty}$ the initial head of the pump where $Q = 0$ is

$$H_{0th\infty} = \frac{u_2^2}{g} \qquad (2-22)$$

12

The theoretical constant speed characteristic curve can then be drawn in the H (Q) graph by entering the value 0 for Q on the ordinate and determining the value for the duty point $H_{th\infty}$ point (Fig. 2.6).

The linking of both duty points represents the theoretical constant speed characteristic curve. The actual constant speed characteristic curve deviates from the theoretical by the amount of the hydraulic losses of the pump in the impeller and the diffuser device and follows a path similar to a parabola.

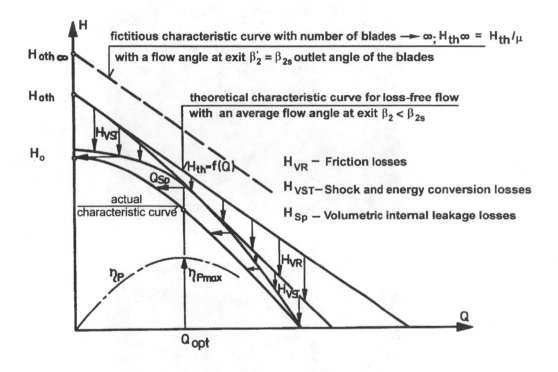

Fig. 2.6 The H (Q) curve (constant speed characteristic curve)

2.3.1 The influence of a finite number of blades on the total head H and further hydraulic losses

The head given for an infinite number of blades $H_{th\infty}$ is reduced due to the incomplete flow guidance where there is a finite number (mainly 7) of blades. In this case the angle of the outlet flow $ß_2$ to the impeller tangent is less than the blade outlet angle $ß_{2s}$. This process is shown in Fig. 2.7.

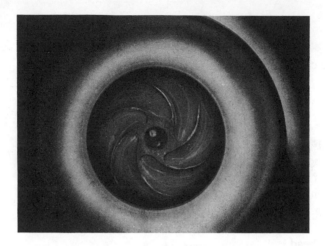

Rate of flow where there is little granulate

Rate of flow where there is much granulate

Fig. 2.7 Deviation of the streamlines from the blade geometry (relative flow pattern)

By using a rotoscope (Section 2.9.6 "Start of cavitation and cavitation pattern in absolute and relative flow pattern") and introducing plastic granulate into the pumped liquid, the relative flow in the impeller can be made visible if there is a transparent cover. The deviation of the relative streamlines leaving the impeller from the blade outlet angle $ß_{2s}$ of the impeller can be clearly seen.

In this case the actual flow angle at the exit β_2 is less than the blade outlet angle β_{2s}. For a fictitious impeller with an infinite number of blades and a thickness tending toward zero the flow angle at exit β'_2 is equal to β_{2s} so that a velocity diagram can be drawn up for this fictitious flow (Fig. 2.8). A comparison of the velocity diagrams for the actual and fictitious flow shows vortex components $c'_{u2} > c_{u2}$, so for a fictitious head defined by the vortex component c'_{u2} we get

$$H_{th\infty} = \frac{u_2 \cdot c'_{u2}}{g} > H_{th} = \frac{u_2 \cdot c_{u2}}{g} \qquad (2 - 23)$$

This difference can be determined by means of a decreased output factor

$$\mu = \frac{H_{th}}{H_{th\infty}} = \frac{c_{u2}}{c'_{u2}} \qquad (2 - 24)$$

Utilisation of the decreased output factor, for example, enables the fictitious vortex component c'_{u2} for the number of blades $\rightarrow \infty$ and therefore the blade angle $\beta_{2s} = \beta'_2$ to be determined from the velocity diagram for the fictitious flow (Fig. 2.8) when designing the impeller blading.

The theoretical head H_{th} is reduced by the losses which occur due to:

- volumetric internal leakage losses at the radial clearance between the impeller and casing;
- friction losses in the blade channels;
- energy conversion losses (velocity in pressure due to changes in direction and cross- section);
- shock losses where the angle of the approach flow to the impeller blades is not vertical.

In this case the actual constant speed characteristic curve is below that of the theoretical constant speed (Fig. 2.6)

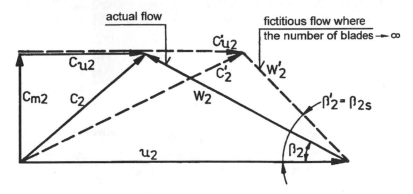

Fig. 2.8 **Velocity diagram for the actual and fictitious flow (dotted) where the number of blades** $\rightarrow \infty$ **; actual average flow angle at exit from the impeller** $\beta_2 <$ $\beta'_2 = \beta_{2s} =$ **outlet angle of the blades**

The deviation of the actual constant speed characteristic curve (H (Q) curve) from the theoretical Q - H_{th} curve is determined by the hydraulic and volumetric efficiency η_h and η_v. It represents the "hydraulic" losses of the pump. The overall efficiency allowing for the mechanical losses is obtained as follows:

$$\eta_{ges} = \eta_h \cdot \eta_v \cdot \eta_m \qquad (2 - 25)$$

From the hydraulic pump power output

$$P_h = Q \cdot H \cdot \rho \cdot g \qquad (2 - 26)$$

and from the overall efficiency of the pump we get the motor power (shaft) output P_w

$$P_w = \frac{Q \cdot H \cdot \rho \cdot g}{\eta_{ges}} \qquad (2 - 27)$$

$$
\begin{aligned}
Q &= \text{rate of flow} \\
H &= \text{head} \\
\rho &= \text{density} \\
g &= \text{gravity acceleration}
\end{aligned}
$$

On the basis of the considerations outlined in Sections 2.1 and 2.2, there are methods of calculation which enable a preliminary calculation of the actual H (Q) curve for the design of an impeller. The variables established in this way are, however, only approximate values. The precise constant speed characteristic curve can only be determined on the test bench. For centrifugal pumps with a constant speed n, there is therefore a specific capacity Q assigned to each head H. Arrangement of these duty points side by side in the H (Q) diagram provides the pump characteristic line or constant speed characteristic curve, because it can be determined on the test bench by throttling using a gate valve. The H (Q) curve also corresponds to the power characteristic curve P with the associated efficiency curve η_p.

The characteristic curve can be steep or flat, stable or unstable depending on the design of the impeller. The steepness of the pump characteristic line is given by the ratio between the difference in the initial head H_0 = head where Q = 0 and the head of best efficiency for this. This is expressed as a percentage; flat curves have a slope of approximately 10%, steep ones can be up to 60% and more. Fluctuating heads cause substantial changes in capacity where the curves are flat, but steep curves on the other hand have substantially smaller effect. Where there are possible fluctuations in head, this fact must therefore be allowed for when choosing the type of centrifugal pump to be used.

2.4 System-head curve

The task of the pump as a machine is to impart energy to a fluid, in order to convey the fluid through a pipe system from a point of low energy content to a point of higher energy content. To do this, static (potential) and kinetic (moving) energy are applied to the fluid. Both forms of energy are equal. They can be interchanged by applying Bernoulli's equation and together represent the head H of the pump. In the steady state, the head H of the pump is equal to the head H_A of a system with the dimension m. The head of the system is derived from the task to be performed.

Fig. 2.9 shows some of the most important system arrangements. For incompressible fluids where the flow conditions are constant, we get the following in accordance with the principle of energy conservation

$$H = H_A \qquad\qquad (2 - 28)$$

$$H_A = \frac{p_a - p_e}{\rho \cdot g} + \frac{v_a^2 - v_e^2}{2g} + z_a - z_e + H_{ves} + H_{vda} \qquad\qquad (2 - 29)$$

H_A = total head of system

p_a = pressure in discharge vessel

p_e = pressure in suction vessel

v_a = velocity in discharge vessel

v_e = velocity in suction vessel

ρ = density of fluid

g = gravity acceleration

z_a = level of surface of discharge liquid relative to any datum plane (line)

z_e = level of surface of suction liquid relative to any datum plane (line)

H_{vse} = sum of losses between points e and s on the inlet side of the system

H_{vda} = sum of losses between points d and a on the pressure side of the system

A uniform reference pressure is to be used for pressures p_a and p_e. For open vessels with barometric pressure p_{amb} on the surface of the liquid, the following is to be used.

$$1 \text{ mmHg} = 1 \text{ Torr} = 133{,}3 \text{ N/m}^2$$

The head H can be determined by measurement at the suction and outlet branches of the pump on a test bench or in the system.

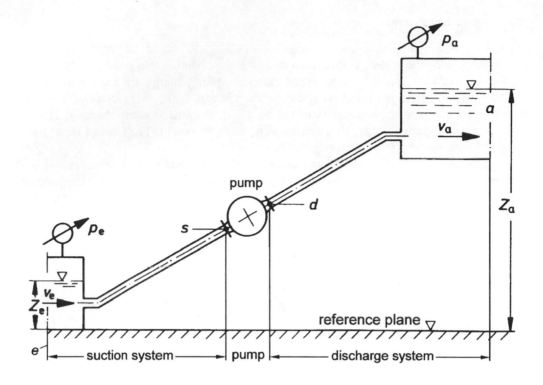

Fig. 2.9   Diagram of a pumping system [2 - 3]

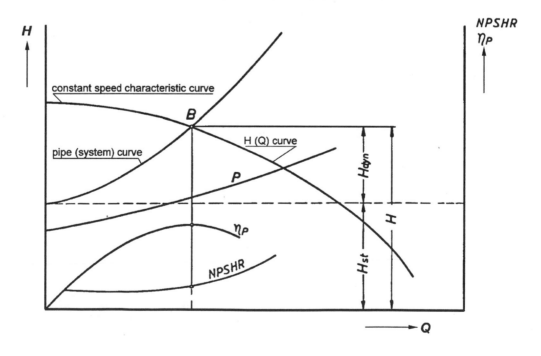

Fig. 2.10 System and H (Q) curve

The duty point B of the centrifugal pump is the intersection point of the piping characteristic curve and H (Q) curve.

As can be seen from Fig. 2.10, the loss curve of the piping added to the static head forms the system head curve in the H (Q) diagram. A corresponding variable for the power input, efficiency of the pump and the NPSHR value is assigned to each duty point. In the design of the operating data of a centrifugal pump, care should be taken to ensure that the pump works as close as possible to the point of best efficiency, i.e. that duty point B and the point of best efficiency coincide. As can also be seen from Fig. 2.10, the power input for self priming centrifugal pumps is lower up to average specific rotational speed with increasing rate of flow and reducing head. When designing the drive motor, care should be taken to ensure that no motor overload occurs with a fluctuating head and there is still adequate reserve. This reserve should be:

> up to 7,5 kW approx 20 %
> from 7,4 up to 40 kW approx 15 %
> above 40 kW approx 10 %

2.4.1 Change in the rate of flow Q for a change in static head H_{st} with the hydrodynamic component remaining constant at the system head curve

As can be seen from Fig. 2.11, the rate of flow Q reduces to Q_1 where there is a rise in the static head. In the reverse case it reduces from Q to Q_2 if there is a reduction in the static component at the system head curve. This means that flat curves produce relatively large fluctuations, and those for steeper ones are smaller, as can be shown from the values Q'_1 and Q'_2 .

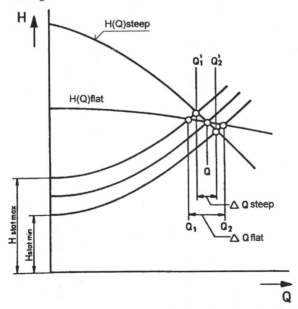

Fig. 2.11 Change in rate of flow with a fluctuating static head

2.4.2 Two centrifugal pumps operating in parallel with the same curve, separate suction pipes and a common pressure pipe. The length of the suction pipes should be short compared with that of the pressure pipe.

Centrifugal pumps in technical process plants frequently pump through a common pressure pipe with the pumped liquid being drawn mainly through separate suction pipes. Usually only one pump operates at a time. Where there is an increased demand, one or more pumps are switched in either automatically or manually, thus producing parallel operations. For this reason, knowledge of the following constructions is important in the planning and design of pumping stations. Assumption: both pumps are exactly the same with regard to their operating behavior, i.e. they have the same constant speed characteristic curve and in fact the same H (Q) curve shown dotted in Fig. 2.12. Assume the static head to be H_{st}. If any one of the two pumps is pumping, the head H_1 is established according to the rate of flow Q_1 because the head increases from H_{st} to H_1 due to the pipe friction losses. If a second pump is now brought on line with the first, the rate of flow doubles at each ordinate level and we therefore get the common H (Q) curve of pumps. An increased rate of flow, however, means higher friction losses and this shifts the duty point of the system on the piping characteristic curve from B_1 to B_2. The friction losses increase with the square of the rate of flow and this means that the piping characteristic curve is always a parabola with the apex at Q = 0. Whilst one pump alone would have had to overcome the total head H_1, during parallel operation the total head H_2 is overcome by both pumps.

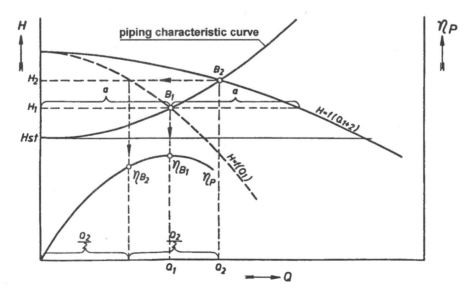

Fig. 2.12 **Parallel operation using two centrifugal pumps with the same characteristic curves**

This leads to a reduction in the rate of flow of the individual pump and therefore also to an impairment of the efficiency of the pumps where one of the pumps alone operates at the point of best efficiency at a head H_1 and rate of flow Q_1.

It therefore follows:

- that with parallel operation the rate of flow is less than the total of the rates of flow of centrifugal pumps operating singly;

- that parallel operation leads to a reduction of the efficiency of the pumps under certain circumstances

2.4.3 Two centrifugal pumps in parallel operation with an unequal characteristic curve, separate suction pipes and a common pressure pipe (Fig. 2.13)

Let us assume the suction pipes to be shorter than the pressure pipe. Both curves of the pumps are shown with I and II. By adding them together we get the curve (I + II). The summarized characteristic curve (I + II) coincides with the curve of pump II from point A to point C

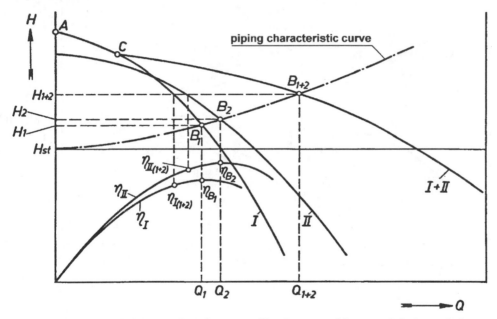

Fig. 2.13 Parallel operation of two centrifugal pumps with unequal characteristic curves

because pump I cannot achieve the head of pump II because of its lower curve. The lift-type check valve of pump II is closed during parallel operation by the head of pump I until pump II can also deliver. This is the case in point C. From this point onwards both pumps deliver together with the rates of flow being added. If any pump is operated singly, the particular duty points B_1 and B_2 occur at the intersection of the system curve and the pump curve, with the associated flow data Q_1, H_1 and/or Q_2, H_2 and η_{B1}, η_{B2}. In parallel operation, the duty point B_{1+2}, rate of flow Q_{1+2} at head H_{1+2} is achieved, which is less than $Q_1 + Q_2$ because the larger rate of flow increases the friction loss. If when operated individually the pumps worked at maximum efficiencies η_{B1} and η_{B2}, the efficiency of the individual pumps in parallel operation would then be reduced to $\eta_{I(1+2)}$ and $\eta_{II(1+2)}$.

2.4.4 Centrifugal pumps with a common suction and common pressure pipe

This situation should be avoided where possible. Sometimes, however, plant conditions mean that it is not possible to have two separate suction pipes. The same considerations apply here in principle as for the common pressure pipe. Fig. 2.14 shows, the same as Fig. 2.12, the interaction of two centrifugal pumps with the same H (Q) curve with a common pressure and common suction pipe. Let the geodetic suction lift be given as H_{Sts}. During single operation a manometric suction lift $H_{s1} = H_{Sts} + H_{vs1}$ is established, resulting from the geodetic suction lift and the friction losses in the pipe. When a second pump is brought on line, the friction losses of the suction pipe are increased to H_{vs1+2}. and therefore the manometric suction lift to H_{s1+2}. During single operation a net positive suction head $NPSHR_1$ was required for rate of flow Q_1. In parallel operation the pumps deliver only $0,5 \cdot Q_2$ each i.e. a lower rate of flow than Q_1. This therefore reduces the NPSHR value for each of the two pumps to $NPSHR_{1+2}$. Whilst the manometric suction lift increases during parallel operation of the pumps, the NPSHR in contrast reduces in the example shown.

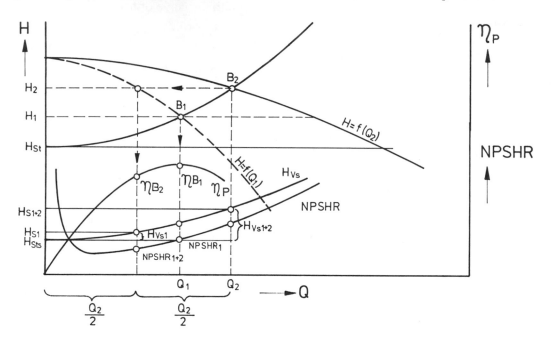

Fig. 2.14 Centrifugal pumps with a common suction and a common pressure pipe

Where it is possible to operate the pumps with a common suction pipe, these factors must therefore be allowed for, e.g. when drawing a fluid in a system where the surface of the inlet fluid is subject to barometric pressure the maximum suction lift for parallel operation is to be set at:

$$Z_e \leq \frac{p_{amb} - p_v}{\rho \cdot g} + \frac{v_e^2}{2g} - NPSHR_{1+2}$$

(For further information on determining the maximum suction lift, refer to Section 2.9 ("Suction capacity - net positive suction head (NPSH) and cavitation"))

22

2.4.5 Centrifugal pumps with an unstable constant speed characteristic curve and short pipe line [2-4]

Fig. 2.15 is a schematic of a centrifugal pump with a very short pipe and a system head curve which is practically horizontal. The head H of the pump therefore approximates to the static head H_{st}. Assume that the pump begins to deliver at duty point 1; the liquid level and therefore the head will increase until the consumption V equals the rate of flow Q. This may be reached at duty point 2 and we therefore have a stable operating condition with delivery adjusting automatically to consumption. If consumption is now less than Q_3, e.g. Q_4, the liquid level first rises up to duty point Q_3, and from there onwards the centrifugal pump cuts out. Where the rate of flow is less than Q_3 the static head in the vessel is greater than the head which can still be generated by the pump. The pump ceases to deliver until the liquid level drops to H_0 and then resumes delivery, i.e. at a rate of flow Q_5 whose associated head H_5 is equal to H_0. Because only Q_4 is consumed, the liquid level again rises and the cycle is repeated.

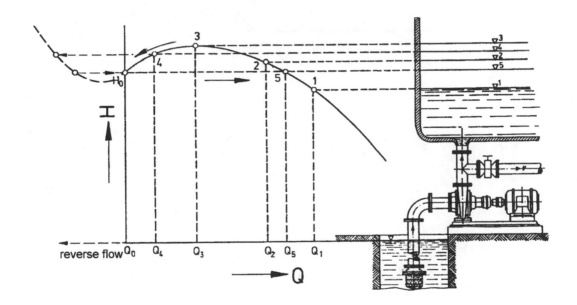

Fig. 2.15 Centrifugal pump with an unstable constant speed characteristic curve and a short pipeline

The constant speed characteristic curve from the critical duty point Q_3 to duty point Q_0 is the unstable branch of the H (Q) curve. For centrifugal pumps which operate with a long pipeline we also have the pipe friction losses in addition to the geodetic altitude to be allowed for. The critical duty point of these pumps is the point of contact of the upwardly-displaced system head curve H_A with the constant speed characteristic curve (Fig. 2.16). It therefore depends on the shape of the piping characteristic curve. Unstable curves should therefore be avoided where pumps are operated in parallel.

23

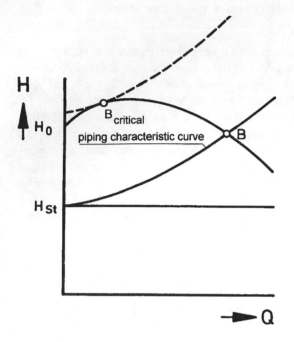

Fig. 2.16
Shortening of the unstable section of the constant speed characteristic curve due to pipe friction losses in a long pipeline

2.5 Control of centrifugal pumps

2.5.1 Throttling control (Fig. 2.17)

The total head of the installation H_A is comprised of the static component H_{st} and dynamic component H_{dyn}

$$H_A = H_{st} + H_{dyn} \tag{2 - 30}$$

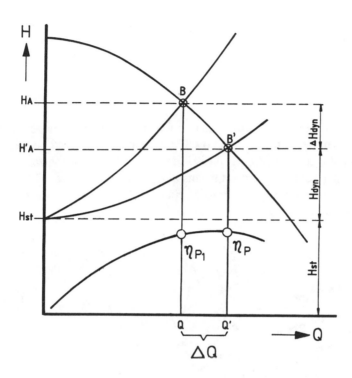

Fig. 2.17 Throttling control pump characteristics

The pump requires this head for a service flow rate Q. If the pump now does not deliver the required service flow rate Q, but instead Q' at a lower total head of the installation H'_A at duty point B', then throttling control (gate valve or orifice) must be used to set duty point B at flow rate Q. This means an additional pipe friction loss ΔH_{dyn}, i.e. an increase in the dynamic component of the system head curve, so that the formula for H_A is now:

$$H_A = H_{st} + H_{dyn} + \Delta H_{dyn} \tag{2 - 31}$$

The additional pipe friction loss ΔH_{dyn} may however be created only in the pipeline because throttling control on the inlet side poses the danger of cavitation. Where this type of control is employed, allowance must of course be made for a deterioration in efficiency because the drop in head is converted to heat in the throttling device. The efficiency deteriorates with the ratio of the unthrottled head to the set head. If η_p is the pump efficiency

25

at a flow rate Q' in the unthrottled condition, we get the following degree of utilization of the system in the normal working range B

$$\eta_{\text{Efficiency(Unit)}} = \eta_p \frac{H_{st} + H_{dyn}}{H_A} \qquad (2\text{-}32)$$

For this reason pumps with a flat H (Q) curve should be used where possible for throttling control.

Pumps of this kind enable a substantial reduction in capacity ΔQ with a low throttled head. Throttling control always means loss control, but is nevertheless the most frequent method of controlling centrifugal pumps, because it is the simplest and cheapest to implement.

2.5.2 Variable speed control

Variable speed control is a type of control which within certain limits is almost completely loss-free but is very expensive. If the rate of flow Q_1 in a pipe system is to be regulated to a rate of flow Q_2 without incurring additional loss due to throttling control, then speed has to be changed from n_1 to n_2. Each new speed means a new throttling curve. The speed n_2 is to be chosen such that the pump characteristic curve for n_2 intersects the system head curve at duty point B_2 (Fig. 2.18).

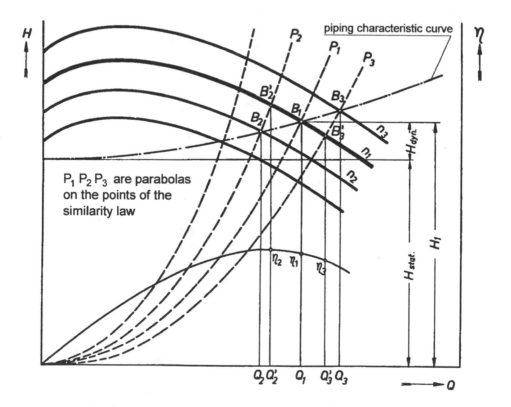

Fig. 2.18 Variable speed control

26

The new throttling curve for n_2 is obtained from the following considerations:
Let for example (2 - A) the new speed n_2 equal 0,8 n_1. To achieve control with the minimum loss, equal shock conditions must pertain at the impeller inlet e.g. a shock-free condition at the inlet (Fig. 2.19). For the velocity triangle at the impeller you then get $n_2 = 0,8$ n_1 thus $u_2 = 0,8$ u_1 and $\beta_{n1} = \beta_{n2}$. The similarity of the inlet triangles therefore means that the inlet velocity c_1 at the same inlet angle β_1 behaves as a function of the velocities and therefore as a function of the speeds. We therefore get:

$$\frac{u_{n1}}{u_{n2}} = \frac{c_{m1n1}}{c_{m1n2}} \qquad (2 - 33)$$

However, because the inlet cross-section for the given impeller does not change, inlet velocity Q is directly proportional to inlet velocity. We therefore get:

$$\frac{u_{1n1}}{u_{1n2}} = \frac{c_{n1}}{c_{n2}} = \frac{Q_1}{Q_2} \qquad (2 - 34)$$

$$\frac{u_{1n1}}{u_{1n2}} = \frac{n_1}{n_2} = \frac{Q_1}{Q_2} \qquad (2 - 35)$$

$$Q_2 = \frac{Q_1 \cdot n_2}{n_1} \qquad (2 - 36)$$

The rate of flow of a pump where the operating conditions are equal is therefore proportional to its speed.
For example:

$$Q_2 = 0,8 \cdot Q_1$$

$$\text{für } n_1/n_2 = 0,8$$

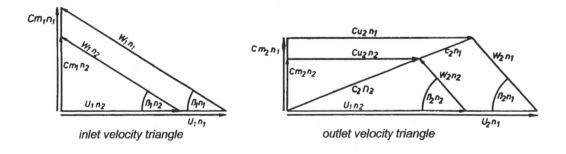

inlet velocity triangle   outlet velocity triangle

Fig. 2.19 Velocity triangles for different speeds e.g. $n_2 = 0.8\ n_1$

27

For a given flow angle at exit β_2 the following is applicable to the impeller outlet conditions

$$\beta_{2n1} = \beta_{2n2}$$

The outlet flow rate Q is equal to the established rate of flow. The meridional velocity c_{m2} thus changes in relation to the flow rates.

$$\frac{c_{m2n1}}{c_{m2n2}} = \frac{Q_1}{Q_2} \qquad (2\text{-}37)$$

$$c_{m2n2} = \frac{c_{m2n1} \cdot Q_2}{Q_1}$$

In the example

$$c_{m2n2} = 0,8 c_{m2n1}$$

$$c_{u2n2} = 0,8 c_{u2n1}$$

From the similarity of the outlet velocity triangles we now get the following example:

$$c_{u2n2} = 0,8 c_{u2n1}$$

The theoretical head H_{th} is obtained as follows from the principal of linear momentum

$$H_{th} = \frac{u_2 \cdot c_{u2}}{g} \qquad (2\text{-}38)$$

In the example:

$$H_{thn2} = \frac{u_2 \cdot c_{u2}}{g} = \frac{0,8 \cdot u_{2n1} \cdot 0,8 c_{u2n1}}{g}$$

$$H_{thn2} = \frac{0,8^2 \cdot u_{2n1} \cdot c_{u2n1}}{g}$$

or $\qquad H_{thn2} = 0,8^2 H_{thn1}$

where $\qquad \dfrac{n_2}{n_1} = 0,8$

The following conditions therefore generally apply.
The heads of the centrifugal pump under equal operating conditions behave as the square of its speeds.

$$\frac{H_1}{H_2} = \left(\frac{n_1}{n_2}\right)^2$$

thus $\qquad H_2 = \dfrac{H_1}{\left(\frac{n_1}{n_2}\right)^2}$ $\hspace{3cm}$ (2 - 39)

In the example: $\qquad H_2 = 0,8^2 H_1$

The motor power output P_w of a centrifugal pump is obtained from the equation

$$P_W = \frac{\rho \cdot g \cdot Q \cdot H}{\eta_{ges}} \hspace{3cm} (2\text{ - }40)$$

In the example:

$$Q_2 = 0,8 \cdot Q_1$$

$$H_2 = 0,8^2 \cdot H_1$$

$$\rho_1 = \rho_1 \qquad \text{, because the same fluids were present}$$

$$\eta_{ges1} \approx \eta_{ges2}$$

because equal shock conditions were present during the change of speed
(equal angle in the velocity triangles)

Therefore we get $\qquad P_{W2} = \dfrac{\rho \cdot g \cdot Q_2 \cdot H_2}{\eta_{ges}}$

$$P_{W2} = \frac{\rho \cdot g \cdot 0,8 \cdot Q_1 \cdot 0,8^2 \cdot H_1}{\eta_{ges}}$$

$$P_{W2} = 0,8^3 \cdot P_{W1}$$

$$\frac{P_{W1}}{P_{W2}} \approx \left(\frac{n_1}{n_2}\right)^3$$

$$P_{W2} \approx \frac{P_1}{\left(\frac{n_1}{n_2}\right)^3} \tag{2-41}$$

The required motor power output of a centrifugal pump therefore changes with the cube of its speed.

What was derived from the special case of a shock-free entry generally applies to all duty points whose velocity triangles are similar. The following therefore applies.

> For two operating states with different speeds the flow rates behave in the same way as the speeds, the heads in the same way as their squares and the powers in the same way as the cube of their speeds. The hydraulic efficiencies are equal, the pump efficiencies reduce slightly with speed.

In the H (Q) graph duty points with equal shock states lie on parabolas whose apexes lie in the datum of the co-ordinates (similarities parabolas Fig. 2.18 and 2.20).

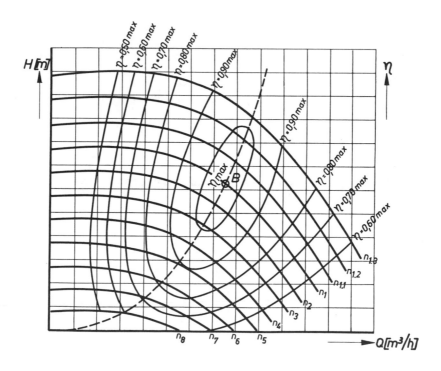

Fig. 2.20 H (Q) curves relative to speed (shell diagram)

2.5.2.1 Efficiency during rotation speed control

The duty points of equal shock states which lie on the affinity parabolas passing through the co-ordinates data have approximately equal efficiency. If n_1 is set as the efficiency curve for the speed, the efficiency for duty points B_2 or B_3 of a new constant speed characteristic curve is obtained e.g. n_2 and n_3 by finding the similar points B'_2 and B'_3 on the original constant speed characteristic curve for n_1 and reading off the efficiency in the given efficiency curve for n_1 at the flow rate of this duty point Q'_2 or Q'_3 (Fig.. 2.18).

If the foregoing is checked on a completed pump on the test bench, it will be seen that the points where the efficiency is equal do not coincide completely with the similarity parabola, but instead form almost closed elliptical curves in the upper and lower range. These diagrams are therefore referred to as shell diagrams. The deviation from this similarity parabola in these ranges is due to the losses which do not precisely follow the square law. It can also be seen that the efficiencies shown in the shell diagram represent the total efficiencies of the pump. However, η_p also includes η_m , the mechanical efficiency which unlike the output does not alter according to the cube, but instead linearly with speed. For this reason the conversion of the pump output using the affinity law is also only an approximation.

For similar operating states we therefore get:

$$\frac{P_{W1}}{P_{W2}} \approx \left(\frac{n_1}{n_2}\right)^3 \qquad\qquad (2 - 42)$$

It is particularly important that the affinity law is valid only if it does not lie within the range of cavitation.

The shell diagram of a centrifugal pump clearly shows the possible applications of this pump
(Fig. 2.20). It has an optimum duty point B where the maximum possible efficiency is achieved. Such shell diagrams are particularly suited to determining the efficiency of pumping installations.

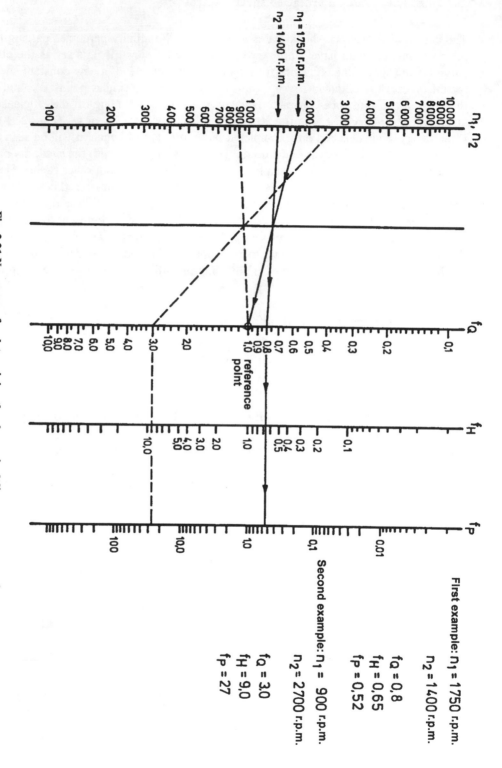

Fig. 2.21 Nomogram for determining the change in delivery parameters of
Q; H ;P w for changes in speed

First example: n_1 = 1750 r.p.m.

n_2 = 1400 r.p.m.

f_Q = 0,8
f_H = 0,65
f_P = 0,52

Second example: n_1 = 900 r.p.m.

n_2 = 2700 r.p.m.

f_Q = 3,0
f_H = 9,0
f_P = 27

2.5.2.2 Energy saving by variable speed control [2-5]

As we have seen, centrifugal pumps follow the affinity law, also referred to as Newton's law of similitude. According to this, the rate of flow Q changes with the square of the head H and the required motor power output with the cube of the speed.

When planning systems, allowances for the head of pumps are frequently given to be on the "safe side". The disadvantage of this is that the required status has to be established by throttling using a gate valve or orifice. Another possibility is to return part of the flow through a bypass into the suction or supply vessel. Both these methods of adjusting the operating parameters are, however, associated with energy losses which under certain circumstances can be of an unacceptable magnitude.

NB! (canned motor pumps and magnetic drive pumps)

> When operating such pumps in the overload range, which can occur when pumping without throttling control, the maximum flow rate specified by the manufacturer may be exceeded, and for self-priming centrifugal pumps this can lead to thermal overload of the motor. In the case of canned motor pumps and magnetic drive pumps where no additional pressure increase of the partial flow is provided for by means of an auxiliary impeller, then an impermissible temperature rise can occur due to the smaller pressure gradient and this results in a reduction in the stator-clearance flow. There is also the danger that the pump will run in a cavitation condition in the NPSHR > NPSHA overload range.

The most energy-saving type of required status setting can be achieved by changing the speed. In this way only sufficient power is fed to the pump as is required to overcome the load. Statistically only 1 - 2% of centrifugal and side channel pumps used in technical process plants are speed controlled, which means that a considerable amount of energy is still lost by wasteful matching to the required operating states.

2.5.3 Frequency converter for matching the delivery parameters

The speed of three-phase induction motors for pump drives can be changed using static frequency converters. These devices change a constant power supply with its associated frequency into a converted voltage and frequency [2-6].

The speed of the motor is directly proportional to the frequency of the supply. Changing the voltage and frequency in an equal ratio means that the torque output at the motor shaft is available as an almost constant nominal torque over the complete speed range. The speed range in the 50 Hz range is 1:10. Fig. 2.22 shows examples of possible power savings on centrifugal pumps.

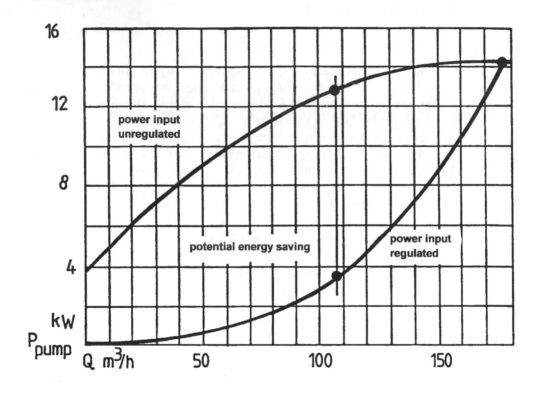

Fig. 2.22 Example of energy saving on centrifugal pumps using variable speed control (Danfoss-GmbH)

There are systems where stoppage of the pumps must not occur on failure of the frequency converter. Therefore it is important to ensure that in the event of malfunction the frequency converters are changed over and the pumps can run directly from the power supply. This is only possible if the frequency converters have an output voltage which the motor can take without changing the terminal connections, including those from the power supply, e.g. 3 x 380 V where a 220/380 V motor is star connected.

Although frequency converters still represent a considerable investment cost they are becoming cheaper. If the cost of the converter is set against the annual saving in energy cost the outlay is recovered very quickly, particularly for pumps with a long service life.

Min. investment differential costs	3500 DM
Max. investment differential costs	7000 DM
In stages	0500 DM
Running hours	7200 h
Energy cost	80 DM/MWh

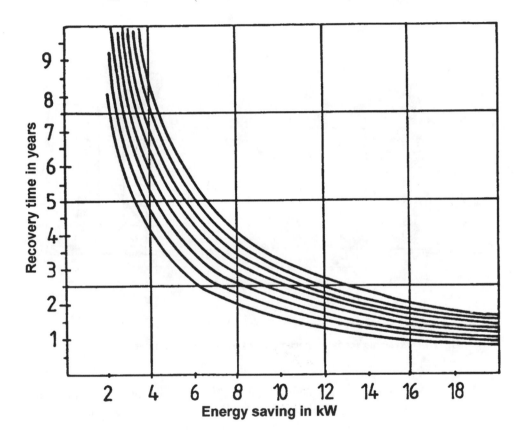

Fig.. 2.23 An example from an installation by the Hoechst company showing the recovery time of investment costs relative to energy saving for variable speed control centrifugal pumps.

Kuchler [2-7] has compiled a computer program for an installation by the Hoechst company which shows the recovery time for investment costs as a function of energy saving (Fig. 2.23).

2.6 Matching the service data at impellers

2.6.1 Correcting the impeller output where the head is too great

If the impeller characteristic curve does not agree with the required flow characteristics for the actual operating conditions, this can be corrected by changing the impeller outlet conditions. If the required duty point is below the characteristic curve of the pump, the required values of Q and H can be matched as required by reducing the impeller diameter. To do this, in the case of volute, i.e. pumps with a blade-less annular space as a diffuser device, the diameter of the impeller is reduced. For centrifugal pumps which have a diffuser for pressure conversion (this type is used mainly for multistage centrifugal pumps), the blades only are reduced and the shrouds of the impeller are left, refer to Fig. 2.24.

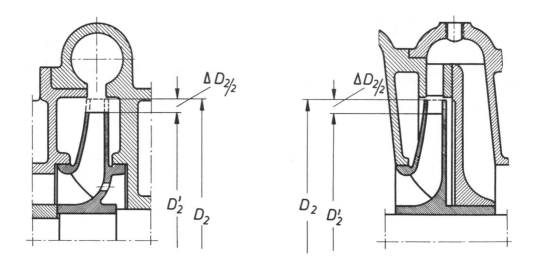

Fig. 2.24 Matching the flow characteristics by reducing the impeller diameter.

In the first case, the reduction of the shrouds can easily be allowed for because the reduction of the impeller diameter reduces disc friction loss by more than the deterioration of the fluid flow due to efficiency losses. In the latter case it is not advisable to turn down the shrouds because the fluid flow will then be inadequate. The change in the operating characteristics obtained by reducing the diameter of the impeller can be approximately calculated. If the original diameter was D_2 , then the new external diameter is D'_2. The same applies to the new flow rate with Q' instead of Q and H' instead of H.

If an approximation is obtained using the affinity laws, then with the same outlet angle β_2 and same disc width b_2 the flow rates change with the diameter and the head with the square of the diameter. From this it follows that the output reduces with the cube of the impeller diameter.

Presented as a formula:

$$\frac{Q'}{Q} = \frac{D_2'}{D_2}$$

(2 - 43)

$$Q' = Q\frac{D_2'}{D_2}$$

$$\frac{H'}{H} = \left(\frac{D_2'}{D_2}\right)^2$$

(2 - 44)

$$H' = H\left(\frac{D_2'}{D_2}\right)^2$$

$$\frac{P'}{P} = \left(\frac{D_2'}{D_2}\right)^3$$

(2 - 45)

$$P' = P\left(\frac{D_2'}{D_2}\right)^3$$

If the impeller is reduced in accordance with this equation, it can be seen that the actual flow parameters achieved are less than previously calculated. This is because the change in the blade angle β_{2s} reduces the blade energy and the reduction in the output of the impeller increases. This appears particularly pronounced for impellers with a large specific speed. The effect of the poor liquid flow is felt in the blade channels. In most cases the impeller blades on fast running pumps no longer overlap so that the flow is less precise. For this reason the impeller should not be turned down to the calculated diameter but corrected to the newly established blade diameter D'_2.

For impellers in the specific range n_q of 10 r.p.m to n_q of 25 r.p.m a correction factor K of 0,75 to 0,8 is preferable. Where n_q is greater a correction factor of 0,75 to 0,7 should be used. For ΔD we then get

$$\Delta D = K \cdot (D_2 - D_2')$$

(2 - 46)

with this still only being an approximation which must be checked by a trial run. The variety of designs of pumps, the shape of the impeller blades and the inlet conditions as well as the type of diffuser mean that this formula cannot be generally applied and, as already mentioned, can only be used as an approximation.

2.6.2 Matching the impeller output where the head is too low

If the specified impeller does not achieve the required head, a correction can be made within certain limits. The head of an impeller is calculated according to the formula

$$H \sim H'_{th\infty} = \frac{u_2 \cdot c'_{u2}}{g} \qquad (2 - 47)$$

whereby $H'_{th\infty}$ is the head of an impeller with a number of blades $\rightarrow \infty$, whose blade angle corresponds to the angle of flow β'_2.

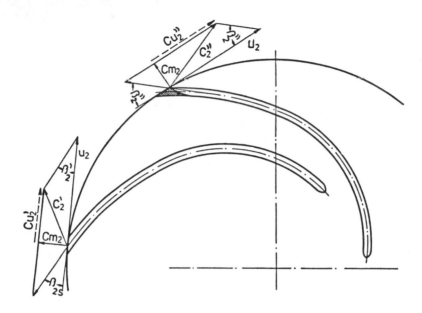

Fig. 2.25 Matching characteristics to service data by sharpening impeller blades

It follows therefore that the head depends not only on the peripheral velocity but also on the component c_{u2} of the absolute outlet velocity c_2. The magnitude of c_{u2} in turn depends on the blade outlet angle β_{2S}. An increase in β_{2S} results in the growing of c_{u2}. If the outlet angle $\beta_{2S} = \beta'_2$ is changed by sharpening the blades at the outlet as shown in Fig.. 2.25, we get β''_2. Because β''_2 is greater than β'_2, this means that c''_{u2} will also be greater than c'_{u2}. We therefore get

$$H''_{th\infty} = \frac{u_2 \cdot c''_{u2}}{g} > H'_{th\infty}. \qquad (2 - 48)$$

The actual increase in head possible in this way at the given flow rate is, however, relatively small and does not exceed 2 - 3%. For multistage centrifugal pumps this head difference can, however, be decisive in achieving satisfactory matching to the required service data.

38

2.7 Influence of viscosity on the characteristic curves.

An increase in the viscosity of the pumped fluid changes the characteristic curves of the pump. Flow rate and head reduce, accompanied by an increase in power input, i.e. efficiency is lowered (Fig.. 2.26). The characteristic curves for pumping viscous fluid can only be accurately determined by trial.

For Newtonian liquids a method recommended by the Hydraulic Institute New York can be used for converting the "water characteristic curves" using factors which depend on viscosity, flow rate and head (Fig.. 2.27).

The correction values k_Q, k_H and k_η relative to flow rate Q_z, head H_z and viscosity are given in Fig.. 2.27.

$$Q_z = k_Q \cdot Q_w$$

$$H_z = k_H \cdot H_w$$

$$\eta_z = k_\eta \cdot \eta_w$$

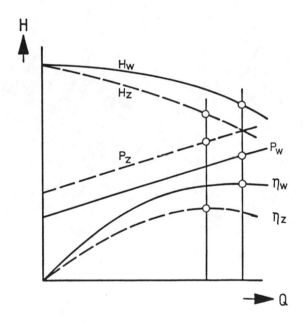

Fig.. 2.26 Reduction in performance when handling viscous liquids.

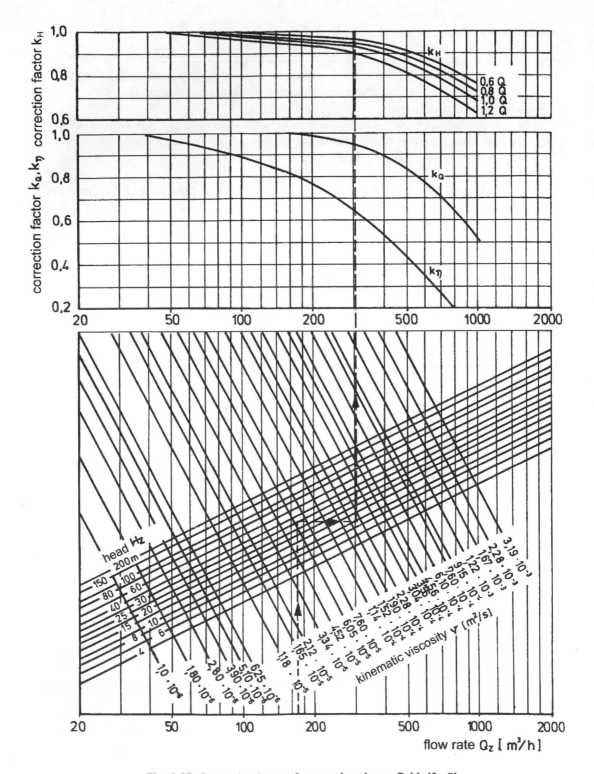

Fig.. 2.27 Correction factors for pumping viscous fluids [2 - 8]

40

2.8 Efficiency and specific speed [2 - 4]

The widespread application range of centrifugal pumps requires a basis for comparing these machines which can be numerically expressed. The term "specific speed" has proved itself in practice for this purpose. This is a variable obtained from the service data which has great practical significance for the design and choice of pumps.

Definition:

The specific speed n_q of a centrifugal pump is the required speed of one of the present pumps which are geometrically similar in all parts, which delivers a flow rate of $1m^3/s$ at a head of $1m$.

This is:

$$n_q = n \cdot \frac{Q^{\frac{1}{2}}}{H^{\frac{3}{4}}}$$

(2 - 49)

where:

$$\begin{aligned} n &= \text{revolutions per minute} \\ Q &= \text{flow rate in } m^3/s \\ H &= \text{head in meters per stage} \end{aligned}$$

Centrifugal pumps are grouped according their specific speed. These groups are as follows.

1. Radial vane	n_q	=	11 - 38 r.p.m
2. Francis screw	n_q	=	38 - 82 r.p.m
3. Mixed flow	n_q	=	82 - 160 r.p.m
4. Axial flow	n_q	=	100 - 500 r.p.m

The type number of a centrifugal pump, for which the specific speed n_q represents a quantity, is therefore determined by n_q but not by the absolute speed n of the machine. As shown in equation (2-49), the specific speed increases with n and q and decreases with H.

Between the three variables (n, Q and H) expressed in n_q and the efficiency of the pump, there is a very close functional relationship which exists for different reasons, as will be shown in the following text.

High speed demands wide blades with a relatively small diameter. This means that the clearance cross sections between the impeller and casing are small relative to the usable width of the blade and therefore also the internal leakage losses relative to Q. The large channel width means that the relative friction losses also remain within limits although the relative speeds are large. Efficiency at high specific speed is therefore generally good. Low speed, however, demands narrow discs (small Q) with a large diameter (large H). In this case the increase in diameter is limited by the disc friction losses occurring on the outside of the impeller, because this loss of power at extremely low specific speed slightly increases the pump power output and therefore greatly reduces efficiency. Multistage pumps must therefore be used if the specific speed formed from the service data is less than a certain magnitude (approximately n_q = 15 r.p.m). This means that a specific speed per stage is greater: for example if for the complete machine n_{qM} = 10 r.p.m, then n_q for the individual stages is as follows

Table 2 - 1

No. of stages i =	1	2	3	4	5	6
$n_q = n_{qM} i^{3/4}$ r.p.m	10	16,8	24,5	28,3	33,5	38,3

Für The following can be used for the disc friction loss P_r

$$P_r = M \cdot \omega \qquad (2 - 50)$$

$$P_r = \frac{1}{2} \cdot c_M \cdot \rho \cdot \omega^3 \cdot R^5 \qquad (2 - 51)$$

where:

P_r = disc friction
ρ = density of fluid
R = impeller radius
ω = angular velocity
c_M = torque coefficient

The factor c_M represents a friction coefficient which takes account of the Reynold's number and therefore the viscosity of the fluid to be pumped. Furthermore c_M depends on the impeller shroud geometry and the roughness of the shroud wall. Fig. 2.28 [2 - 9] shows over c_M over Re.

Disc friction loss therefore increases with the diameter of the disc, i.e. to the 5th power. It can therefore be clearly seen that impellers with a lower specific speed are subject to a relatively higher disc friction, which of course represents a loss, than impellers with a large n_q. Clearly this also leads to a requirement for a multistage design if n_q drops below a minimum limit, to be more closely defined. By setting a suitable impeller blade angle the head per stage can be changed within certain limits without changing the impeller diameter. Increasing the outlet angle β_{2s} also increases the head but care must be taken to ensure that the outlet angle must not be set to any steep angle to achieve the necessary head at the smallest possible impeller diameter, because otherwise the exchange losses (vortex formation at the impeller outlet, reverse flow) become too great. There is an optimum at which the hydraulic efficiency can be optimized at the particular impeller diameter and outlet angle. For radial impellers this outlet angle β_{2s} is approximately 25° to 30°.

By means of double curved impeller blades, where the start of the blades is drawn forward into the suction orifice, it is also possible to increase the head per stage without increasing the diameter of the impeller. This is because of the reduced specific impeller loading of the lengthened impeller blades.

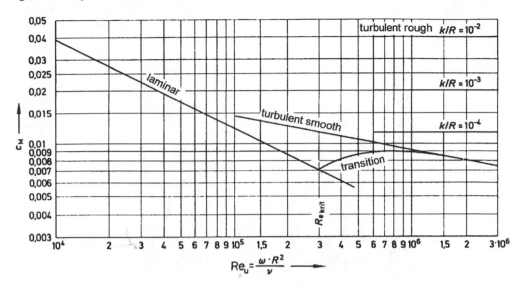

Fig.. 2.28 Torque coefficient c_M of free rotating discs.

A very limited possibility of reducing the impeller diameter where the head per stage remains constant consists of increasing the number of impeller blades. This arrangement does however cause an increase in the internal friction surfaces which very quickly counteracts the reduction in the disc friction losses.

The most satisfactory number of impeller blades is obtained from :

$$z = 6,5 \, \frac{D_2 + D_1}{D_2 - D_1} \cdot \sin \frac{\beta_{2s} + \beta_{1s}}{2} \qquad (2 - 52)$$

where:

z	=	number of impeller blades
D_2	=	external diameter of impeller
D_1	=	inlet diameter of impeller
β_{2s}	=	outlet angle of impeller
β_{1s}	=	inlet angle of impeller

If a diffuser fitted with blades is arranged downstream of the impeller for pressure conversion (for almost all multistage pumps), then to avoid sympathetic oscillations the number of diffuser blades must not be the same as the number of impeller blades and their number must also not be a common multiple of the impeller number.

If the limits of the above possibilities for minimizing the impeller diameter are reached, this means that the impeller diameter in conjunction with a particular speed is a measure of the achievable head.

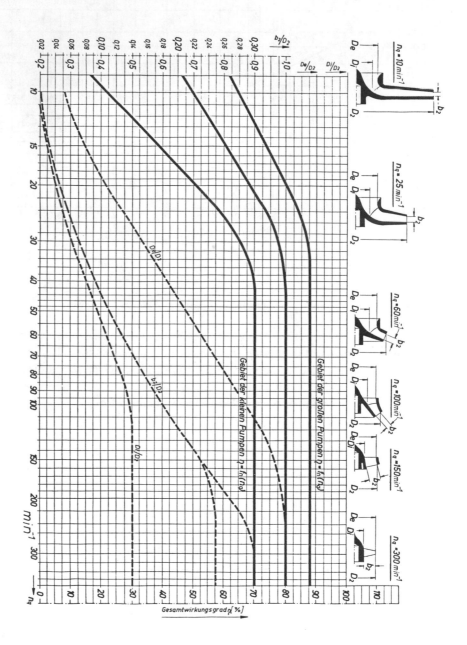

Fig. 2.29 Impeller shapes and efficiency as a function of specific speed n_q, plotted as a mean value of a band

44

As already described, it is possible to raise the type number of a centrifugal pump by increasing the number of stages and therefore improving the efficiency. Fig. 2.29 gives an overview of the path of the pump efficiency to be expected relative to the specific speed from centrifugal pumps which are well designed with regard to the normal working range of best efficiency and associated dimensions of the impeller. As the Reynold's number increases with the same n_q, i.e. larger pump dimensions, the value for efficiency increases because the viscosity forces reduce compared with the mass forces.

The Reynold's number Re_u for a centrifugal pump is as follows:

$$Re_u = \frac{\omega \cdot r_2^2}{\upsilon} \qquad\qquad (2 - 53)$$

where:

ω = angular viscosity in s^{-1}
r_2 = radius of impeller in cm
ν = liquid kinematic viscosity in cm²/s

Re_u is dimensionless and increases with ω and r. Centrifugal pumps are therefore more economical than smaller pumps with the same specific speed as dimensions increase.

Fig.. 2.30 Impeller with double curvature blades n_q = 30 r.p.m.

If one considers the efficiency curve, the steep rise of the curve from $n_q = 5$ r.p.m to $n_q = 40$ r.p.m is obvious. An increase in the type number within these limits by using multistage construction therefore results in a worthwhile improvement in efficiency, whilst above $n_q = 40$ r.p.m the increase in efficiency can no longer be considered worthwhile because of the substantially increased construction costs of the pumps. The normal working range of centrifugal pumps for municipal and local water utilities is mainly between $n_q = 15$ r.p.m and $n_q = 30$ r.p.m. This range is shifted downwards to $n_q = 10$ r.p.m for centrifugal pumps for industrial applications, due in part to the higher pressures required (e.g. boiler feed pumps). In special applications where only a single stage type can be used, e.g. brewery and dairy pumps, pumps with a specific speed up to $n_q = 7$ r.p.m are also built. Fig.. 2.30 shows an impeller with double curvature blades where $n_q = 30$ r.p.m.

2.9 Suction capacity - net positive suction head (NPSHA) and cavitation [2-10]

Large sums have to be paid out each year for the repair of centrifugal pumps because a large number of these machines are not correctly installed. A key factor in this is cavitation, a type of malfunction in pumps due to insufficient vapor pressure in the flowing fluid because the pump has not been correctly installed.

An explanation of the phenomenon of cavitation in centrifugal pumps is difficult if one considers that it is necessary to observe and measure physical, chemical and thermodynamic processes within a short period ($t = 0,5 \cdot 10^{-4}$ to $2 \cdot 10^{-4}$ s) on a very small area ($A = 10^{-3}$ to 10^{-4} mm^2) at high pressures ($p = 10^3$ to 10^5 bar) and at high frequencies ($f = 2500$ s^{-1}) [2-11]. It is generally the case that the mechanics of cavitation are based on Bernoulli's equation.

This states that in the stationary flow of an ideal fluid in a homogenous gravitational field, the sum of all energies of one and the same thread of flow is always constant with regard to a fluid level of random characteristics. It is therefore in accordance with Fig. 2.31:

$$z_1 + H_1 + \frac{v_1^2}{2g} = z_2 + H_2 + \frac{v_2^2}{2g} \qquad (2 - 54)$$

$$
\begin{array}{ll}
g & = \text{gravity acceleration} \\
H_1 ; H_2 & = \text{potential energy of a flowing liquid} \\
V_1 ; V_2 & = \text{velocity of flow} \\
Z_1 ; Z_2 & = \text{static elevations of the fluid above the datum point}
\end{array}
$$

With a liquid flow horizontal to the reference plane $Z_1 = Z_2$ we get:

$$H_1 + \frac{v_1^2}{2g} = H_2 + \frac{v_2^2}{2g} \qquad (2 - 55)$$

This equation states the internal pressure H of the fluid decreases with an increase in speed v. The limit to the possible pressure drop of a flowing fluid is set by the vapor pressure p_v. Besides the type of fluid, this depends mainly on its temperature.

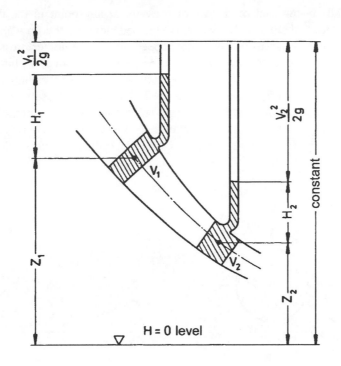

Fig. 2.31 The Bernoulli equation.

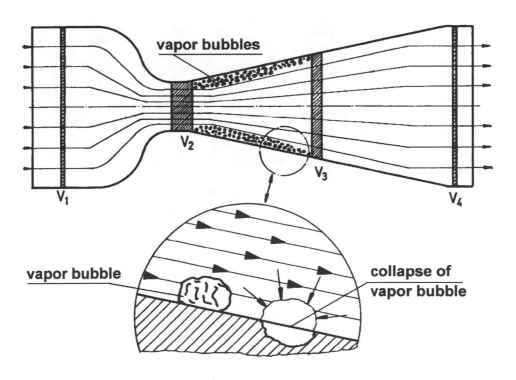

Fig. 2.32 Cavitation in a venturi.

47

If the pressure falls below that of the vapor pressure at any point in the fluid flow, gas vapor bubbles are formed (Fig.. 2.32) which are carried along by the flow and then collapse (implosion) suddenly in areas of lower flow speed due to the higher pressure prevailing in these areas. The collapse of these bubbles takes place, as stated, within an extremely small area at high pressures and at high frequencies. At the start of the implosion the bubbles, which are mainly spherical, become indented. This occurs close to the wall, on the side turned away from the wall, and on the side subject to the higher pressure in the case of bubbles in the free flow.   As the indentation of the bubbles increases, microjets are formed which burst into two or more parts (Fig.. 2.33). If the bubbles are close to the wall this microjet impinges the surface of the wall at high velocity [2-12].

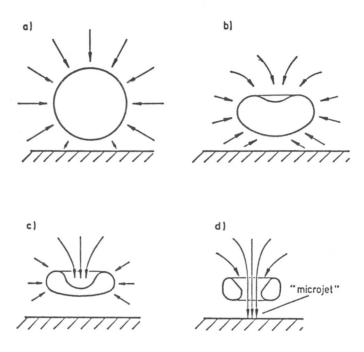

Fig.. 2.33 Asymmetric collapse of bubble close to wall.

The result is that the surface of the machine part affected by the implosion process is subjected to constantly changing loads which can lead to material fatigue and fatigue fractures. This generation and collapse of gas vapor bubbles is called cavitation. At first only small uneven spots are caused on the surface, but these soon become large rough patches making the surface resemble a "moon landscape". Cavitation erosion increases with increasing unevenness and not infrequently leads to complete piercing of the affected part, for example the impeller blade.

Fig.. 2.34 shows a disc ring of an annular nozzle vacuum pump with severe cavitational erosion and Fig.. 2.35 is a close-up view of this ring. These illustrations clearly show the destructive effect of cavitation. Other secondary effects are noise, vibration and a reduction in the parameters in the H(Q) graph. The degree of cavitation can increase up to complete breakdown of the flow. Immediately cavitation erosion occurs and the first unevenness in the surface of the material forms, an automatic self-destructive process is set up resulting in secondary cavitation. Fluid flowing over recesses where material has broken away forms vortices which follow the law $v_u \cdot r = const$.

Fig. 2.34 Disc ring of an annular nozzle vacuum pump damaged by cavitation.

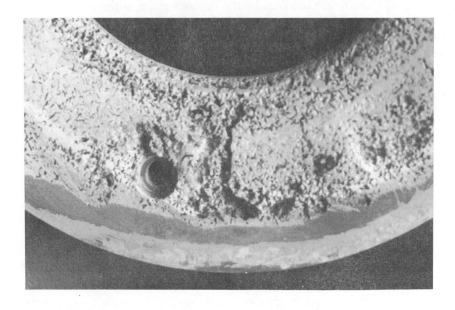

Fig. 2.3.4.1 Close-up view of the disc ring shown in Fig. 2.34.

In this case v_u is the velocity of the vortex in the direction of the circumference with r being its radius. Again, we therefore get, according to Bernoulli, an increase in peripheral velocity with a reduction in radius. A vapor bubble then forms in the centre of the vortex. If this is swept away by the flow a collapse occurs at a point of higher pressure. This generates further cavitation erosion if the collapse occurs on the surface of the material and not in the flow away from this. For this reason it is important that areas subject to cavitation be as smooth as possible.

Fig. 2.35 Section through the disc ring shown in Fig. 2.34

2.9.1 The NPSH value for centrifugal pumps

Malfunction-free operation of a centrifugal pump in accordance with the characteristic curve can only be guaranteed if it is ensured that the fluid cannot vaporize at any point in the pump. This applies particularly to the area of the impeller inlet. To achieve this, the energy available from the system present at the inlet cross section of the pump must be equal to or greater than that required by the impeller to bring the liquid to the maximum flow velocity predominating in the start of the impeller channel and to overcome the friction losses from the inlet cross section to the cascade. This total head is called the net positive suction head (NPSH).

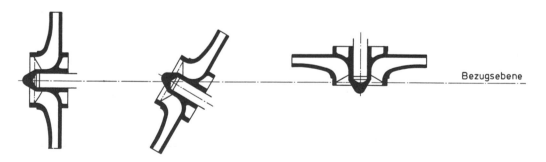

Bezugsebene = reference plan

Fig. 2.36 Datum planes for various impeller positions

A uniform datum plane and reference cross section must be assumed for a clear and unambiguous definition of this term. The difference between the total head at any point of a flowing liquid and the vapor pressure level is called the net total head.

For centrifugal pumps the horizontal plane which passes through the centre of the impeller inlet (Fig. 2.36) is specified as the datum plane in accordance with ISO 2548 and the total head relative to this plane is designated the NPSH. The referent cross section is the inlet cross section (inlet connection flange). To understand the term NPSH a distinction must be made between the NPSHA provided by the system and the NPSHR required by the pump to avoid cavitation (NPSHA is the net positive suction head available and NPSHR is the net positive suction head required).

2.9.1.1   Net positive suction head available (NPSHA)

The NPSHA is the net total head provided by the system at the inlet cross section (reference cross section) of the pump (not of the impeller) at point S, the centre of the suction connection (Fig. 2.37).

It consists of the absolute pressure p_s predominating at this point less the vapor pressure of the fluid in the inlet cross section, plus the total head from the mean flow velocity in the reference cross section. In this case all variables are to be given in meters. NPSHA can be defined as follows on the basis of the measurements on a running pump.

$$\mathrm{NPSHA} = \pm\, z_s + \frac{p_s - p_v}{\rho g} + \frac{v_s^2}{2g} \qquad\qquad (2\text{-}56)$$

If the centre of the reference cross section does not coincide with the reference plane then it is necessary to introduce the height difference z_s into the equation (Fig. 2.37 and 2.38). When the centre of the inlet cross section is higher than the reference plane, z_s is positive and negative when it is lower.

The NPSHA can be determined from the net total head at the inlet of the system for plants which are in the planning stage or are shut down. In this case this is reduced by the value of the losses in the pump connecting pipe.

$$\mathrm{NPSHA} = \pm z_e + \frac{p_e + p_{amb} - p_v}{\rho g} + \frac{v_e^2}{2g} - H_{vs} \qquad\qquad (2\text{-}57)$$

The magnitude $v_e^2/2g$ represents the flow energy occurring in the system, but this is usually negligibly small. The difference in level between the inlet liquid level and the reference plane is represented by z_e. When the liquid level is higher than the reference plane z_e is positive.

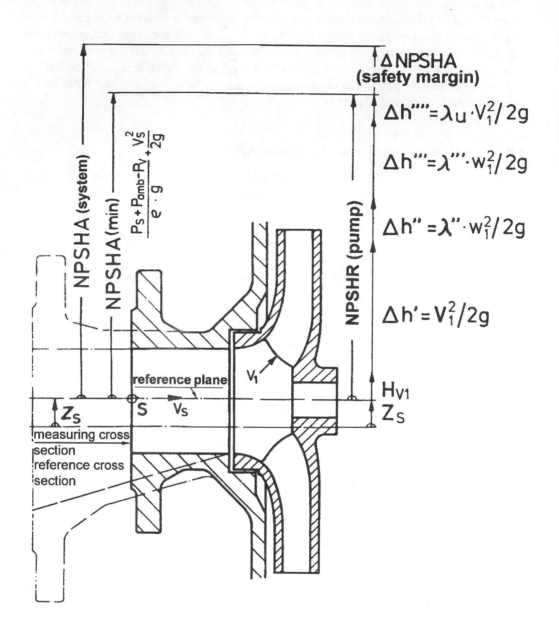

Fig. 2.37 Influence variables on NPSHR

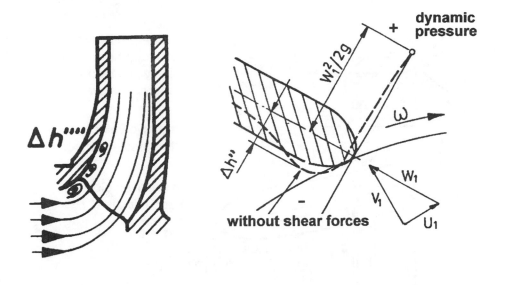

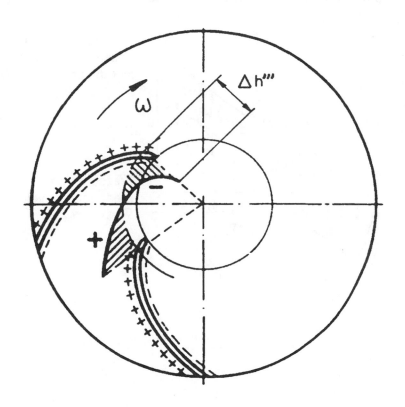

Fig. 2.37.1 Distribution of velocity and pressure over the blade contour

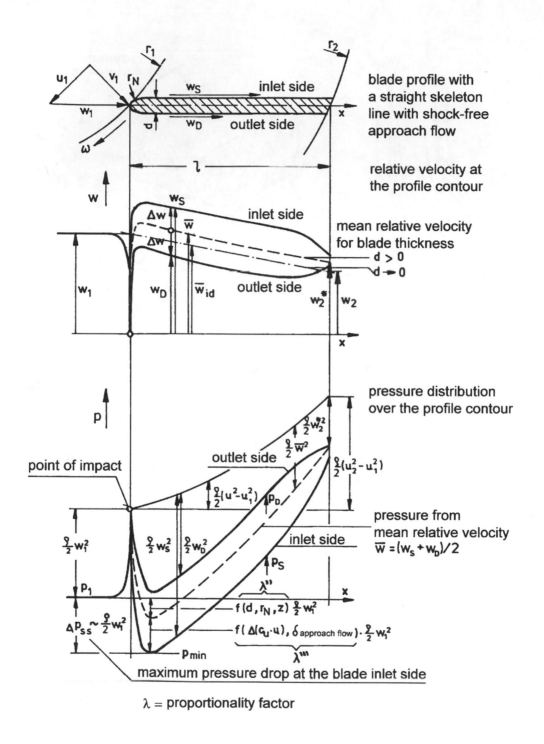

λ = proportionality factor

Fig. 2.37.2 Distribution of velocity and pressure over the blade contour

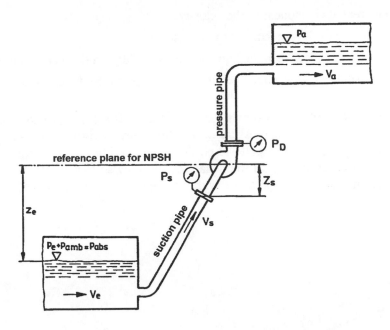

Fig. 2.38 Centrifugal pump system

Formula symbols (Fig. 2.37, 2.37.1, 2.37.2 and 2.38)

g	=	normal gravity acceleration
H_{vs}	=	losses in the inlet pipe including inlet losses and losses in valves, filters etc.
H_{vl}	=	losses in the suction branch
NPSHA	=	the NPSH produced by the system
NPSHR	=	the NPSH required for the pump relative to the permitted degree of cavitation
ΔNPSHA	=	excess of NPSHA over NPSHR (safety allowance)
p_{amb}	=	atmospheric pressure
p_v	=	vapor pressure at the inlet temperature on the impeller inlet
p_e	=	pressure in the inlet of the system (normally on the surface of the liquid on the inlet side)
p_s	=	absolute pressure in the inlet of the suction branch
Q	=	rate of flow
u	=	peripheral velocity
v_e	=	mean velocity of flow in the inlet of the system (normally in the vessel on the inlet side)
v_s	=	mean velocity of flow in the inlet cross section of the pump

55

v_u = circumferential component of the velocity

v_1 = mean velocity of flow on entry into the cascade

z_e = difference in level between the liquid on the inlet side and the datum plane

z_s = difference in level between the centre of the pump and the datum plane

Equation (2-57) for the NPSHA curve shows that where there is a constant geodetic suction lift or suction head and constant pressure on the fluid surface on the inlet side, ignoring the inlet energy $v_e/2g$, the existing NPSH changes only with the level of loss of the inlet pipe. But this in turn changes with the square of the flow so that the NPSHA curve in the H(Q) diagram represents a parabola with its open end downwards and the apex at Q = 0 (Fig. 2.39).

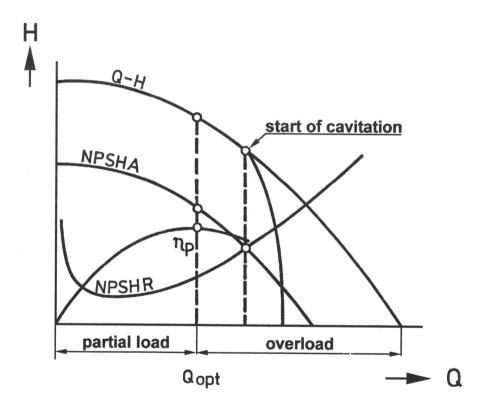

Fig. 2.39 NPSHA and NPSHR in the H(Q) diagram

Example (2-B)

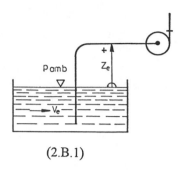

(2.B.1)

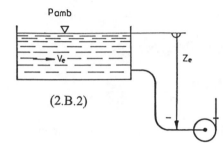

(2.B.2)

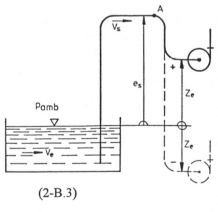

(2-B.3)

Examples for determining the NPSHR or the max. suction lift +Ze or min. suction head -Ze.

1. Pump delivers from open suction tank.

Equation

$$NPSHA = -Z_e + \frac{(p_{amb} - p_v)}{\varrho \cdot g} + \frac{V_e^2}{2g} - H_{vs}$$

$$Z_e \leqq \frac{(p_{amb} - p_v)}{\varrho \cdot g} + \frac{V_e^2}{2g} - H_{vs} - NPSHA$$

2. Pump delivers from open inlet tank with a suction head - Ze.

Equation

$$NPSHA = Z_e + \frac{(p_{amb} - p_v)}{\varrho \cdot g} + \frac{V_e^2}{2g} - H_{vs}$$

$$Z_e \geqq NPSHR + H_{vs} - \frac{(p_{amb} - p_v)}{\varrho \cdot g} - \frac{V_e^2}{2g}$$

With a further reduction in the value of -Ze the amount of -Ze increases.
This means an increase in the required suction head.

3. Pump delivers from an open tank with a rising and falling suction line (avoid this arrangement if possible). Calculation of the suction lift or suction head using examples 1 or 2.
Condition for the non evaporation at point A.

Equation

$$e_s \leqq \frac{(p_{amb} - p_v)}{\varrho \cdot g} + \frac{V_e^2}{2g} - \frac{V_s^2}{2g} - H_{vA}$$

H_{vA} is the coefficient of friction of the suction line from the suction line inlet up to point A.

Example for determining the NPSHA

57

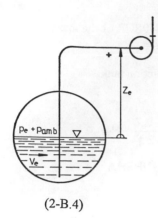

(2-B.4)

4. Pump delivers from closed suction tank.

Equation

$$NPSHA = -Z_e + \frac{(P_e + P_{amb} - p_v)}{\varrho \cdot g} + \frac{V_e^2}{2g} - H_{vs}$$

$$Z_e \leqq \frac{(P_e + P_{amb} - P_v)}{\varrho \cdot g} + \frac{V_e^2}{2g} - H_{vs} - NPSHR$$

5. Pump delivers from closed inlet tank with a suction head -Ze.

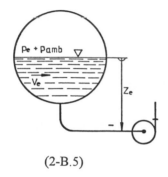

(2-B.5)

Equation

$$NPSHA = Z_e + \frac{(p_e + P_{amb} - p_v)}{\varrho \cdot g} + \frac{V_e^2}{2g} - H_{vs}$$

$$Z_e \geqq NPSHR + H_{vs} - \frac{(P_e + P_{amb} - p_v)}{\varrho \cdot g} - \frac{V_e^2}{2g}$$

A further reduction in the -Ze value increases the amount of -Ze.
This means an increase in the required suction head.

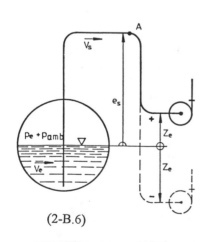

(2-B.6)

6. Pump delivers from closed tank with rising and falling suction lines (avoid this arrangement if possible).
Calculation of the suction lift or suction head using examples 4 or 5.
Condition for non evaporation at point A.

Equation

$$e_s \leqq \frac{(p_e + P_{amb} - p_v)}{\varrho \cdot g} + \frac{V_e^2}{2g} - \frac{V_s^2}{2g} - H_{vA}$$

H_{vA} is the coefficient of friction of the suction line from the suction line inlet up to point A.

Examples for determining the NPSHA

58

2.9.1.2 Required NPSHR

The NPSHR depends on the geometry of the impeller, rate of flow, speed of the pump and, as new research shows, also on the physical properties of the fluid to be pumped. The latter may be initially ignored for the purposes of this approach. The NPSHR represents the total head at the inlet cross section of the pump which is necessary in order to avoid, or reduce to an acceptable level, vapor formation and subsequent cavitation. The mathematical determination of NPSHR serves only as an approximation for the range of best efficiency and should be regarded by the planner as an initial approximation, with precise values obtainable only by a test bench trial. The value of NPSHR can be influenced by design measures such as the shape of the impeller blades (double curvature blades which project into the suction orifice of the impeller provide a lower specific blade loading and therefore smaller NPSHR value), the thickness of the blades, the number of blades and the inlet angle of the blades β_{ls}, the inlet diameter D_1 and any alignment of the flow using a diffuser in the inlet connection.

The duty point Q_{opt} can then be taken as:

in accordance with VDMA

$$NPSHR = (0,3 \div 0,5) \cdot n \cdot \sqrt{Q} \qquad (2 - 58)$$

with n in s^{-1} and Q in $m^3 \cdot s^{-1}$

in accordance with Thoma

$$NPSHR = \sigma \cdot H \qquad (2 - 59)$$

with

$$\sigma = k \cdot n_q^{\frac{4}{3}}$$

and n_q in accordance with formula (2 - 49)

$$n_q = \frac{n \cdot \sqrt{Q}}{H^{\frac{3}{4}}}$$

$$n = min^{-1}$$

$$Q = m^3 \cdot s^{-1}$$

$$k \approx 0,0014$$

Observation of the NPSHR curve reveals that usually in the partial load range there is first a slow but steady rise in the curve with a contrasting sharp increase in the overload range. The difference in the magnitude of the increase in the overload range compared with the partial load range can be explained by the fact that in the overload range with a proportional increase in Q and an associated NPSHR which increases approximately with the square, the shock losses when the flow is passing over the impeller blade tips are very important. Although these losses are also present in the partial load range, they have substantially less impact because Q is correspondingly smaller and the velocity of the approach flow v to the cascade is less. An interesting but very important phenomenon which is frequently ignored in practice is the renewed increase in the NPSHR curve in the vicinity of the ordinates in the H(Q) diagram. This curve path, which at first glance is somewhat peculiar, can be explained by the fact that where there is severe throttling control there is a correspondingly large reverse flow of fluid from the pressure side, i.e. from the diffuser, into the impeller and therefore an increased energy loss. As a result, the temperature of the fluid rises. The liquid thus heated passes through the narrow radial gap between the neck of the impeller and casing wear ring and suffers an increase in velocity, which according to Bernoulli can result in a drop in pressure and therefore in the vapor pressure being too low. To prevent this requires an additional NPSH which is reflected in a rise in the NPSHR curve. The point at which the NPSHR and the NPSHA curves intersect is the point of entry into the cavitational range. The H(Q) curve and the curve of the recorded output begin to drop (Fig. 2.39). Increasing cavitation leads to noise and vibration. The destructive forces of the bubble mechanism also begin to take effect.

2.9.2 Cavitation-free operation of centrifugal pumps

Cavitation-free operation is present if the following condition is met:

$$NPSHA \geq NPSHR \qquad (2 - 60)$$

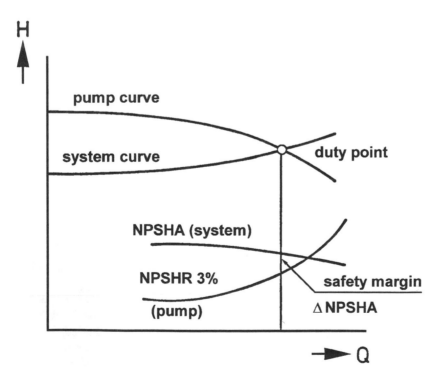

Fig. 2.40 NPSHA safety margin with respect to NPSHR in the H(Q) diagram

To be able to operate a system without cavitation it is best to select an NPSHA which is as large as possible, provided this is economically acceptable. In any case a safety allowance of at least 0.5 m should be provided to avoid possible damage to the pump or a decrease in output (Fig. 2.40). It should also be noted that the planned duty point is not always precisely obtained in practice. Even a relatively small movement of the duty point towards the overload range can cause a substantial increase in the NPSHR on the H(Q) curves under certain circumstances.

The previous types showed the change in the energy pattern within the pump only under normal circumstances in a flow state which corresponded to, or was at least close to, the particular given efficiency. Unwanted surface roughness such as casting pimples or inaccurate casting can still lead to an increase in the NPSHR and, for example, cause detachment turbulence at such points which then cause an additional reduction in the internal pressure. Because of this it is also necessary to increase the inlet energy provided by the operator at the inlet cross section by a certain amount ΔNPSHA to allow for these inaccuracies.
The values of the NPSHR specified by the pump manufacturer represents good average values and do not allow for possible deviations.

2.9.3 Development of the NPSHR curve

The NPSHA must always be greater than the NPSHR if cavitation-free operation is to be guaranteed. The operator and manufacturer must therefore agree on the reference point "S", i.e. the intersection of the reference and measuring cross section with the datum plane, to which all energy balances are referenced. Therefore all the influences from "S" which result in the NPSHR are to be added (Fig. 2.37). In detail these are as follows:

- The friction loss from the inlet cross section to the start of the impeller blade H_{v1}. This value is mainly very small and can be ignored in many cases.

- The reduction in internal pressure due to the increase in velocity of the pumped liquid on the absolute flow v_1 predominating just before the cascade.
  ($\Delta h' = v_1^2 / 2 \cdot g$)

- Further increase in the relative velocity due to the displacement effect of the finitely thick blades and the flow around the leading edges of the blades;
  ($\Delta h'' = \lambda'' \cdot w_1^2 / 2 \cdot g$)

- The effect of the inertia forces due to the deflection of the relative flow in the rotating impeller cascade, which has the effect of a reduced relative velocity and increased pressure on the front side of the blade (pressure side) and an increased relative velocity with a pressure decrease on the rear side of the blades (inlet side).

The distribution of the pressure increase at the blade "leading edge" (Fig. 2.37.1) corresponds to an approach flow without shear forces.

If blade shear forces are acting the so-called "circulatory additional velocities" are overlaid as shown in Fig. 2.37.2. ($\Delta h''' = \lambda''' \cdot w_1^2 / 2 \cdot g$)

This is shown very clearly by the photographs (Fig. 2.52.1 and 2.53.1) showing the relative flow of an impeller. The light spots in the impeller blade channel are vapor bubbles along the inlet side of the blade. On the pressure side of the blade (dark spots) the pumped medium is still in a state of liquid aggregation. The inertia forces which act on the fluid clearly contribute substantially to the formation of the NPSHR.

- The deflection of the flow up to 90° for radial flow impellers also contributes to a lowering of the internal pressure. The laws of centrifugal movement can be seen on the pumped medium, resulting in a centrifugal effect so that the flow does not run in concentric paths. A higher pressure predominates at the outer wall because that is where the larger centrifugal force acts. There is on the other hand a drop in pressure at the inner wall so that the streamlines detach. ($\Delta h'''' = \lambda_u \cdot v_1^2 / 2 \cdot g$)

  All these influences together give the NPSHR a total head which has to withstand an equal or greater level from the system side to avoid vaporization of the fluid and subsequent cavitation.

If the individual influence variables acting on the NPSHR (Fig. 2.41) are added to form a common parabola, we get an NPSHR curve similar to Fig. 2.42 (dotted line) which starts approximately above the coordinate origin (the curves c and d in Fig. 2.41 do not end in the coordinate origin. If the suction flow is free of vortex torque $w_1^2 = v_1^2 + u_1^2$ then only v_1 is proportionally dependent on the volume flow Q, with the peripheral velocity u_1 in contrast having no dependency on the volume flow).

It is, however, generally known that the NPSHR curve shows no such path in any case and that in the further course from duty point Q_{opt} in the direction of the overload range there is a departure from the theoretical parabola formula. In fact the derived NPSHR parabola reduces to a point in the duty point Q_{opt} with the NPSHR increasing on the left and the right (partial load range and overload range). This can be explained by the behavior of the approach flow to the cascade. In the partial load range the impeller blades receive a "frontal thrust" and in the overload range a "back thrust". This causes detachment of the flow with the formation of vortices in the detachment zone and increased flow velocity due to the displacement effect caused by the detachment. Up to the core of the detachment vortex the vortex component v_u increases according to the vortex rule $v_u \cdot r = $ constant, so that critically-low pressures can occur in the area of the vortex core.

The equations of VDMA (2-58) and Thoma (2-59) can be used for approximate calculation of NPSHR.

These formulae can be used with sufficient accuracy for rough calculation within a ±10% range of the best efficiency point. In this case the Thoma formula provides more reliable values and is therefore preferable. The precise course of the NPSHR curve can only be determined by a test bench run at the appropriate operating temperature using the relevant fluid.

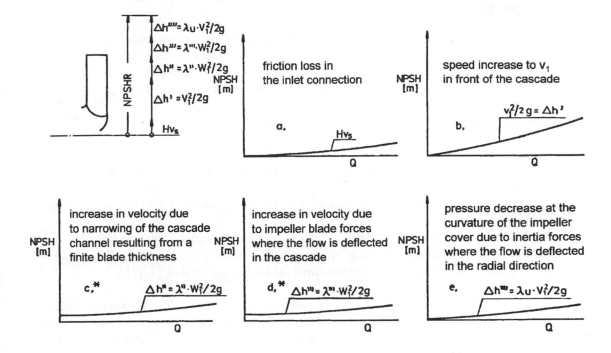

Fig. 2.41 Development of the NPSHR curve

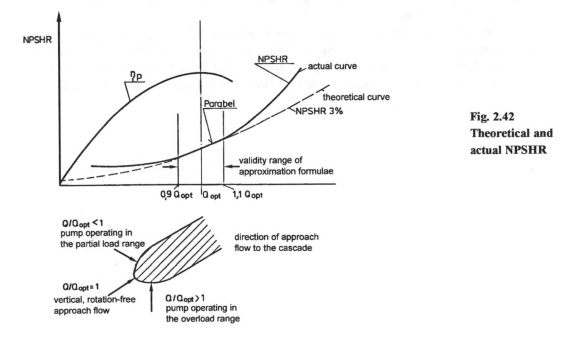

Fig. 2.42
Theoretical and
actual NPSHR

2.9.4 The influence of the physical properties of the fluid and temperature on the NPSHR

In the H(Q) diagram the NPSHR curves are drawn independent of temperature and physical properties of the fluid, in other words they are equally valid for all fluids and temperatures. Observations in day-to-day practice, however, show that there can be considerable deviations in the NPSHR values for different fluids and temperatures compared to the values obtained on a test bench using water at a temperature of approximately 20 °C. The deviation in the NPSHR response can be observed relative to vapor pressure and fluid density.

The phenomenon described above is contained in the literature under the heading "Thermal cavitation criterion". The thermal cavitation criterion allows for the amount of gas formation in the negative pressure zone subject to cavitation, which is dependent on fluid physical properties and temperature. In general, this results in improved NPSHR values compared to water at 20 °C.

Fig. 2.43 gives the correction values of the NPSHR for various fluids compared to the corresponding values for transport of water at 20 °C. The standard characteristic curve for NPSHR is reduced by the respective measured values for the NPSHR [2-13].

This phenomenon can be explained by the way in which the beginning of cavitation is determined. In purely physical terms the points at which cavitation starts is always the same (provided flow rates and rated speeds are the same). In practice, the beginning of cavitation is set at 3% head drop of the H(Q) curve (Fig. 2.44) due to the difficulty of observing the processes at the start of the blade. At this point in time, however, cavitation is already taking place and in fact at an advanced stage so that the vapor bubbles begin to restrict the cross section of the blade channels [2-14].

If the internal pressure falls below the vapor pressure, vapor is realeased in amounts which vary with different fluids.

The same applies to identical fluids at different temperatures. Water at 20 °C, for example, has a specific volume $v = 57.85$ m^3/kg, but this falls to as low as 0.39 m^3/kg at 150 °C. In other words the vapor volume of water at 20 °C is 148 times that at 150 °C. The same applies to the relationship between the vapor volumes of hydrocarbon and cold water.

Note:

> If the NPSHR values according to Fig. 2.43 are reduced by more than 50% of the values for water at 20 °C, the NPSHR value used is only half that for water (20 °C).

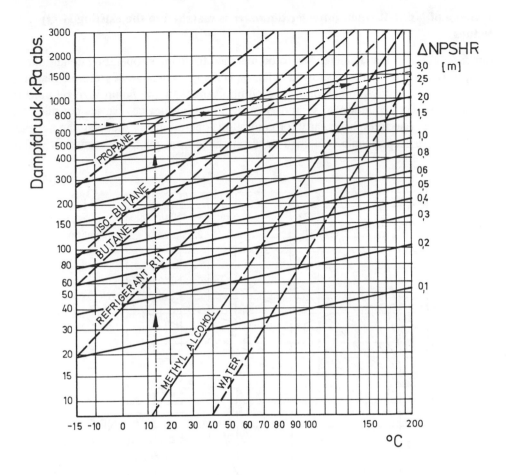

Fig. 2.43 Reduction of the NPSHR 3% values of various fluids relative to temperature [2 - 13]

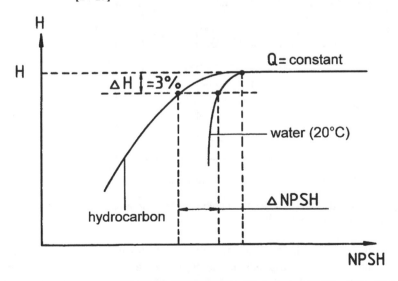

Fig. 2.44 NPSHR 3% with water at 20 °C and hydrocarbons

2.9.5 Change of NPSHR when impeller diameter is matched to the existing H(Q) values

In many cases the impeller diameter has to be matched to changed operating conditions, if for example the safety margins allowed when specifying system losses during planning were too great, the geodetic position of the pump changes or the pump is to be used in some other way. The reduction of Q and H using throttling devices causes considerable losses and is to be rejected, particularly for continuous operation. By reducing the impeller diameter the pump output can be matched to system conditions relatively accurately within certain limits.

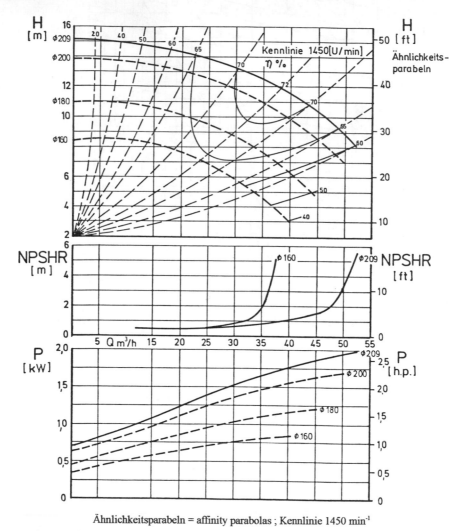

Ähnlichkeitsparabeln = affinity parabolas ; Kennlinie 1450 min⁻¹

Fig. 2.45 Change to the NPSHR curve in the overload range for a standard chemical centrifugal pump, size 32-160

Reducing the impeller diameter reduces the capacity in proportion to, and the head approximately with the square of, the impeller diameter. However, the pump reacts completely differently with regard to its NPSHR characteristics. Duty points with equal shock conditions lie on the affinity parabolas which pass through the coordinate origins. They have approximately equal efficiencies and in the H(Q) graph they produce, as shown, the so-called shell diagram. In the duty point of best efficiency the individual impeller diameter is therefore to be calculated with an NPSHR which is the same or similar. To the left of the Q optimum in the partial load range an NPSHR path which is mainly the same or deviates little from that of the full impeller diameter is achieved due to the reduction of the flow rate and therefore the levels of the absolute and relative velocities at the cascade. To the right of the Q optimum, however, the NPSHR sharply increases because due to the shortened blade lengths the specific blade loading is greater and thus the pressure difference between the front and rear sides of the blade increases. The course of the changed NPSHR curve can only be determined on a test bench. It is however important to note the following:

When the impeller diameter is reduced there is to some extent a substantial increase in the NPSHR in the overload range (Fig. 2.45).

2.9.6 Start of cavitation and cavitation pattern in the absolute and relative flow patterns

To be able to observe the flow patterns in the impeller and diffuser and therefore to further explain what has already been stated, the absolute and relative flows were photographically recorded using a special test bench designed by Prof. Dr. -Ing. W. Stieß. In this way stroboscopic images of the momentary states in the impeller and diffuser, changing with the rotating frequency of the impeller and relative flow in the impeller, can be studied by continuous observation and also photographically recorded. This method of observation is such that the observer or camera appear to be rotating with the impeller.

Construction of test bench (Fig. 2.46 and 2.46.1)
A centrifugal pump as shown in Fig. 2.47 is installed in the system in such a way that a transparent Plexiglas panel enables the flow patterns in the impeller and diffuser to be observed from the pressure side of the impeller.

Fig. 2.46 View of the test bench shown in Fig. 2.46.1

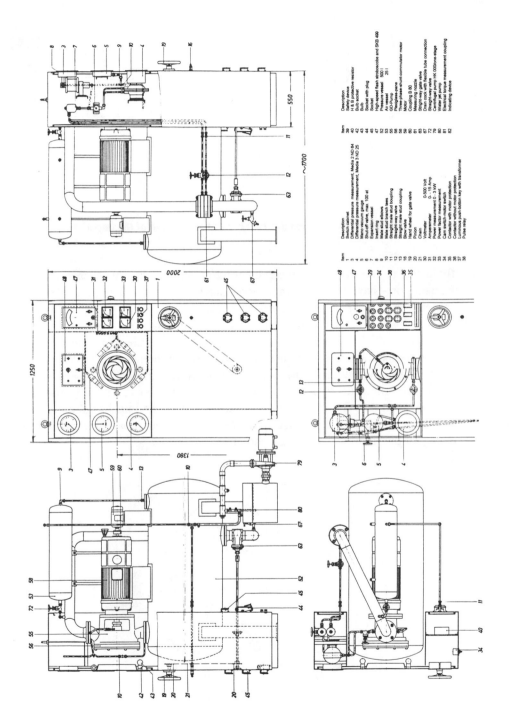

Fig. 2.46.1 The Sieß centrifugal pump test bench for research and teaching purpose:

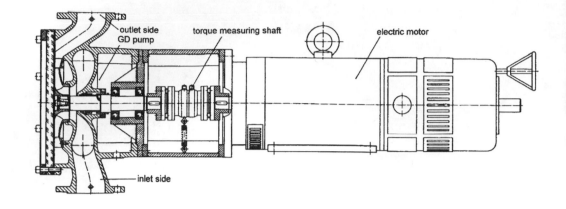

outlet side
GD pump

torque measuring shaft

electric motor

inlet side

Fig. 2.47 Test pump with a Plexiglas cover over the rear end of the impeller

Absolute flow (Fig. 2.48)
This is directly visible to the stationary observer using a continuous light. It can be recorded by camera using time exposures. When stroboscopically observed using a strobe light, the rotating impeller is illuminated in the same position on each rotation. This makes the impeller appear stationary if the rotational speed is greater than approximately 1000 r.p.m. The use of this method of illumination makes it possible for example to see whether the blade channels are completely filled, whether and where vortex detachments occur on the impeller and also if cavitation occurs where this happens in the impeller and whether any vapor bubbles are present. The stroboscopic picture does not however show the flow pattern, in other words it also doesn't show the relative flow in the impeller although it is viewed as stationary, because in each case only the instantaneous images which change with the frequency of the rotational speed are shown.

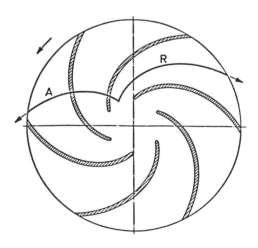

Fig. 2.48 Absolute (A) and relative (R) flow in the impeller

Relative flow in the impeller (Fig. 48)

To be able to view the relative flow, the observer or camera must rotate about the axis of the impeller at the same speed. But the same effect can also be obtained in a simpler manner by viewing through a rotating prism which rotates in the impeller axis at half the speed but in the same direction of rotation as the impeller and which enables the impeller to appear as though it is stationary in the picture.

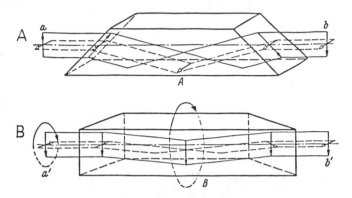

Fig. 2.49 Beam path in the prism

A   starting position of the complete setup
B   prism after rotation through 90° in the direction of the arrow with the object rotated through 180° in the same direction
a   object
a'   object after rotation through 180° in the direction of the arrow
b   picture
b'   picture with spatial position unchanged

The path of the beam in the prism at the initial position A and after a rotation of the impeller through 180° with rotation of the prism in the same direction through 90° (Fig. 2.49) shows that the arrows chosen as the picture remain frozen in the initial position [2-15].

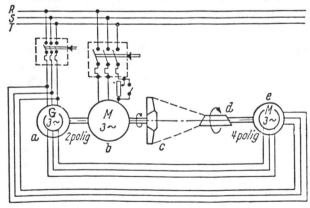

Fig. 2.50 Diagram of the turboprism [2-15]

a transmitter generator
b drive motor $\left.\right\}$ $n = n_M$ d prism
c impeller   e receiver motor $\left.\right\}$ $n_M / 2$

71

The rotoscope by Prof. Stieß constructed on this principle is synchronized with the pump impeller by an electric shaft (Fig. 2.50).

Viewing through the rotating prism shows the impeller to be always at rest whilst the diffuser and casing appear to be rotating. The relative streamlines can be made visible by adding granular particles to the flow (Fig. 2.7).

The relative flow pattern shows the extent to which the flow actually follows the impeller blades, where vortex detachments occur and during cavitation where and how the vapor bubbles move and collapse. It also shows the positions of potential cavitation erosion.
By placing a television camera in front of the prism, the relative flow pattern can be transmitted to a screen thus affording the possibility of observing the absolute and relative flows simultaneously.

The following photographs (Fig. 2.51 to Fig. 2.54.1) illustrate these processes very clearly.

Fig. 2.51 and 2.51.1.
This shows cavitation-free operation

Fig. 2.52 and 2.52.1
This shows the start of cavitation with the first bubble cores forming. No change can as yet be detected in the H(Q) diagram.

Fig. 2.53 and 2.53.1
Here the cavitation is extending and the H(Q) curve begins to fall.

Fig. 2.54 and 2.54.1
We now have full cavitation, the curve drops sharply, vapor bubbles enter the diffuser and delivery stops completely.

Impellers which run in the cavitation range show typical signs of wear at the blade tips and as operation in the cavitation range continues the damage becomes self generating. An interesting observation can be made on multistage pumps which are operated under conditions of severe cavitation over a long period. The blade tips of the diffuser following the first impeller are attacked although these are in fact on the pressure side. The explanation for this can be seen in Fig. 2.54. Where cavitation is very strong the vapor bubbles move from the impeller blade channel into the diffuser. Each time an impeller blade inlet side (rear side of the blade) sweeps past the tip of a diffuser blade, vapor bubbles move from the blade inlet side to the tip of the diffuser where they condense. The vapor bubbles collapse as they approach the next blade pressure side (front side of impeller blade). The vapor bubbles at the diffuser blades form and decay periodically with the frequency of the rotational speed times the number of impeller blades. The cavitation on the impeller, in contrast, is constant even though certain fluctuations due to an unsymmetrical approach flow and reactions from the diffuser back into the impeller (rotational frequency times the number of diffuser blades) can be observed. Fig. 2.55 and 2.55.1 show the destructive

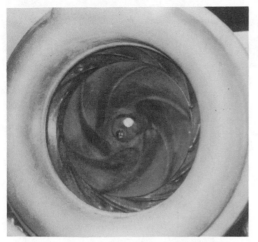

Fig. 2.51 Absolute flow pattern

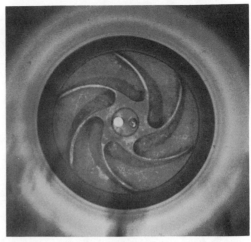

Fig. 2.51.1 Relative flow pattern

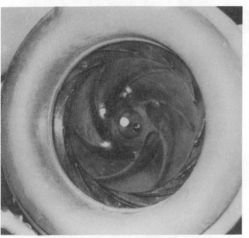

Fig. 2.52 Absolute flow pattern

Fig. 2.52.1 Relative flow pattern

Fig. 2.53 Absolute flow pattern

Fig. 2.53.1 Relative flow pattern

Fig. 2.54 Absolute flow pattern   Fig. 2.54.1 Relative flow pattern

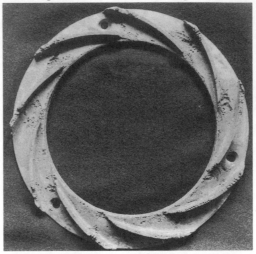

Fig. 2.55 A diffusor damaged by cavitation

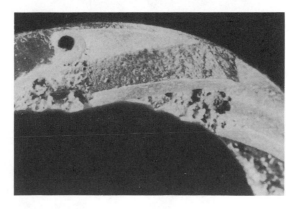

Fig. 2.55.1 Diffusor blade tips damaged by cavitation of the kind shown in Fig. 2.55

effect of cavitation, as previously described, on a diffuser. Fig. 2.56 shows particularly pronounced cavitation erosion on a side-channel pump.

Fig. 2.56 Cavitation erosion on the impeller of a side-channel pump

Advanced cavitation can not only damage the impeller of the first stage of a centrifugal pump but also the first diffusor. Cavitation wear in the impeller and diffuser occurs mainly at the points shown in Fig. 2.57. This is shown clearly in Fig. 2.54, 2.55 and 2.55.1.

The bubble size or bubble drag length L_{SCH} can be used to determine a 3% NPSHR (Fig. 2.58) [2.12]. The test bench shown in Fig. 2.46 is suitable for this purpose. The criteria for NPSHR, shown in Fig. 2.59, are given in Table 2-II.

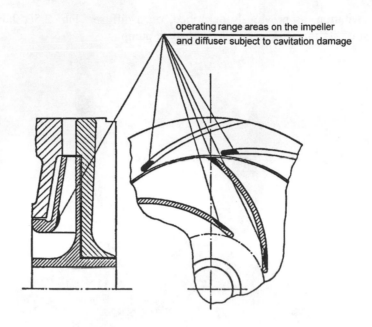

Fig. 2.57 Main areas of cavitation on the impeller and diffuser

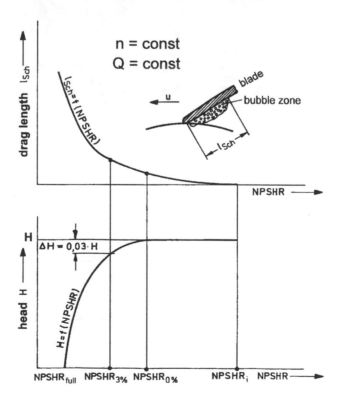

Fig. 2.58 Pattern of collapse of characteristics due to cavitation in the impeller of a
centrifugal pump

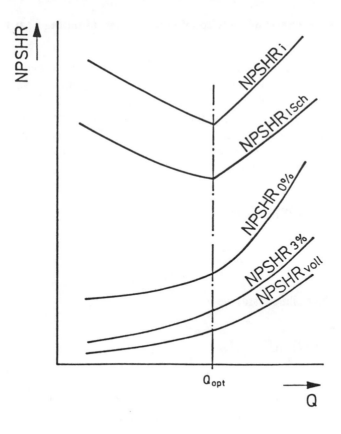

Fig. 2.59 NPSHR values of a centrifugal pump specified according to various criteria [2-12]

Table 2 - II Criteria for (NPSHR) of centrifugal pumps [2-12]

Criterion	Designation	Determination
Start of cavitation	$(NPSHR)_i$	Visual or acoustic
Defined expansion of cavitation	$(NPSHR)_i$ $(z.\ B.\ (NPSHR)_{l\ =\ 20\ mm})$	Visual
Defined drop in head	$(NPSHR)_{\Delta H}$ $(z.B.(NPSHR)_{3\ \%})$	Cavitation change curves
Avoidance of impermissible material damage	$(NPSHR)_{Erosion}$	Long term tests, evaluation of system experience

77

2.97 The behavior of the NPSHR of centrifugal pumps when pumping near boiling fluids (liquid gas pumping)

Liquid gases, such as ammonia, propane, butane, pentane, ethane, freons and methane are stored mainly in the boiling state, in other words in a state of aggregation at which in a unit of time the same amount of molecules change from the gaseous to the liquid state and vice versa. This means, however, that if there is the slightest drop in internal pressure at any point in the pump (for example due an increase in velocity) gas forms immediately. For this reason, pumps which pump such fluids must always have a geodetic suction head. The required suction head z_e is obtained from:

$$z_e \geq NPSHR + H_{sv} - \frac{p_e + p_{amb} - p_v}{\rho g} - \frac{v_e^2}{2g} \qquad (2-61)$$

where $p_e + p_{amb} = P_v$

(because the fluid is in the boiling state) we get

$$z_e = NPSHR + H_{sv} - \frac{v_e^2}{2g} \qquad (2-62)$$

Where v_e is negligible, the necessary suction head reduces to the level of the required NPSHR value plus the friction losses in the suction pipe.

Cavitation-free operation requires, as stated, that the NPSHA be greater than NPSHR. Provided the liquid and gas phase in the supply vessel are in thermodynamic equilibrium this can be easily achieved by choosing an adequate suction head or reducing the losses on the inlet side (appropriate dimensioning of the inlet pipe). If liquid gases with lower temperatures are pumped, difficulties arise due to possible heating of the pumped fluid in the inlet pipe. The rise in vapor pressure caused by this corressponds a loss of NPSHA. In addition to good insulation of the suction pipe, the dwell time of the liquid in the suction pipe also plays a part in this case. The longer the dwell time, i.e. the slower the velocity of the flow, the greater the possible temperature rise in the suction pipe and therefore the loss of NPSHA. The suction pipe must therefore not be too large. Experience has shown that the optimum velocity speed range is 0,8 to 1 m/s. The pre-rotation of the flow in the suction pipe also reduces the NPSHA. Where the flow is subject to rotation high velocities are produced towards the centre of the pipe and therefore, according to Bernoulli, lower static pressures. Avoidance of rotation of the liquid should therefore already begin at the inlet into the suction pipe.

Observations using glass pipes confirm that vapor bubbles produced at the inlet of the pipe are not destroyed despite a rise in the internal pressure where there is an increase in the geodetic liquid column to the pump, and these pass into the cascade of the impeller. To avoid the generation of vortices, the pipeline to the pump should be tangential where there is a vertical boiler or arranged laterally for a horizontal boiler. If this is not possible, the pipeline must be provided with an anti-vortex cross which prevents rotation of the inflow.

The inflow conditions become more difficult if the temperature equilibrium between the liquid and gas phases in the suction vessel are disturbed, perhaps by connecting compressors in refrigeration plants or emptying liquid gas vessels, e.g. tankers, which have no separate gas balancing line or other pressure equalization. If the discharge is faster than the corresponding temperature equalization with the environment, the volume of liquid removed must be continuously replaced by vaporization of the liquid gas. This vaporization process leads to cooling of the liquid and therefore a reduction in the vapor pressure in the vessel. If one assumes that the liquid particles passing through the supply pipe retain their temperature up to the pump and that during the dwell time in the supply pipe the pressure in the gas phase drops, e.g. 1 m. Fl. S., this has the end effect of a loss of 1 m NPSHA. The requirement that the NPSHA must always be larger than the NPSHR produces a maximum permissible rate of pressure drop in the vessel, i.e. for physical reasons the rate at which a liquid gas vessel can be emptied is limited. The following applies for the maximum rate of pressure drop on the surface of the liquid at the inlet side:

$$\text{Max}\left(\frac{dpe}{dt}\right) = (\text{NPSHA} - \text{NPSHR}) \cdot \rho \cdot g \cdot \frac{Q}{V_L} \qquad (2 - 63)$$

$Pe =$ positive pressure (above air pressure on the surface of the liquid on the inlet side)
$Q =$ rate of flow
$V_L =$ volume of suction pipe
$\rho =$ density
$t =$ time
$g =$ gravity acceleration

A similar phenomenon, which can lead to a noticeable deterioration in the NPSHA, occurs when emptying liquid gas vessels which are situated outdoors. Where the vessel is heated by solar radiation it can be severely chilled by sudden rain thus disturbing the thermodynamic equilibrium between the liquid and gas phases. The gas phase in this case cools down more quickly than the liquid phase so that the pressure in the gas phase can drop below that of the corresponding vapor pressure of the liquid. This pressure drop in the gas phase results in a loss of NPSHA which can be calculated by converting formula (2-63). Sprinkling vessels in mid-summer can also have the same effect and can lead to considerable difficulties with the NPSH [2-16].

2.9.8 Inducers, their purposes and operating characteristics

2.9.8.1 Purpose

Inducers are axial impellers placed on the same shaft immediately before the first impeller of a centrifugal pump (Fig. 2.60 and 2.61) which generate additional static pressure before the cascade of the impeller. They are used particularly where the available NPSHA does not exceed or is not equal to the NPSHR required by the pump. In many cases inducers are also used as a precautionary measure if the anticipated losses of the feed or suction pipe cannot be precisely determined or fluctuations of the NPSHA due to changes in the geodetic altitude of the liquid level on the inlet side, or application of pressure to it, are expected.

Inducers are also particularly suitable where liquids are being pumped which contain dissolved gases. In both cases the inducer can serve to prevent cavitation or decreased output provided it is correctly calculated and matched to the delivery output of the impeller which it supplies.

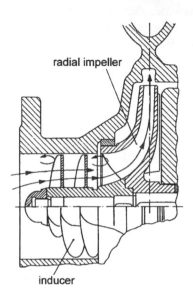

radial impeller

inducer

Fig. 2.60 Inducer positioned before the impeller [2-17]

If the system energy and the NPSH are related directly to the impeller inlet (normally the energy balance takes place at the inlet cross section - inlet connection of the pump), the energy increase of the fluid produced by the inducer is added to the system energy. The inducer therefore provides additional safety against liquid cavitation whilst also preventing the dissolved gas entrained in the liquid changing to a gaseous state of aggregation due to an increase in velocity.

2.9.8.2 Improvement of the NPSHR by use of an inducer

In addition to the increase in pressure before the first impeller of the pump, the inducer also performs a further task in that it ensures that the inlet conditions to the cascade are such that positive pre-rotation conditions are produced thus improving the NPSHR. The output of the inducer must however be matched to that of the impeller. If this is done, very good results can be achieved and the NPSHR values can be halved or even further reduced in the duty point (design point) of the pump (Fig. 2.62).

Fig. 2.61 Inducer and impeller on the shaft of a canned motor pump

The NPSHR curve of the impeller and inducer must be known for the use of an inducer to be assessed. Due to the defined suction capacity of the inducer its NPSHR curve increases relatively sharply to the left and right of the design point and can, as shown in Fig. 2.63, intersect the NPSHR curve of the impeller at one or two points. This means that above this range the inducer causes a deterioration in the NPSHR value of the pump.

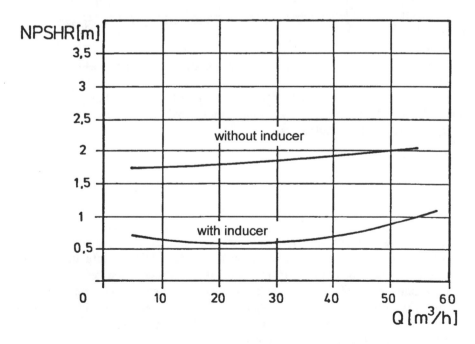

Fig. 2.62 Impact of an inducer on NPSHR

In the overload range (to the right of the design point) more liquid is supplied to the inducer than its suction capacity can cope with. The inducer deflects excess volume rate of flow, thus resulting in vortices forming in the liquid and detachment of the flow at the blade inlet side. For rates of flow which are less than those calculated for the design point, insufficient liquid is fed to the inducer (partial load range) and this causes detachments and vortices on the pressure side of the inducer blades. Both phenomena cause a noticeable increase in the NPSHR of the inducer. The vapor bubble forming on the pressure side of the blades due to the vortices have a greater detrimental effect on the NPSHR characteristics of the first impeller of the pump than those which form on the inlet side of the impeller. The inducer should therefore be designed for a larger than necessary flow rate, approximately 2 to 2,5 times the rated volume rate of flow, so that it always runs within the partial load range, thus providing a safety margin against the NPSHR increasing into the overload range.

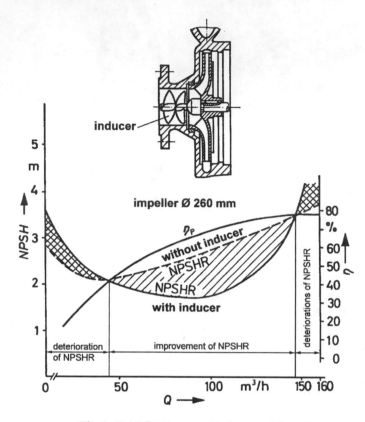

Fig. 2.63 NPSHR curve of inducer and impeller

2.9.8.3 Inducers used for two-phase mixtures

The delivery of two-phase mixing is a further field of application of the inducer. Normally, centrifugal pumps with a small to medium specific speed (n_q) can convey a maximum of 6 to 7% by volume of gas in the undissolved, gaseous state of aggregation at the same time. Because of pre-rotation, larger gas proportions lead to segregation before the cascade. The specific lighter proportions collect in the core of the rotation and restrict the inlet cross section so that the flow is first reduced and then finally stops. Use on an inducer causes the flow to be force fed into the impeller channels and from there it is delivered in the flow to the impeller outlet. By using inducers, therefore, undissolved gas proportions of up to half the volume rate of flow can be conveyed.

2.9.8.4 Use of inducers to counteract liquid-gas cavitation

The cavitation process already described is based on the formation of bubbles by the local vapor pressure being undershot and implosion in the area of higher pressure due to the vapor bubbles being carried along by the flow.

Cavitation caused by dissolved gases entrained in the liquid, however, behaves quite differently. The gases in the solution can then change to a gaseous state of aggregation (expansion instead of implosion) if the velocity of the liquid increases (e.g. at the impeller inlet) whereby according to Bernoulli the internal pressure is lower. This is referred to as gas cavitation although the designation is not quite correct. The gas bubbles emerging from the liquid begin to gather on the cascade at the inlet side of the blades and, because they are specifically lighter than the fluid, they are not carried through to the pressure side. The continuing formation of gas bubbles causes restriction in the impeller inlet and finally the complete stoppage of the flow. The process for liquid cavitation is basically similar but with the difference that in that case it is not the gas but the vapor bubbles which restrict the cross section. Whilst the damage is caused by the imploding vapor bubbles during liquid cavitation, the gas bubbling out during gas cavitation is completely harmless in this respect. If liquid and gas cavitation occur simultaneously, the gas bubbles lie over the vapor bubbles and act as hydrophores so that the collapse of the vapor bubbles is much more resilient, which is indicated by a distinct reduction in noise. The material damage is therefore correspondingly lower or is completely absent. The pump power output is certainly affected by this so that this operating state cannot be tolerated for long.

With the aid of inducers, both the gas and liquid cavitation can be counteracted by suppressing recirculation on suction side of the impeller.

2.9.9 Increase in NPSHR values in the partial load range due to heating-up of the fluid on the inlet side; effect of pre-rotation and recirculation [2-16]

2.9.9.1 Dependency on the physical fluid property and temperature particularly for canned motor pumps and magnetically-coupled pumps

Special attention should be paid to the NPSHR path in the area of the ordinates. Here the fluid physical property dependency (Section 2.9.4 "Effect of physical properties of fluid and temperature on the NPSHR") is particularly noticeable and, contrary to what has been said above, it begins to have a negative effect.

Using a hermetic centrifugal pump with standard hydraulics acc. DIN 24256/ISO 2858 of the type 32 - 250 (Fig. 2.64) as an example (2 - C) , we will attempt to demonstrate the

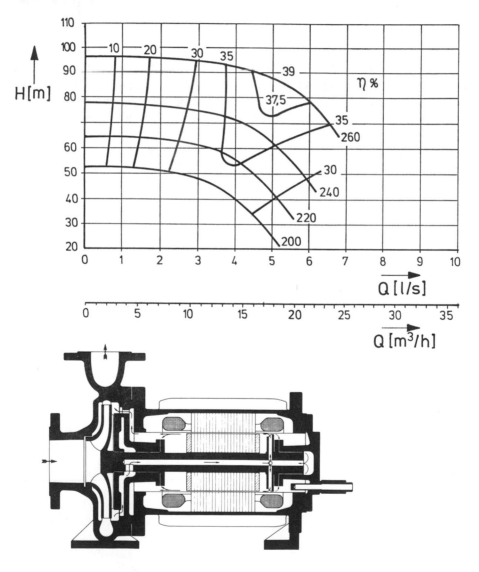

Fig. 2.64 Characteristic curves and functional diagram of a canned motor pump with type 32-250 standard hydraulics acc. DIN 24256/ISO 2858

different influences on NPSHR and the minimum flow Q_{min} for an impeller diameter of D_2 = 260 mm. In addition the following should also be noted:

Due to the severely throttled delivery, recirculation flows out of and into the impeller (Fig. 2.65) occur at the impeller inlet and outlet in the partial load range and severe pre-rotation occurs. This produces shear layers between the normal and reverse flow thus forming vortices and these in turn form vapor bubbles which implode at the pressure side of the impeller blades.

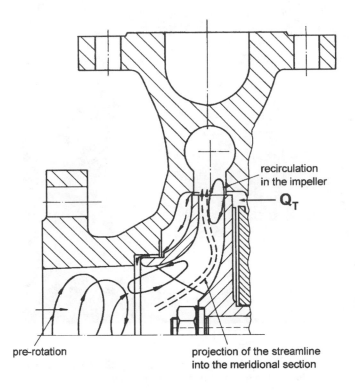

recirculation
in the impeller

Q_T

pre-rotation

projection of the streamline
into the meridional section

Fig. 2.65 Pre-rotation and recirculation in an impeller operating in the partial load range

Fig. 2.66 plots the relative flow pattern of an impeller in the partial load range close to the ordinates in the H(Q) diagram which clearly shows the recirculation flow at the impeller inlet and outlet. The shear forces naturally lead to the generation of "bubble clouds" which cause increased cavitation erosion.

The pump is equipped with a pressure-increasing secondary impeller located in the rotor chamber, which takes off the partial flow at an annular filter, sucks it through the hollow shaft and back into an annular chamber at the outer periphery of the main impeller, in other words back into the pressure chamber of the pump.

Drive is effected by a 15 kW canned motor. Under severely throttled operating conditions, there is a high level of heat generation resulting from friction heat of the energy given off to the impeller by the drive. In addition, there is also mixed temperature at the pressure side of the impeller due to the return of the partial flow heated over the gap between the rotor and stator and the mixing of this flow with the main flow of the pump. The fluid heated in this way travels to the cascade of the impeller in the form of leakage via the sealing ring gap. This results in a temperature increase at the impeller inlet and thus in an adequate increase in the vapor pressure. This increase, however, must be countered by a corresponding static pressure increase on the suction side in order to prevent vapor bubble formation, i.e. cavitation. Put another way, the NPSHR characteristic curve rises towards the ordinates once again.

Fig. 2.66 Recirculation on the suction and pressure side of an impeller operating in the partial load area (close to the ordinates) (relative flow patterns)

Fig. 2.67 shows the computed NPSHR characteristic curves for R 22, NH_3 and water, the latter at 20 °C and 100 °C (refer also to Section 3.2.7 "Heat balance of a canned motor pump"). The calculation takes account of the measured NPSHR curve for water at 20 °C as well as the influences of the fluid physical properties ρ, c, p_v, the heat loss of pump and motor and the leakage flows from the pressure to the suction side. In this respect, the steep rise of the characteristic curves in the vicinity of the ordinates compared to the values for water at 20 °C is conspicuous.

In this example the friction loss characteristic line of the suction pipeline should exhibit the path of the H_{vs} (Q) curve entered in the diagram.

Assuming fluids R 22 and NH_3 are stored in the boiling condition, we get

$$Ze_{min} = NPSHR + H_{vs} \ , \ \text{where} \ \frac{v_e^2}{2} = 0.$$

The respective friction losses from the pipeline characteristic curve must be added to the values of NPSHR in order to obtain the corresponding value Ze_{min}. Ze_{min} is the minimum suction head from the inlet fluid level to the reference level for fluids which are in boiling

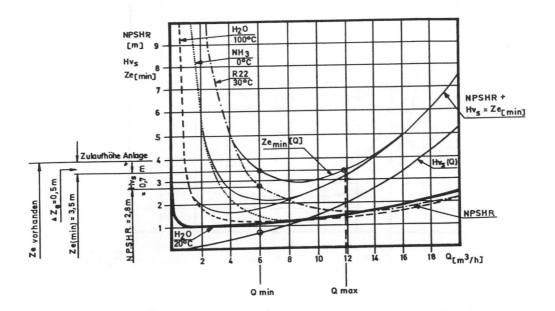

Fig. 2.67 Fluid property dependency of the NPSHR -(Q) characteristic curve for type 32-250 standard hydraulics acc. DIN 24256 and minimum suction heads in boiling condition

condition necessary to prevent cavitation. In the case of horizontally mounted centrifugal pumps, this point is the centre of the pump axis. Further Ze_{min} (Q) curves are thus created for the minimum required suction head. When transporting R22 at a temperature of 30 °C, for example, a suction head

$$Ze_{min} = 2,8 \text{ m} + 0,7 \text{ m} = 3,5 \text{ m}$$

above the vapor pressure head H_v is necessary with a corresponding flow Q of 6 m³/h. However, this is already Q_{min}, the minimum flow rate at which the pump may be operated at this suction head. On the basis of the Ze_{min} (Q) curve, it is possible to read off the necessary minimum flow to be taken off at the pump outlet connection to avoid cavitation. At the same time, however, the maximum flow rate Q_{max} is fixed on the basis of the Ze_{min} -Q curve. In the above example, Q_{max} is 11.8 m³/h.
For reasons of safety, Ze_{min} is to be increased by 0.5 m, with the result that the recommended suction head in this case is 4 m. If Q_{min} cannot always be guaranteed from the process side, it must be secured by means of a bypass line fitted with an orifice. The bypass flow must under no circumstances be returned to the suction line of the pump, but must be transported to the inlet reservoir; otherwise the desired effect of minimum flow is lost.
Fig. 2.68 shows the influence of the flow temperature on the NPSHR (Q)-curve for Freon R22. The lower the temperature, the more favorable the NPSHR values.

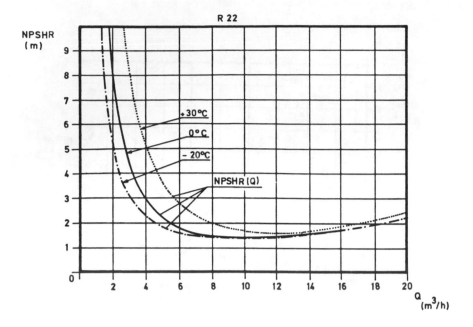

Fig. 2.68 Temperature dependence of the NPSHR (Q) characteristic curve for
Freon R 22

The considerably more rapid rise of the curve for 30 °C is explained by the progression of
the vapor pressure curve of R 22; see Fig. 2.69.

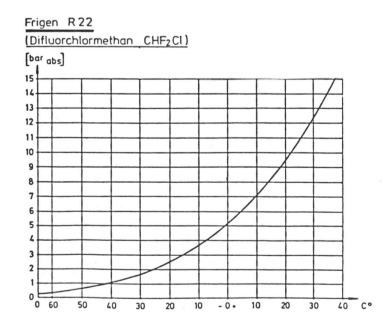

Fig. 2.69 Temperature - vapor pressure curve for Freon R 22

2.9.9.2   Influence of wearing clearance

The clearance geometry of the necessary sealing clearance between the pump casing and impeller also exercises considerable influence on NPSHR. To use the same example, the NPSHR characteristic curves are listed in Fig. 2.70 for R 22 at +30 °C   relative to the radial clearance "S" and thus to the clearance leakage losses Q_v at a gap length L = 12 mm. As can be seen from the characteristic curves, narrower clearances cause lower flow losses Qv and thus result in better NPSHR values. Wider clearances on the other hand impair suction capability. This is particularly the case if the wear rings are worn. As the smallest gap width to be realized on the gap ring is necessarily larger than the existing gap width on the plain bearings, the bearing gap width has an indirect influence on the suction capacity of the pump.

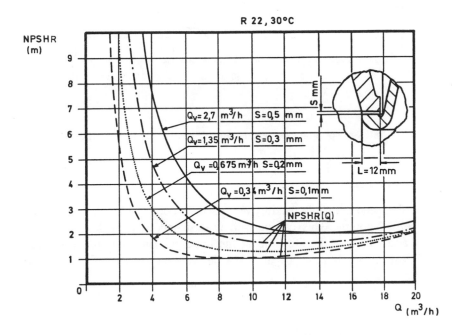

Fig. 2.70 Influence of the clearance leakage Qv on NPSHR for Freon R 22 at 30 °C

2.9.9.3   Influence of motor power loss

Fig. 2.71 shows the influence of the motor power loss on NPSHR (Q) for R 22 at 30 °C. As the motor heat loss is forwarded into the pressure chamber at the impeller discharge of the pump and mixes with the flow, the leakage flow arrives at the suction side at a correspondingly higher temperature/vapor pressure. The NPSHR curve for $P_0 = 0$ is a suction pressure curve such as would be found in a conventional centrifugal pump without the influence of motor heat loss. It can be seen from this curve that the influence of pump heat loss is dominant compared to the motor heat loss.

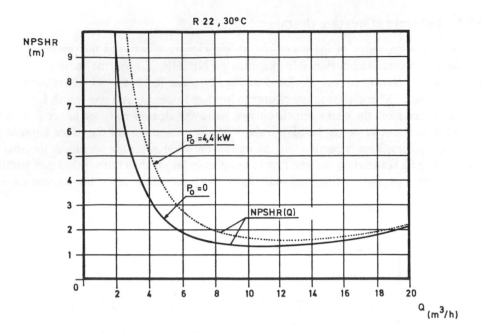

Fig. 2.71 Influence of motor power loss on NPSHR for Freon R 22 at 30 °C

2.9.10 Suction specific speed

The suction specific speed is mainly used by end users to compare the cavitation behavior of various centrifugal pumps. This comparison is designed to provide evidence of the suction capacity and therefore the life time of the impeller. The suction specific speed is defined for the best efficiency point by at a 3% decrease in head simular to the known specific speed, as:

$$S_{qs3\%} = n \cdot \frac{Q_{opt}^{\frac{1}{2}}}{(NPSHR_{opt3\%})^{\frac{3}{4}}} \qquad (2 - 64)$$

where:

Q_{opt}	=	flow rate at best efficiency point
n	=	rated speed of pump
$NPSHR_{opt\ 3\%}$	=	NPSH of pump at operating point Q_{opt} at a 3% head drop

It should be noted that there is not necessarily a relationship to the specific speed n_q, but rather more that $S_{qs3\%}$ is to a particular degree influenced by the design of the impeller inlet geometry. This was confirmed by investigations of Surek upon the suction specific speeds of actual centrifugal pumps relative to specific speeds (Fig. 2.72). From the point of view of the operator, however, high suction specific speeds are preferred.

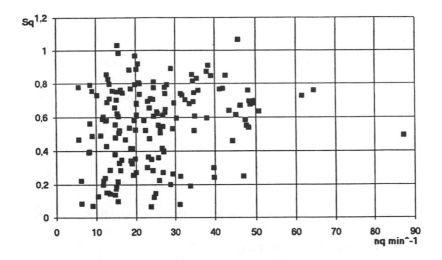

Fig. 2.72 Suction specific speeds of actual single-stage centrifugal pumps relative to specific speed [2-18]

These, however, are not always reliable indicators of the cavitation resistance of an impeller [2-19]. When running under partial load, the design of the blade, particularly the blade inlet angle which has to be correspondingly reduced where the impeller inlet diameter is increased, causes different cavitation processes. Due to reverse flows and the associated formation of vortices, various cavitation processes occur corresponding to the particular blade configuration. This therefore gives rise to the possibility that, despite a lower suction specific speed, lower cavitation wear occurs during the operation of the pump compared to a similar pump with a higher suction specific speed.

2.9.11 Cavitation and materials

At the outset it must be mentioned that there are no materials which are resistant to cavitation damage. There are of course substantial differences in resistance to cavitation between individual materials but in the end they are all attacked by its great stress. The more brittle the material, such as gray cast iron, Si casting glass etc. the more they are damaged by cavitation because brittle materials fatigue more rapidly. It is sometimes necessary to operate pumps close to or above the cavitation limit for reasons of design or price, so that the parts subject to cavitation must be made of materials which withstand cavitation attack for as long as possible. These are mainly high grade steels, laminar ferritic cast steel with a high chrome content and various bronzes and stellites. It is important that the parts exposed to cavitation are kept as smooth as possible, preferably polished. Drawn materials are then usually more resistant than castings. It is also important that for impellers the cascade profiles have a good aerodynamic shape.

2.9.12   Measures for the avoidance of cavitation

A distinction must be made here between measures which the operator can implement and those which can be taken by the manufacturer.

Measures by the <u>operator</u>

- Reduction of geodetic suction lift or increase in suction head.
- Short suction line with the largest possible cross section.
- Valves, bends, curves avoided where possible or the maximum radii used.
- The temperature of the fluid to be kept to a minimum.
- Application of a gas pressure to the surface of the liquid in closed suction or supply vessels.

Measures by <u>pump manufacturers</u>

- Impellers with double curvature blades drawn well forward into the suction orifice
- Avoidance of short deflection radii at the blade cover.
- Reduction of the thickness of the impeller blades.
- Use of a smaller blade inlet angle β_{1s}.
- Reduction in speed.
- In the case of centrifugal pumps setting a larger type number nq for the impeller of the first stage even though it will operate outside the range of best efficiency but will be within the partial load range on the NPSHR curve.
- Fitting an inducer.
- Aligning the flow to the impeller by fitting a guide vane in the inlet connection.

2.10 Examples from conventional pump construction

Fig. 2.73 Standard chemical pump to DIN 24256/ISO 2858 (Lederle)

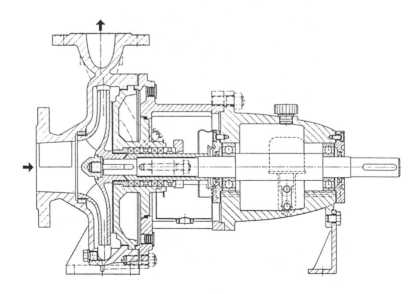

Fig. 2.74 Section view of standard chemical pump to DIN 24256/ISO 2858 (KSB)

Fig. 2.75 Public water supply installation (Lederle)

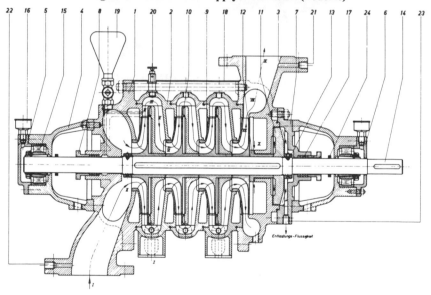

1	Inlet casing	7	Stuffing box housing	13	Round nut	19	Priming funnel
2	Stage	8	Stuffing box	14	Shaft	20	Bleed screw
3	Delivery casing	9	Impeller	15	Roller bearing with washer	21	Gauge connection
4	Bearing housing	10	Diffuser	16	Stauffer lubricator	22	Vacuum gauge connection
5	Bearing housing cover, closed	11	Balance disc	17	Packing	23	Connection for drain valve
6	Bearing housing cover, open	12	Spacer sleeve	18	Cover	24	Thrower

Fig. 2.76 Section view of pump shown in Fig. 2.75 (Lederle)

References for Chapter 2

[2-1]   Collektiv.:   Technisches Handbuch Pumpen (Technical Manual, Pumps),
                          Verlag Technik Berlin, 5th edition 1976

[2-2]   Ackert, I.:   Eulers Arbeiten über Turbinen und Pumpen (Eulers work on turbines and
                          pumps),
                          Special edition Zürich; Ovell Füssli, 1957

[2-3]   Diezel, F.:   Turbinen, Pumpen und Verdichter (Turbines, pumps and compressors),
                          Vogel- Verlag Würzburg 1980

[2-4]   Neumaier, R.:   Handbuch neuzeitlicher Pumpenanlagen (Manual of modern pump systems),
                          Verlag Alfred Schütz Lahr/Schwarzwald 1966

[2-5]   Neumaier, R.:   Seitenkanal-Strömungsmaschinen (Kollektiv) (Side channel flow machines
                          (collective)
                          Verlag und Bildarchiv H.W. Faragallah, Sulzbach/Ts
                          Publisher H.W. Faragallah 1992

[2-6]   Callis, D.:   Ständige Weiterentwicklung-stufenlose Drehzahlregelung von Kreiselpumpen mit
                          Frequenzumformern (Continuous development of stepless speed control of
                          centrifugal pumps using frequency converters) Pumpen (Pumps);
                          Vulkan-Verlag, Essen 1987

[2-7]   Kuchler, G.:   Energieeinsparung bei der Regelung von Kreiselpumpen (Energy saving in the
                          control of centrifugal pumps). Pumpen; Bauelemente der Anlagentechnik
                          (Pumps; Structural elements of system engineering)
                          Vulkan- Verlag, Essen 1987

[2-8]   Standards of Hydraulic Institute New York USA 1955

[2-9]   Bohl, W.:   Technische Strömungslehre (Technical fluid dynamics) Vogel
                          Verlag Würzburg 1980

[2-10]   Neumaier, R.:   Kavitation und Gesamthaltedruckhöhe von Kreiselpumpen mittlerer
                          Schnelläufigkeit (Cavitation and total NPSH of medium-speed centrifugal
                          pumps)
                          Maschinenmarkt Vogel Verlag Würzburg, year 83, vol. 86, October 1977

[2-11]   Schönberger, W.: Untersuchung über Kavitation an radialen Kreiselrädern (Investigation of
                          cavitation on radial impellers)
                          Dissertation Technical University of Darmstadt 1966

[2-12]   Stoffel, B.:   Pumpen und Pumpenanlagen (Pumps and pump systems),
                          Technical Academy Esslingen
                          Course No.: 14535/64.068 1991
[2-13]   Hydraulic Institute, 14th edition 1983

[2-14]   Schiele O.:   Pumpen zur Förderung von Flüssiggasen und Flüssiggasgemischen bei tiefen
                          Temperaturen (Pumps for delivery of liquid gasses and liquid gas mixtures at
                          low temperatures)
                          KSB - Technische Berichte (Technical Reports) 8 (1964)

                          Sichtbarmachung der Strömung in Radialverdichterstufen, besonders der
[2-15]   Fister, W.:   Relativströmung in rotierenden Laufrädern, durch Funkenblitze (Rendering the

flow in radial compressor stations, particularly the relative flow in rotating impellers, visible by flash lighting).
Brennstoffe - Wärme - Kraft (Fuels, Heat, Power) (BWK) vol. 18 (1966) No. 9

[2-16] Krämmer, R.;   Kreiselpumpen und rotierende Verdrängerpumpen hermetischer Bauart (Hermetic
        Neumaier, R.:   centrifugal and rotating displacement pumps)
                            Hermetic-Pumpen GmbH, Company Literature 1982

[2-17] Ochsner, K.:   Der Universal Inducer Entwicklungsstand bei Inducern Für Kreiselpumpen (The
                            Universal Inducer. Development state of inducers for centrifugal pumps)
                            Special edition C 339. D Fa. Ochsner Linz

[2-18] Surek, D.:   Kavitation bei rückwärtslaufenden Kreiselpumpen im Turbinenbetrieb
                            (Cavitation in reverse running centrifugal pumps used as turbines)
                            Aus Pumpen als Turbinen (Kollektiv) (Pumps as turbines) (collective)
                            Publisher: W. H. Faragallah;
                            Verlag und Bildarchiv W. H. Faragallah 1993

[2-19] Stoffel, B.;   Zur Problematik der spezifischen Saugzahl als Beurteilungsmaßstab für die
        Hergt, P.:   Betriebssicherheit einer Kreiselpumpe (The problem of suction specific speed as
                            an assessment criterium for the operating safety of a centrifugal pump)
                            Pumpen (Pumps). 2nd edition.
                            Publisher: G. Vetter Vulkan Verlag Essen 1992

Part II

Hermetic centrifugal pumps

Basic Principles

3. Hermetic centrifugal pumps, machines of the new generation

In the history of modern centrifugal pump design, which spans more than 100 years, the pumps necessary for public utilities and industry have been brought to a high state of technical development. From the point of view of manufacturer and operator the problems of flow, strength and materials have been largely solved. Comprehensive literature on centrifugal pumps provides information on the successful efforts of research and training in conjunction with the pump manufacturing industry to meet all the requirements of pumping applications. One particular problem area for manufacturers is the steadily increasing demand from space technology, the chemical industry, medicine and reactor engineering for fully leak-free hermetic pumps. Further factors are the population's growing awareness of the need for environmental protection and the steadily increasing requirements of legislation.

The original stuffing boxes as shaft seals, which are still largely in use today, have been mainly superseded in problem applications by the use of mechanical sealings. These enable the unavoidable leak losses to be reduced to an acceptable minimum.

But the mechanical seal alone could not, and cannot, guarantee zero leak losses because it depends on a cool lubricating flow and therefore a defined leak flow for its seal. Not until the invention of the canned motor pump (by the Swiss B. Grämiger 1913) (Fig.3.1) and the permanent magnet central coupling (by the brothers Charles and Geoffrey Howard 1949) were the conditions created for the hermetic, leak-free operation of centrifugal pumps.

Although the canned motor pump was invented relatively early, its practical use mainly for the construction of leak-free centrifugal pumps was not recognized by the chemical and nuclear industries until the beginning of the fifties. This was paralleled by the development of the magnetic coupling for leak-free power transmission from the drive motor to the pump.

3.1 Hermetic torque transfer by the canned motor pump/permanent magnet central coupling

What both systems have in common is the magnetic transmission of torque through a rigid thin-walled coaxial casing (Fig. 3.2 and 3.3) (can) of non-magnetic, product-resistant material. The can forms the separation between the fluid and atmosphere and thus provides a hermetic sealing element of the pump unit. In both systems eddy currents are induced when using metallic materials, which leads to a rise in temperature of the can. A partial flow drawn from the pump pressure side provides a cooling flow through the rotor-stator-can gap and is then returned to the main flow for cooling the motor/coupling and removal of heat loss. Where non-metallic materials such as ceramics or carbon fiber - reinforced plastics are used, the heating due to induction is absent but not that caused by rotor friction in the fluid.

KAISERLICHES PATENTAMT

PATENTSCHRIFT
— № 278406 —
KLASSE 21 *d*. GRUPPE 39.

BENJAMIN GRÄMIGER in ZÜRICH, Schweiz.

Vorrichtung zum Antrieb einer in einem Gehäuse vollkommen gasdicht
eingeschlossenen Arbeitsmaschine.

Patentiert im Deutschen Reiche vom 5. März 1913 ab.

AUSGEGEBEN DEN 28. SEPTEMBER 1914.

Die Erfindung bezieht sich auf die an sich bekannten Vorrichtungen zum Antrieb einer in einem Gehäuse eingeschlossenen Arbeitsmaschine durch Gleichstrom, wobei zwecks Vermeidung einer Stopfbüchse der Läufer des Motors mit der anzutreibenden Arbeitsmaschine in das Gehäuse gasdicht eingeschlossen ist, und Läufer und Ständer des Motors durch eine einen Teil des Gehäuses bildende Wandung aus geeignetem Material voneinander getrennt sind. Gemäß der Erfindung ist der Läufer als Kurzschlußanker ausgebildet, während in an sich bekannter Weise im geeignet gewickelten Ständer vermittels umlaufender Bürsten, die ihren Antrieb von einer kleinen Hilfsmaschine erhalten, und eines feststehenden Kollektors durch Gleichstrom ein Drehfeld erzeugt wird, das den Läufer mitnimmt und in Drehung versetzt. Eine solche Vorrichtung eignet sich insbesondere zum Antrieb des Verdichters einer kleinen Eismaschine, wenn dieser Verdichter in einem Gehäuse vollkommen gasdicht eingeschlossen ist.

In allen Fällen, in denen es sich für notwendig oder wünschenswert erweist, eine in einem Gehäuse eingeschlossene Arbeitsmaschine unter Vermeidung einer Stopfbüchse durch einen Elektromotor mit gasdicht voneinander getrenntem Rotor und Stator anzutreiben, ist ein Kurzschlußanker deshalb besonders zweckmäßig, weil ein solcher gar keiner Stromzuführung von außen bedarf und keine Isolation enthalten muß. In der Regel wird nämlich der Raum, in welchem der Läufer

sich bewegen soll, von Gasen, Dämpfen oder Flüssigkeiten erfüllt sein, und diese üben schädliche und zerstörende Wirkungen auf die Isolationsmaterialien aus. Zudem wäre ein Läufer mit Bürsten und Kollektor in dem vollkommen unzugänglichen, abgeschlossenen Raume praktisch nicht zu verwenden, selbst wenn jene schädlichen chemischen Wirkungen fehlen würden. Im weiteren wäre es schwierig, die den Strom zuführenden und unter Betriebsspannung stehenden Drähte gasdicht in das Gehäuse einzuführen. Die Verwendung eines üblichen Gleichstrommotors ist also für solche Fälle praktisch ausgeschlossen. Es soll nun aber gemäß der Erfindung auch bei Gleichstrombetrieb als Läufer ein Kurzschlußanker verwendet und im Ständer durch Gleichstrom ein Drehfeld erzeugt werden.

Zur näheren Erläuterung diene ein Ausführungsbeispiel, welches in den Fig. 1 und 2 dargestellt ist. Fig. 2 insbesondere zeigt schematisch die Wicklung des Ständers und die Verbindung derselben mit den Kollektorlamellen und den umlaufenden Bürsten. Eine in einem Gehäuse *g* eingeschlossene, nicht eingezeichnete Arbeitsmaschine soll durch Gleichstrom angetrieben werden. Der als Kurzschlußanker ausgebildete Läufer *r* ist mit der Welle *W* der anzutreibenden Maschine verkeilt. Zwischen ihn und den Ständer *s* ist eine glockenförmig geschlossene Hülse *h* geschoben. Diese ist bei *A* mit dem Ständer *s* gasdicht verbunden, und dieser selbst ist bei *B* gasdicht an das Gehäuse *g* angeschlossen.

Die Wicklung des Ständers ist hier als Ringwicklung angenommen. Der feststehende Kollektor *k*, von dessen Lamellen aus die Stromzuführungen zu den aufeinanderfolgenden Spulen der Wicklung des Ständers führen, steht von innen in Kontakt mit zwei um 180° versetzten Kohlenbürsten b_1, b_2, welche in Taschen eines aus nicht leitendem Material hergestellten und auf der Welle w_1 sitzenden Stückes *i* stecken und durch die Wirkung der Federn f_1, f_2 und durch ihre eigene Fliehkraft angepreßt werden. Die Bürste b_1 ist durch das biegsame Kabel d_1 mit dem Schleifringe c_1, die Bürste b_2 durch das biegsame Kabel d_2 mit dem Schleifringe c_2 leitend verbunden. Den beiden Schleifringen c_1, c_2 wird Gleichstrom durch die feststehenden Bürsten b_3, b_4 zugeführt bzw. abgenommen. Die Welle w_1 erhält ihren Antrieb von dem kleinen Nebenschlußmotor *m*. Die Umdrehungszahl des im Ständer erzeugten Drehfeldes ist abhängig von der Drehzahl dieses Motors.

Der Strom nimmt beispielsweise folgenden Weg. Von der Bürste b_3 geht der aus irgendeiner Quelle stammende Gleichstrom an den Ring c_1, sodann durch das Kabel d_1 zur umlaufenden Bürste b_1 und von dieser zu derjenigen Lamelle des feststehenden Kollektors *k*, auf der sie augenblicklich gleitet. Von dieser Lamelle fließt der Strom in die Wicklung des Ständers *s*, verteilt sich dort in zwei Teilströme und kehrt bei der erstgenannten, gegenüberliegenden Lamelle des feststehenden Kollektors in die zweite umlaufende Bürste b_2, sodann durch das Kabel d_2 in den zweiten Schleif-

ring c_2, von diesem in die zweite feststehende Bürste b_4 und endlich wieder zur Stromquelle zurück. Weil von der umlaufenden, stets gleichpoligen Bürste b_1 der Strom der Reihe nach in die aufeinanderfolgenden Lamellen des feststehenden Kollektors *k* tritt, wandern Stromeintritt und gleichermaßen der Stromaustritt fortlaufend auf der Ständerwicklung im Kreise herum, und damit dreht sich auch das durch diesen Gleichstrom erzeugte magnetische Feld, welches den Kurzschlußanker unter Schlüpfung mitnimmt.

PATENT-ANSPRUCH:

Vorrichtung zum Antrieb einer in einem Gehäuse vollkommen gasdicht eingeschlossenen Arbeitsmaschine, z. B. des Kompressors einer Eismaschine, durch Gleichstrom, wobei zwecks Vermeidung einer Stopfbüchse der Läufer des Elektromotors mit der Arbeitsmaschine in das Gehäuse eingeschlossen ist und Läufer und Ständer durch eine einen Teil des Gehäuses bildende Wandung aus geeignetem Material gasdicht voneinander getrennt sind, dadurch gekennzeichnet, daß der Läufer als Kurzschlußanker ausgebildet ist, welcher von einem Drehfelde mitgenommen wird, das in an sich bekannter Weise vermittels umlaufender, von einer kleinen Hilfsmaschine angetriebener Bürsten und eines feststehenden Kollektors im geeignet gewickelten Ständer durch Gleichstrom erzeugt wird, der den umlaufenden Bürsten zugeführt wird.

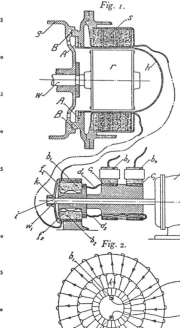

Fig. 1.

Fig. 2.

Fig. 3.1 First patent No. 278406 of a canned motor pump from the Imperial German Patent Office, by Benjamin Grämiger from the year 1913

In contrast to the magnetic clutch, there is a higher generation of heat in the canned motor because of the added electrical losses. What both types have in common is that the bearings are mounted in the product space and must therefore be able to withstand the conditions imposed by the fluid passing through. The bearings are normally designed as plain bearings although in a few cases they consist of roller bearings.

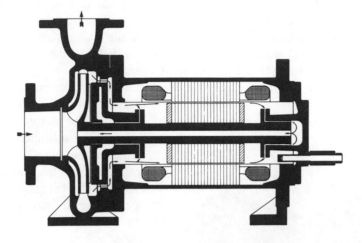

Fig. 3.2 Diagram of a canned motor pump

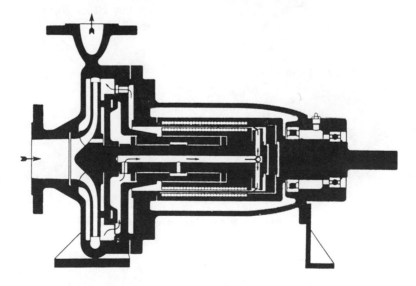

Fig. 3.3 Diagram of a magnetic coupling pump

These days the majority of bearing materials are ceramic. Extreme hardness and corrosion resistance, good running properties together with the resistance to cyclic thermal stress mean these materials are used almost universally. The disadvantage, however, is their susceptibility to shock and impact. The problem of mixed friction which occurs during the startup and run-down phases is readily solved by use of such materials.

3.2 The three phase canned motor [3-1]

Pumps with a rotating drive system which operate without a shaft seal, i.e. leak-free and needing a direct drive, must form a common unit with the drive motor so that no rotating parts pass through the common casing to the outside (Fig. 3.2), because technically there is no seal between two spaces under different pressures if a rotating shaft passes through. This means that parts under voltage are close to the operating fluid. The problems posed by this are solved by the canned motor. A normal three-phase squirrel cage motor with a squirrel cage rotor has a thin, tubular lining of high grade stainless steel or nickel alloy in its stator bore (Fig. 3.11) with the ends sealed or welded into the relevant end bearing covers. Fig. 3.4 to 3.11 show the development of a canned motor.

The can is the decisive part of the machine which turns the pump motor into a hermetic

unit and gives the complete assembly its name. The motor has to fulfill two tasks. It must impart the necessary torque to the pump and also form the sealing unit of the complete machine. Provided the corrosion characteristics of the pumped liquid permit, the rotor can rotate in the liquid-filled rotor space without a protective casing. In most cases, however, a corrosion proof casing is also provided to protect against attack from the liquid (Fig. 3.8 and 3.9).

Fig. 3.4 Rotor blank

Fig. 3.5 Rotor blank pressed onto shaft

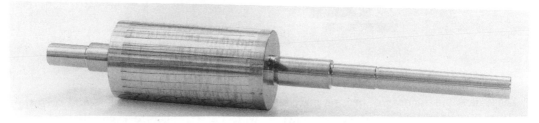

Fig. 3.6 Machined rotor and shaft

Physically, the can represents a stationary conductor in the rotating field in which an electric current is induced, whose strength depends on the field strength, intersection speed, wall thickness and conductivity. Magnetically, the can and sheathing of the rotor represent an increase in the air gap which increases the no-load current. The rotating field of the machine generates eddy-current losses in the can. The rotating magnetic field of the three-phase induction motor intersects the rotating and the fixed can and exerts an EMF on this. Metallic sleeves of this kind are, however, good current conductors so that a current is generated in them which increases their temperature. In a stator sleeve of non-magnetic high grade steel with a wall thickness of 1 mm, heat losses of 10 to 15% of the useful output power can be expected due to eddy currents. This is the main reason why canned motors have a lower efficiency than normal motors. If reinforcing rings are fitted for stability reasons to the free ends outside the stator lamination of the can, further eddy currents are generated which can,

Fig. 3.7 Rotating can with endplates for rotor prodection

however, be substantially reduced by providing an insulting layer between the stator sleeve and reinforcing ring [3-2].

Fig. 3.8 Rotor, encased (corrosion- proof)

Stainless steels, codes 1.4571 and 1.4506, or molybdenum alloys with a high grade nickel content (Hastelloy B or C) are the materials used mainly for cans.
The latter is particularly suitable because it not only has a very low iron content of not more than 3% but it is also very resistant to corrosion. Hastelloy alloys also have a higher electrical resistance than stainless steels of the aforementioned material groups.

Table 3.1 Electrical resistances

Material	Resistivity $\left[\frac{\Omega \cdot mm^2}{m}\right]$
Stainless steel 4571 und 4506	0,75
Hastelloy B	1,35
Hastelloy C	1,39

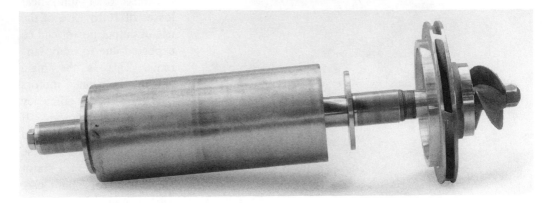

Fig. 3.9 Complete rotor

The thickness of the can and of the rotor jacket determine the size of the magnetic gap and therefore of the losses. Every effort is therefore made to keep the wall thickness to a minimum. The reduction of this wall thickness is, however, governed by the following two factors:

      I The internal pressure of the liquid
      II The resistance to corrosion

Re I

Cans are exposed to the internal pressure (~ P_2 = outlet pressure of pump) of the operating fluid. The critical stress from the pressure is felt in the free part between the stator lamination and the support in the casing. Although the resistance to hydraulic pressure can be increased using support rings, this makes for a more expensive machine and increased eddy current losses.

Re II

Canned pumps are used mainly to convey difficult, aggressive liquids and the cans used must be chosen to suit the pumped fluid. However, they do not always guarantee complete corrosion resistance, and wear must be allowed for. This is worthy of note because the highest temperature in a pump unit (for non externally-cooled motors) occurs on the can. The thickness of the can walls is usually between 0.5 and 1 mm.

Fig. 3.10 Stator with steel casing

Fig. 3.11 Stator assembly fitted with can

3.2.1 Operating chracteristics of the canned motor

The fitting of the can substantially affects the operating characteristics of the motor. To understand this requires an explanation of the functioning of electrical machines. The asynchronous machine consists , if the description is limited to the so-called "active part", of the magnet yoke and the primary and secondary windings. This arrangement can be compared with the construction of a transformer. In contrast to the transformer, the magnet yoke of the asynchronous motor is in two parts, namely, the fixed part which is the stator and rotating part which is the rotor. The stator normally contains the primary winding. On asynchronous motors the secondary winding in the rotor is usually a short-circuited squirrel-cage winding.

magn. Fluß = magnetic flux

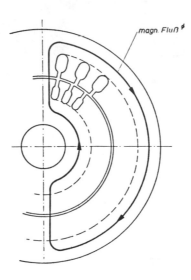

Fig. 3.12 Magnetic flux in a three-phase squirrel-cage motor

There has to be an air gap between the stator and rotor, in contrast to the transformer, so that the rotor can turn. The magnetic flux necessary to form the torque of the machine "flows" through the magnetic circuit as shown schematically in Fig. 3.12. To maintain this flux (which is necessary to develop the torque) the motor draws reactive power from the supply. The level of this reactive power depends on the strength of the flux and the magnitude of the reluctance of the motor. The reluctance of the air is a multiple of the reluctance of the iron laminations. Because a high reactive power consumption in the machine causes electric heat and therefore a deterioration in operation it is advisable to keep the reactive power demand to a minimum. Apart from careful dimensioning of the path in the iron laminations, the air gap of the machine should also be kept to a minimum. The minimum air gap is essentially determined by mechanical factors (installation and measuring tolerances, flexing of the shaft), but nevertheless should be extremely small for electrical reasons (harmonic moments) [3-3].

106

In practice the air gap of two-pole machines ranges between 0,25 and 0,8 mm in the 0,1 to 30 kW performance range.

If cans are now arranged between the stator and rotor (Fig. 3.13.2), the diameter of the rotor has to be decreased corresponding to the thickness of the can compared with a normal machine and the stator bore diameter increased. The material of the cans (as already explained) must meet the following requirements:

I   High mechanical stability
II High chemical resistance
III High magnetic and electrical resistance

The requirements of I and II are derived from the function of the pump, whilst the property required under III depends on the electrical data of the machine. This requirement is further explained in the following.

If the can had been made of a material with good magnetic conducting properties (ferromagnetic), this would have acted as a "magnetic shield" on the stator and rotor. The magnetic circuit would no longer close over the rotor, as shown in Fig. 3.12, but instead over the can.

This increased scatter reduces the breakdown torque and the rated speed leading to poor operating properties and degraded operation of the machine. To avoid, this the cans are made of non-magnetic steel.

Eddy current losses in the cans are unavoidable, particularly in the fixed cans. The rotary field of the machine causes a voltage to be induced in the can, the level of which depends on the density of the magnetic field (induction). So-called eddy currents which cause heat losses are induced in the can due to these voltages. To minimize the eddy current losses, the can should be thin and the magnetic density kept to an absolute minimum. The can material must have a high resistivity. The can losses depend on the following variables:

$$V_{can} \sim \frac{n_s^2 \cdot L \cdot D^3 \cdot s \cdot B_{\delta m}^2}{\rho}$$

where

n_s	=	synchronous speed
L	=	length of stator
D	=	diameter of can
s	=	thickness of can wall
$B_{\delta m}$	=	air gap induction
ρ	=	resistivity of can

If one considers the can losses in the design of the machine, it would be seen that with the same "active volume" of the machine a rotor diameter which is as small as possible has to be chosen and therefore the length of the machine must be increased.

This tendency (small diameter and correspondingly long machine) also has a beneficial effect on the magnitude of the liquid friction of the rotor as this rises approximately to the fifth power of the rotor diameter (at a given speed), whilst the machine length has only a linear effect on the friction losses.

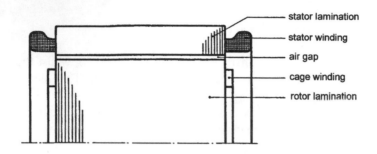

— stator lamination
— stator winding
— air gap
— cage winding
— rotor lamination

Fig. 3.13 Normal three-phase squirrel-cage motor

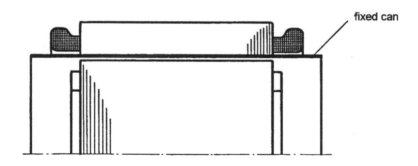

fixed can

Fig. 3.13.1 Three-phase squirrel-cage motor with fixed can

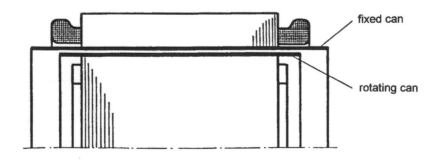

fixed can

rotating can

Fig. 3.13.2 Three-phase squirrel-cage motor with fixed and rotating can

108

3.2.2 Design of the canned motor

When dimensioning the canned motor, the peculiar features of this type must be taken into account. Whilst the limit of utilization of normal machines is determined by the permissible current density of the conductors and the still-acceptable magnetic flow density in the back and teeth of the magnetic circuit, the limit in the case of the canned motor is in the can losses which can be dissipated. This means that magnetically the canned motor gives less yield than a normal motor. A certain measure for the utilization of a machine is the flux density in the magnet gap, the so-called air gap induction. For normal, correctly operated machines this ranges from 0,5 to 0,7 Tesla and for canned motors from 0,4 to 0,5 Tesla. In the following example (3-A) the dimensions and losses of normal two-pole motor and a canned motor, both of explosion-protected design (Ex d), are compared (Table 3-II). Due to the higher friction and eddy current losses, the efficiency of the canned motor is only 81.3% compared with 88.4% for the normal motor.

Table 3-II for example 3-A

			280M-2 Ex-d motor	N80V-2h Canned motor
Measurements and distribution of losses	Nominal output	kW	90	81
	Nominal current at 380V	A	170	170
	Iron losses	kW	4,2	1,1
	Can losses	kW	--	9
	Friction losses	kW	1,6	3,3
	Copper losses, stator	kW	3,4	3,4
	Copper losses, rotor	kW	1,8	1,8
	Total losses	kW	11	18,6
	Power factor		0,91	0,89
Dimensions	External diameter of casing	mm	322	300
	Stator bore	mm	187	158,4
	Air gap	mm	1	2,2
	Length of iron	mm	210	380
	Active volume	dm³	21,5	34,2
	Air gap induction	T	0,64	0,47

The larger magnet gap on the canned motor means a reduction in the power factor from 0,91 to 0,89 and therefore a slightly higher reactive-power absorption. Where the current levels are the same and the electrical consumption is almost the same, the canned motor requires an active iron volume which is approximately 60% higher than that of the normal motor, (i.e. 34,2 dm³ instead of 21,5 dm³).

The essential changes with the canned motor compared with the normal motor relative to the load curves are investigated in the following.

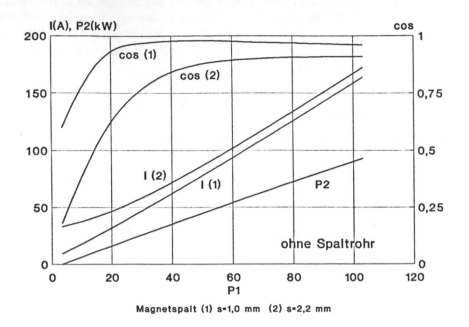

Magnetspalt (1) s=1,0 mm   (2) s=2,2 mm

ohne Spaltrohr = without can ; Magnetspalt = magnet gap

Fig. 3.14 Load curve of a canned motor in accordance with Table 3-II without a can and liquid, with a normal (1) and increased magnet gap (2)

Fig. 3.14 shows the effect on the load curve of the canned motor shown in Table 3-II from an increase in the magnet gap from 1 mm to 2,2 mm, as whereby it is assumed that the motor is first operated without a can or liquid. Essentially the no-load current of the motor increases, its power factor corresfoningly decreases. Under no-load conditions the motor requires a higher reactive power to magnetize the increased magnet gap. At the nominal duty point, however, this influence is masked by the higher active-power input of the motor and the power factor then reaches 0,89 which is almost equal to the value for the normal motor, i.e. 0,91.

The change in the load curve brought about by introducing the can and the influence of the liquid friction is shown in Fig. 3.14.1. The eddy currents induced in the can and the rotor liquid friction shifts the shaft power curve P_2 almost parallel downwards, i.e. the output power P_2 of the motor reduces by a constant amount independent of motor load. The reason why the canned motor has almost constant losses over the complete load curve is explained by the fact that the eddy current and friction losses together normally make up the main proportion of the overall losses of the canned motor. The losses of the motor are essentially determined by the no-load losses. The no-load state of the motor in Fig. 3.14.1 is determined by the point at which the downwards-shifted "power lines" intersect the abscissa.

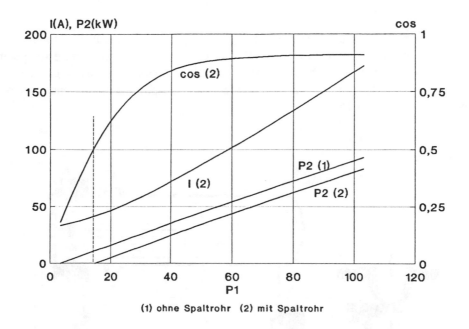

(1) ohne Spaltrohr (2) mit Spaltrohr

ohne Spaltrohr = without can ; Magnetspalt = magnet gap

Fig. 3.14.1 Load curve of motor in accordance with Fig. 3.14 with a can and rotor in liquid

The foregoing example should serve as a guide for the most common case where a canned motor is developed from a standard series of machines. It should be noted that in this case the nominal power is determined by the motor heating. The moments (starting and stalling torques) can be ignored during dimensioning because motors this large are generously sized. The limit on temperature rise is set by the insulating material, the permissible winding temperatures are specified in VDE 0530. The temperature rise depends on the cooling conditions if a specific power loss is used as a basis, which in the case of the canned motor means it depends particularly on the temperature and type of the pumped liquid because the largest part of the power loss is dissipated through this. Therefore, the temperature of the pumped liquid flowing through the rotor-stator gap must be lower than that which the winding can take. To ensure this where the liquid temperatures are higher, the insulation class of the winding of the canned motor is frequently H or C (silicon-ceramic), whereas normal machines are usually manufactured to insulation class F.

111

VDE 0530 specifies the following permissible winding temperatures for these insulation classes:

Insulation class

F 155 °C
H 165 °C
C 220 °C
C 400 °C

If a pumped medium is used whose temperature is not substantially less than the permissible temperature limit of insulation class H or C, or even above these temperature limits, the motor losses must be dissipated through a separate cooling circuit so that a permissible winding temperature can still be maintained despite the high temperature of the pumped medium (Fig. 3.15).

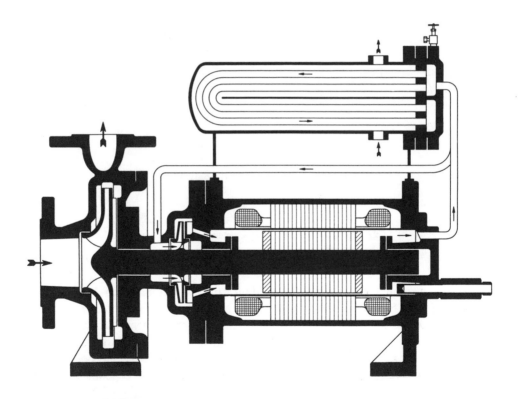

Fig. 3.15 Canned motor pump with externally-cooled motor

3.2.3 The use of canned motors in hazardous locations. Explosion protection according to European Standard "EN"

In cases in which canned motors must be operated in hazardous locations, the statutory provisions as expressed in national and European directives apply; supervision in such cases is carried out in Germany by the Physikalisch-Technische Bundesanstalt (PTB; Federal Institute for Physics and Technology) in Braunschweig. The German and European explosion protection standards stipulated for these motors are contained in the following DIN standards:

- DIN EN 50 014:   Electrical equipment for
                                      hazardous locations,
                                      general regulations

- DIN EN 50 018:   Electrical equipment for
                                      hazardous locations,
                                      flameproof enclosure "d"

- DIN EN 50 019:   Electrical equipment for
                                      hazardous locations,
                                      increased safety "e"

These standards are also included in the directives as a VDE regulation in terms of VDE 0022 under the numbers VDE 0170/0171, Section 1/5.78, VDE 0170/0171, Section 5/5.78 and VDE 0170/0171, Section 6/5.78. The national regulations formerly valid according to VDE 0170/0171, Section 1/12.70, were made ineffective as of 01.05.78.

Canned motors already certified in accordance with the provisions of VDE 0171 may continue to be manufactured and operated without restrictions.

European Standard EN 50014 to 50020
EEx de II C T 1 6
Key:

- E   Motor complies with European standard
- Ex   Explosion protection
- de   Combined type of protection "flameproof enclosure" and
               "increased safety"
- II   Group of electrical equipment which cannot be used
               in mines susceptible to firedamp
- C   The highest class regarding the the maximum
               experimental safe gap (MESG)for type of protection Ex d;
               this is suitable for all gases and vapors;
               limit safe gap width MESG < 0,5 mm
- T1...6   Temperature class for the maximum surface temperature
               for the respective application

Table 3 - III

Temperature class	Maximum surface temperature °C
T1	450
T2	300
T3	200
T4	135
T5	100
T6	85

The lowest ignition temperature of the explosive atmosphere in question must be higher than the maximum surface temperature of the motor.

According to the new type of protection Ex e for the rotor chamber, the temperature increase of the rotor must be taken into account. The most hazardous accident in this context occurs when the impeller is blocked at operating temperature. In this case, the rotor reaches its maximum permissible limit temperature as stipulated by the applicable temperature class within the so-called t_E time.

The t_E time of standard Ex e motors is known. It stipulates that the motor must be protected by a thermally delayed excess-current release which releases within the t_E- time specified on the nameplate.

Therefore, depending on the starting current ratio I_A/I_N , minimum values as defined in EN 50015 must be observed (Fig. 3.16).

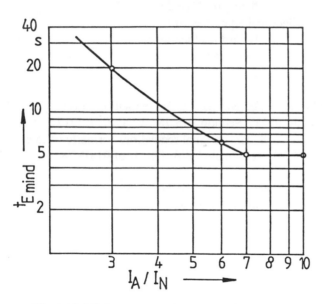

Fig. 3.16 Minimum values for time t_E of motors relative to starting current ratio I_A/I_N

The surface temperature of the rotor must therefore be measured when the rotor is blocked, and the temperature of the pumped medium added to this value. To ensure the observance of the regulations for explosion protection and for supervisory purposes, the legislature has enforced additional safety guidelines for canned-motor pumps. In the test certificates issued by the Physikalisch-Technische Bundesanstalt PTB (Federal Institute for Physics and Technology), conditions are stipulated which must be met for the operation of such pumps in hazardous locations.

These are as follows:

- For reasons pertaining to safety, the rotor chamber must always remain filled with pumped liquid. For this purpose, liquid-level sensors or different measures of at least equal effectiveness must be installed in the system to ensure that the motor is operated only at sufficient liquid levels.
- The coolant/lubricant flow must be monitored by temperature sensors to prevent maximum temperatures from being exceeded. It must be ensured that the temperature of the coolant/lubricant does not exceed ....°C.

3.2.3.1 Level monitoring

To fulfill the requirements of PTB (Federal Institute for Physics and Technology), a level temperature monitoring device was specially developed (Fig. 3.17) which consists of a thermostat, a level sensor and a switching amplifier. A liquid expansion or resistance thermometer is used as a thermostat. The limit temperatures stipulated in the technical specifications of the corresponding canned-motor pumps can be set in the terminal box of the unit using a thermometric scale. Fig. 3.18 shows the circuit diagram for the level/temperature monitoring device and a so-called sensor-controlled double can monitoring system, which may also be installed.

The Niveaustat (level monitoring device) consists of a float equipped with a magnet which moves on a tube in which a reed contact is installed. When the liquid level rises or falls, the reed contact is activated by a magnet. The switching amplifier is designed to provide "intrinsic safety" (EExia) IIC/IIB protection for ambient temperatures of up to 50 °C.

An opto-electronic monitoring system (Fig. 3.19 and 3.20) may also be installed instead of the level monitoring device described above. Opto-electronic sensors are used primarily for low liquid densities ($\rho < 0.5$ kg/dm^3) and high operating pressures ($p > 25$ bar).

The level and temperature monitoring devices for canned-motor pumps are also highly recommended for applications in which no regulations are enforced with regard to explosion protection. Level monitoring ensures that the pumps can be operated only when the liquid level is adequate. For liquids which are pumped at temperatures near the boiling point (such as liquid gas) the level monitor should be installed at Ze_{min} + 0,5 m in the suction line in order to ensure sufficient pressure at the impeller inlet and to protect the pump against damage (cavitation)

where

Ze_{min}	=	NPSHR + Hvs
NPSHR	=	Net positive suction head required [m]
Hvs	=	Frictional resistance in suction line [m]

Level Monitor N 30

up to 100°C, Type N 30.1

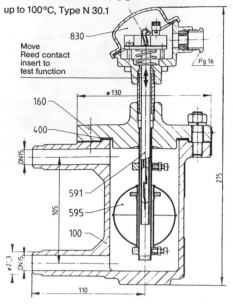

Move
Reed contact
insert to
test function

for 100°C to 370°C
Type N 30.2

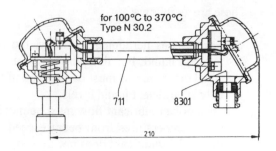

100	Float Housing
160	Cover with guide tube
400	Gasket ⌀ 73 ⌀ 87 x 2
591	Reed Contact
595	Float
830	Connection head N 30.1
830.1	Connection head N 30.2
711	Extension piece

Thermostat T 30

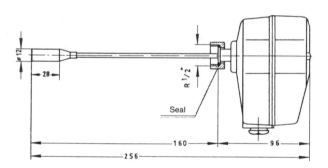

Switching Amplifier S 30

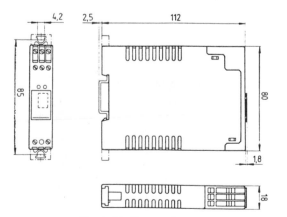

Mechanical data

Dimensions:	18 x 18 x 112 mm
Fixation:	Connection bar or installation plate
Weight:	~ 120 g
Position:	vertical, ex-terminals below
Place of installation:	outside the ex-range
Connection terminals:	self-opening for conductor cross section 1 x 0,1 – 1,5 mm² fine-wired with endsleeves for conductors

Protection acc. to DIN 40050 (IEC 529)	
Casing	IP 30
Connection terminals	IP 20
Range of operating temperature	– 20 ... + 60°C
Range of bearing temperature	– 40 ... + 70°C
Relative humidity in the liquid (no wetting)	95%

Fig. 3.17 Electronic level and temperature monitoring device (Hermetic)

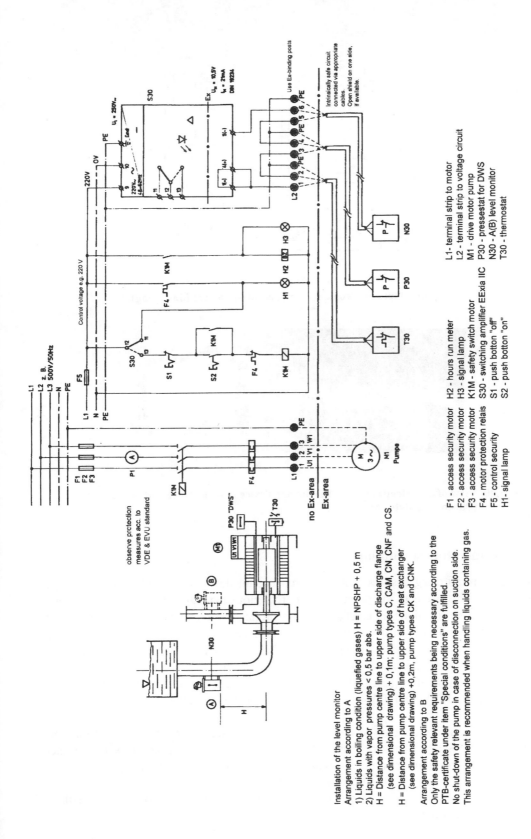

Installation of the level monitor
Arrangement according to A
1) Liquids in boiling condition (liquefied gases) H = NPSHP + 0,5 m
2) Liquids with vapor pressures < 0,5 bar abs.
H = Distance from pump centre line to upper side of discharge flange
    (see dimensional drawing) + 0,1m; pump types C, CAM, CN, CNF and CS.
H = Distance from pump centre line to upper side of heat exchanger
    (see dimensional drawing) +0,2m, pump types CK and CNK.

Arrangement according to B
Only the safety relevant requirements being necessary according to the
PTB-certificate under item "Special conditions" are fulfilled.
No shut-down of the pump in case of disconnection on suction side.
This arrangement is recommended when handling liquids containing gas.

F1 - access security motor   H2 - hours run meter   L1- terminal strip to motor
F2 - access security motor   H3 - signal lamp   L2 - terminal strip to voltage circuit
F3 - access security motor   K1M - safety switch motor   M1 - drive motor pump
F4 - motor protection relais   S30 - switching amplifier EExia IIC   P30 - pressestat for DWS
F5 - control security   S1 - push botton "off"   N30 - A(B) level monitor
H1- signal lamp   S2 - push botton "on"   T30 - thermostat

Fig. 3.18 Installation of the integrated electronic pump monitoring device (Hermetic)

117

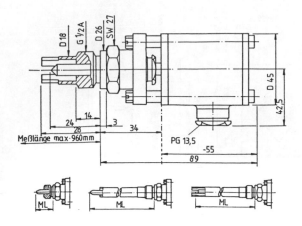

Fig. 3.19 Opto-electronic sensor (Phoenix-Analytec Marketing)

Fig. 3.20 View of opto-electronic level monitoring device in accordance with Fig. 3.19
(Phoenix-Analytec Marketing)

3.2.3.2 Temperature monitoring

Temperature monitoring ensures that the pump is shut down in case of excessively high temperatures. This protects the pump from more serious damage. The temperature monitoring device also functions as a reliable indicator that the pump is operating within the permissible range. Temperature monitoring can also be provided in the form of a maximum-minimum circuit to prevent the pump from being started before the fluid in the pump has reached its correct viscosity, e.g. when the solidification point of the fluids pumped is above ambient temperature (e.g. fatty acids). Motor protection may also be provided by thermistors (PTC 170) or a PT 100 resistance thermometer installed in the windings.

Before any type of motor can be supplied as an "explosion-proofed" motor, it must be tested by the PTB (Physikalisch-Technische Bundesantalt - Federal Institute for Physics and Technology) to ensure compliance with the regulations. The motors are provided with a test certificate after the test has been successfully passed (Fig. 3.21). The manufacturer is obliged to construct each unit according to the tested unit and to carry out exactly defined routine check tests on each unit.

Physikalisch-Technische Bundesanstalt

Certificate of Conformity

PTB No.

This certificate is issued for the electrical apparatus:

Canned motor type

manufactured and submitted for certification by:

Hermetic-Pumpen GmbH
D-7803 Gundelfingen

This electrical apparatus and any acceptable variation thereto is specified in the Schedule to this Certificate and the documents therein referred to.

Physikalisch-Technische Bundesanstalt being an Approved Certification Body in accordance with Article 14 of the Council Directive of the European Communities of 18 Dec. 1975 (76/117/EEC) confirms that the apparatus has been found to comply with the harmonized European Standards.

Electrical Apparatus for Potentially Explosive Atmospheres

EN 50 014:1977 (VDE 0170/0171 Part 1/	) General Requirements
EN 50 018:1977 (VDE 0170/0171 Part 5/	) Flameproof Enclosure "d"
EN 50 019:1977 (VDE 0170/0171 Part 6/	) Increased Safety "e"

and has successfully met the examination and test requirements that are recorded in the confidential test report.

The apparatus marking shall include the code:

EEx de IIC T

The supplier of the electrical apparatus referred to in this certificate has the responsibility to ensure that the apparatus conforms to the specification laid down in the Schedule to this certificate and has satisfied routine verifications and tests specified therein.

This apparatus may be marked with the Distinctive Community Mark specified in Annex II to the Council Directive of 6 Febr. 1979 (79/196/EEC). A facsimile of this mark is printed on this certificate.

Physikalisch-Technische Bundesanstalt   Braunschweig,
Original signed by

(Dr.-Ing. U. Engel)
Regierungsdirektor

PHYSIKALISCH-TECHNISCHE BUNDESANSTALT 11

Prufbescheinigungen ohne Unterschrift und ohne Dienststempel haben keine Gultigkeit.
Die Bescheinigungen durfen nur unverandert weiterverbreitet werden.
Auszuge oder Änderungen bedurfen der Genehmigung der Physikalisch-Technischen Bundesanstalt, Bundesallee 100, Postfach 33 45, D-3300 Braunschweig.

Fig. 3.21 Unofficial translation of Certificate of Conformity (test certificate for canned motors) (Hermetic)

3.2.4 Test rig for canned motors

In order to apply a load test to the motor independent of pump the canned motor is provided with a shaft gland and coupled to a fluid-friction dynamometer (Fig. 3.22). The heat dissipating partial flow Q_T which is passed through the stator rotor gap is simulated by an auxiliary pump and control valves in the inlet and outlet lines. The torque M at the shaft can be determined by the indication on the pendulum level and length of the lever arm and the speed n (r.p.m) can be determined using an inductive pickup. The motor output power can be determined from this as follows:

$$P_2 = M \cdot n \cdot \frac{\pi}{30} \qquad (3-1)$$

The power P_1 taken by the canned motor, and also the current and voltage, can be determined in a known manner using electric power meters. The values of the efficiency (η_{mo}) and the power factor ($\cos\varphi$) are then obtained from these. At the same time, thermocouples are used to determine the temperatures on the surface of the motor, in the partial inlet and outlet flows and in the winding. The motor limit rating is determined either by the maximum permissible winding temperature or, in the case of explosion-proof canned motors, by the maximum permissible surface temperature of the motor.

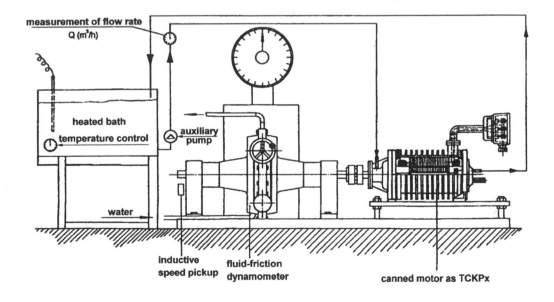

Fig. 3.22 Test rig for canned motors

3.2.5 Heat loss and its dissipation

Losses occur in the complete pump unit which have to be borne by the canned motor in addition to the nominal output. These power losses are converted to heat and must be dissipated. In the pump these losses occur as flow and conversion losses, disc friction losses and internal leakage losses and also as losses necessary to convey partial flow for lubrication and cooling the motor bearings and to remove the greater part of the motor heat losses as well as maintaining the balance of the axial thrust. All these losses are expressed by the pump efficiency η_p. In the motor, mechanical and hydraulic losses occur in addition to electrical losses. The electric losses are made up of the specific total and eddy current losses in the stator, rotor and can together with the copper losses in the stator and rotor. The mechanical losses are due to bearing friction and the hydraulic losses due to the friction losses of the rotor in the liquid. All the losses that are borne by the electric motor are expressed by the efficiency of the motor η_{mo}. The overall efficiency η_{ges} of a unit is therefore calculated from:

$$\eta_{ges} = \eta_p \cdot \eta_{mo} \qquad (3 - 2)$$

The efficiency of a canned motor is determined on a special brake test rig (refer to Section 3.2.4 "Test rigs for canned motors").

If this is known, the pump efficiency can be determined using a plotted calibration curve. The heat generated in the motor due to its losses passes mainly into the lubricating cooling flow Q_T and only a small proportion is dissipated by radiation through the motor casing. The amount of heat dissipated by radiation can be determined by the brake test rig. It represents the difference between the heated dissipation throughout the cooling circuit and the total amount of heat loss. The amount of radiated heat $Q_{WStr.}$ is greater the higher the temperature difference between the cooling flow and room temperature and the higher the permitted temperature rise in the cooling flow. Limits are set in this case to allow for regulations for explosion-proof protection which specify the temperature rise (refer to 3.2.3 "The use of canned motors in hazardous locations").
The radiated heat determined from brake tests amounts to $5 \div 20\%$ of the overall losses of the motor, depending on the operating conditions. It therefore represents a specific factor of the overall heat losses of the motor. If this factor is taken to be equal to f, the heat losses of the motor are therefore:

$$Q_{WVM} = \frac{Q_{WT}}{f} \qquad (3 - 3)$$

where:

Q_{WVM} = overall heat losses of motor
Q_{WT} = amount of heat in partial flow Q_T

121

For canned motors the f values fluctuate between 0,8 and 0,95.

The amount of heat Q_{WT} dissipated in the cooling flow Q_T is equivalent to one part of the overall performance loss. One kWh corresponds to a heat amount of 3600 kJ. If the amount of heat dissipated from the partial flow each hour is now divided by 3600, we get the power loss N_{QT} taken from the partial flow.

(In contrast to the power P the power loss should be designated N).

$$N_{QT} = \frac{Q_{WT}}{3600\frac{s}{h}} \text{ kW} \tag{3 - 4}$$

Q_{WT} is obtained from

$$Q_{WT} = G_T \cdot c \cdot \Delta T_T \text{ kJ/h} \tag{3 - 5}$$

where

G_T	=	volume rate of flow kg/h
c	=	the mean specific heat of the pumped liquid kJ/kg $\cdot$ K and
ΔT_T	=	temperature difference between the inlet and outlet cooling flow K

Example (3-B) should make this clear:

The following values are measured on the motor brake test rig on a canned motor with an active power of 1 kW:

$$P_{output} = 1 \text{ kW} = P_2$$
$$P_{input} = 1,61 \text{ kW} = P_1$$
$$\eta_{mo} = 0,62$$

Temperature of pumped liquid	T	=	14,5	°C		
Room temperature	T_R	=	18,5	°C		
Pumped liquid		=	water			
Density	ρ	=	1000	kg/m³		
Mean specific heat	c	=	4,18	kJ/kg $\cdot$ K		
Partial flow	Q_T	=	2	l/min	=	0,12 m³/h
Temperature difference between the inlet and outlet of the partial flow	ΔT_T	=	4	K		

122

We therefore get:
power loss of motor

$$N_{mo} = P_1 - P_2 \qquad (3 - 6)$$
$$N_{mo} = 1,612kW - 1kW$$
$$N_{mo} = 0,612kW$$

This power loss is dissipated through the cooling flow Q_T and the radiation of the motor. The total amount of heat to be dissipated is therefore:

$$Q_{WM} = Q_{WT} + Q_{WStr} \qquad (3 - 7)$$
$$Q_{WM} = N_{mo} \cdot 3600s/h$$
$$Q_{WM} = 0,612kW \cdot 3600s/h$$
$$Q_{WM} = 2203,2 \ kJ/h$$

The amount of heat in the partial flow Q_T is:

$$Q_{WT} = Q_T \cdot \rho \cdot c \cdot \Delta T_T \qquad (3 - 8)$$
$$Q_{WT} = 0,12m^3/s \cdot 1000kg/m^3 \cdot 4,18kJ/kgK \cdot 4K$$
$$Q_{WM} = 2006,4kJ/h$$

and therefore the power losses of the partial flow is:

$$N_{QT} = Q_{WT}/3600s/h \qquad (3 - 9)$$
$$N_{QT} = 2006,4kJ/h/3600s/h$$
$$N_{QT} = 0,557kW$$

The difference between the total amount of heat to be dissipated and the amount of heat of the partial flow is the amount of radiated heat Q_{WStr}.

therefore:
$$Q_{WStr} = Q_{WM} - Q_{WT} \qquad (3 - 10)$$
$$Q_{WStr} = 2203,2 \ kJ/h - 2006,4 \ kJ/h$$
$$Q_{WStr} = 196,8kJ/h$$

The radiated power loss N_{VStr} can be determined from this.

We get this:

$$N_{Str} = N_{mo} - N_{QT}$$

$$N_{Str} = 0,612 - 0,557$$ (3 - 11)

$$N_{Str} = 0,055 kW$$

The amount of radiated heat therefore corresponds to 9 % of the overall heat loss of the motor.

3.2.6 Pumping liquid near to boiling point

As we have seen in the previous section, the partial flow Q_T is heated by the motor losses. Within its circuit it expands from the head H_d to the suction pressure H_s. To avoid cavitation with boiling liquids, the pressure in the orifice of the impeller must be greater than the vapor pressure of the pumped liquid by the amount of the pressure corresponding to NPSH. A specific temperature is assigned to each vapor pressure. Therefore the temperature of the cooling flow cannot be increased regardless, because vapor which can lead to decreased output of the pump and finally to complete stoppage of delivery sometimes occurs at the inlet into the suction chamber which is in the area of the lower pressure. Example (3-C) shows the particular characteristic of pumping boiling liquids. Take, for example, a canned motor pump used to pump butane with the following data:

Example (3-C)

Head	H	=	60 m	=	3,48 bar
Density	ρ	=	580 kg/m³		
Specific heat	c	=	2,35 kJ/kg·K		
Temperature of air	Te_{air}	=	20 °C		
Temperature of butane	Te_{Butan}	=	20 °C		
Number of stages		=	3		
Partial flow	Q_T	=	5 l/min	=	0,3 m³/h
Factor	f	=	0,9		
Input power	P_1	=	6,25 kW		
Motor efficiency	η_{mo}	=	0,76		
Vapor pressure	p_V	=	2,4 bar abs		

Assuming that the pump is installed so that it has available a suction head H_{zul} = NPSHR plus the friction losses H_{vs}.

124

The rise in temperature of the partial flow Q_T is calculated from the power loss of the motor N_{mo}. This is the difference between the input and output motor power.

$$P_2 = P_{input} \cdot \eta_{mo} = 6,25kW \cdot 0,76 = 4,75 \quad kW \qquad (3 - 12)$$

and therefore the power loss

$$N_{mo} = P_1 - P_2 = 6,25kW - 4,75kW = 1,5 \quad kW \qquad (3 - 13)$$

The power loss dissipated from the partial flow Q_T is lower than the overall power loss by the amount of radiated heat.

Thus

$$N_{QT} = N_{mo} \cdot f = 1,5kW \cdot 0,9 = 1,35 \qquad kW \qquad (3 - 14)$$

therefore the heat loss of the partial flow

$$Q_{WT} = N_{QT} \cdot 3600\frac{s}{h} = 1,35kW \cdot 3600\frac{s}{h} = 4860 \quad kJ/h \qquad (3 - 15)$$

Thus the rise in temperature of the partial flow

$$\Delta T_T = \frac{Q_{WT}}{Q_T \cdot c \cdot \rho} = \frac{4860}{0,3 \cdot 2,35 \cdot 580} = 11,9 \quad K \qquad (3 - 16)$$

If the partial flow were to be fed back to the suction chamber of the first impeller it would evaporate because its temperature would rise from 20 °C to 32 °C because the boiling pressure in the inlet orifice of the pump is 2,4 bar abs + NPSHR · ρ · g . The partial flow must therefore be fed into the space with the higher pressure, or at least into a space in which there is a pressure corresponding to the vapor pressure p_v bei 32 °C + NPSHR · ρ · g . Fig. 3.23 shows the boiling curve for butane relative to temperature and vapor pressure. If the partial flow is now returned to the pressure space behind the first impeller, i.e. into the suction chamber of the second impeller, evaporation would also occur there because the delivery pressure after the first impeller is equal to the vapor pressure of the liquid and the second impeller must also have an inlet pressure increased by NPSHR · ρ · g in its inlet orifice. It cannot be assumed that on the short path from the outlet from the pump shaft to the impeller blade tips a correspondingly lower mixed temperature will be established at the high flow velocities, which would preclude the formation of vapor.

Only the return of the partial flow into the pressure chamber behind the second impeller guarantees adequate safety against the occurrence of vaporization. In this case it is even possible to still allow an increase in the temperature of the partial flow. If, for example, the pump requires an NPSHR 5 m = 0,29 bar, the temperature of the partial flow Q_T could be increased to 37,5 °C.

If the temperature is raised from T_{T2} to T'_{T2} = 37.5 °C, the partial flow could be reduced from Q_T to Q'_T, which would improve the overall efficiency.

For single-stage types of this pump the partial flow must be drawn off after passing the bearing furthest from the pump and returned to the inlet tank via a separate pipe. Pumps for liquids which boil easily should therefore always be of multistage construction unless a design in accordance with Fig. 2.64 (Section 2.9.9.1 "Dependency on the physical properties of the fluid and on the temperature, particularly for canned motor pumps and magnetically-coupled pumps"). In this case a secondary impeller in the rotor chamber again returns the partial flow Q_T into this from the pressure space via the stator-rotor gap.

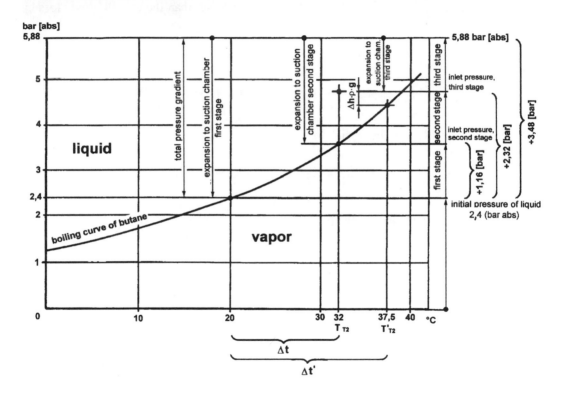

Fig. 3.23 Vapor pressure-temperature curve for butane

126

3.2.7 Heat balance of a canned motor pump

If the power loss N is applied to a liquid flow Q, it causes a temperature rise

$$\Delta T = \frac{N}{\rho \cdot c \cdot Q} \quad K \qquad (3 - 17)$$

In this case the following dimensions are to be allowed for:

$$
\begin{aligned}
\rho &= kg \cdot m^{-3} \\
c &= J \cdot kg^{-1} \cdot K^{-1} \\
N &= W \\
Q &= m^3 \cdot s^{-1}
\end{aligned}
$$

and the relationship

$$
\begin{aligned}
1\ W &= 1\ J \cdot s^{-1} \\
1\ J &= 1\ kg \cdot m^2 \cdot s^{-2}
\end{aligned}
$$

3.2.7.1 Temperature rise of the main delivery flow Q_{ges}

The main delivery flow Q_{ges} must allow for both the pump power loss and the motor heat loss. The rise in temperature of the delivery flow is then

$$\Delta T_{ges} = \frac{N_p + N_{mo}}{\rho \cdot c \cdot Q_{ges}} \qquad (3 - 18)$$

The pump power loss is obtained from

$$N_p = P_p - P_{hydr} = P_p - \eta_p \cdot P_p \qquad (3 - 19)$$

$$N_p = [1 - \eta_p] \cdot P_p$$

In this case P_p is the input power of the pump at the shaft.

The power loss of the motor is obtained as follows:
Firstly the motor output power P_2 is equal to the pump input power.

$$P_2 = P_p$$

and the motor input power.

$$P_1 = \frac{1}{\eta_{mo}} \cdot P_2$$

we therefore get the following for the motor power loss

$$N_{mo} = P_1 - P_2 = P_p \cdot \left[\frac{1}{\eta_{mo}} - 1 \right] \qquad (3 - 20)$$

Therefore by substituting (3-19) and (3-20) in (3-18) we get the temperature rise of the main delivery flow

$$\Delta T_{ges} = \frac{P_p}{[\rho \cdot c \cdot Q_{ges}]} \cdot \left[[1 - \eta_p] + \left[\frac{1}{\eta_{mo}} - 1 \right] \right] \qquad (3 - 21)$$

3.2.7.2 Temperature rise of motor partial flow Q_T

Because the motor partial flow is branched off on the pressure side of the pump it is already heated up by ΔT_{ges} compared with the suction side. In passing through the motor it suffers a further temperature rise by

$$\Delta T = \frac{N_{mo}}{[\rho \cdot c \cdot Q_T]} = \frac{P_p}{[\rho \cdot c \cdot Q_T]} \cdot \left[\frac{1}{\eta_{mo}} - 1 \right] \qquad (3 - 22)$$

i.e. in total compared with the inlet temperature on the suction side

$$\Delta T_{QT} = \Delta T_{ges} + \Delta T \qquad (3 - 23)$$

$$\Delta T_{QT} = \frac{P_p}{[\rho \cdot c \cdot Q_{ges}]} \cdot \left[[1 - \eta_p] + \left[\frac{1}{\eta_{mo}} - 1 \right] \right] + \frac{P_p}{[\rho \cdot c \cdot Q_T]} \left[\frac{1}{\eta_{mo}} - 1 \right]$$

$$\Delta T_{QT} = \frac{P_p}{[\rho \cdot c \cdot Q_{ges}]} \cdot \left[[1 - \eta_p] + \left[\frac{1}{\eta_{mo}} - 1 \right] \right] + \frac{Q_{ges}}{Q_T} \cdot \left[\frac{1}{\eta_{mo}} - 1 \right]$$

If the relationship is defined as $n = \frac{Q_{ges}}{Q_T}$, we get

$$\Delta T_{QT} = \frac{P_p}{[\rho \cdot c \cdot Q_{ges}]} \cdot \left[[1 - \eta_p] + [n + 1] \cdot \left[\frac{1}{\eta_{mo}} - 1 \right] \right] \qquad (3 - 24)$$

The pump output P_p can be expressed as follows by the pump delivery data

$$P_p = \frac{[\rho \cdot g \cdot H \cdot Q_{ges}]}{\eta_p} \qquad (3 - 25)$$

with the dimensions

$$\rho \;\; = \;\; kg \cdot m^{-3}$$
$$g \;\; = \;\; 9,81 \; m \cdot s^{-2}$$
$$H \;\; = \;\; m$$
$$Q \;\; = \;\; m^3 \cdot s^{-1}$$
$$P_p \;\; = \;\; W$$

If P_p is used in accordance with this relationship in (3-21), we get:

$$\Delta T_{ges} = \frac{g \cdot H}{[c \cdot \eta_p]} \cdot \left[[1 - \eta_p] + \left[\frac{1}{\eta_{mo}} - 1 \right] \right] \qquad (3 - 26)$$

The total temperature rise between the suction and pressure side is therefore:

$$\Delta T_{ges} = \frac{g \cdot H}{c} \cdot \left[\left[\frac{1}{\eta_p} - 1 \right] + \frac{1}{\eta_p} \left[\frac{1}{\eta_{mo}} - 1 \right] \right] \qquad (3 - 27)$$

Thus when P_p is substituted in equation (3-24) we get the following for the temperature rise of the partial flow of the motor

$$\Delta T_{QT} = \frac{g \cdot H}{[c \cdot \eta_p]} \cdot \left[[1 - \eta_p] + [n + 1] \cdot \left[\frac{1}{\eta_{mo}} - 1 \right] \right] \qquad (3 - 28)$$

The maximum temperature rise of the motor partial flow relative to the inlet temperature on the suction side is therefore:

$$\Delta T_{QT} = \frac{g \cdot H}{c} \cdot \left[\left[\frac{1}{\eta_p} - 1 \right] + \frac{1}{\eta_p} \cdot \left[\frac{1}{\eta_{mo}} - 1 \right] \cdot [n + 1] \right] \qquad (3 - 29)$$

The variables g, H and c are to be substituted in formulae (3-27) and (3-29) with the following values:

$$g \;\; = \;\; 9,81 \; m \cdot s^{-2}$$
$$H \;\; = \;\; m$$
$$c \;\; = \;\; J \cdot kg^{-1} \cdot K^{-1}$$

Example (3-D):

The following values may be used as an example for a canned motor pump:

$$Q \quad = \quad 40 \quad m^3 \cdot h^{-1}$$
$$H \quad = \quad 46,4 \quad m$$
$$Q_T \quad = \quad 0,6 \quad m^3 \cdot h^{-1}$$
$$c \quad = \quad 4,99 \cdot 10^3 \quad J \cdot kg^{-1} \cdot K^{-1}$$
$$\eta_p \quad = \quad 0,59$$
$$\eta_{mo} \quad = \quad 0,75$$

For this we get for $\quad n = \dfrac{Q_{ges}}{Q_T} = 66,66$

and therefore the temperature rise of the motor partial flow relative to that on the suction side in accordance with (3-29).

$$\Delta T_{QT} = \frac{9,81 \cdot 46,6}{4,99 \cdot 10^3} \cdot \left[\left[\frac{1}{0,59} - 1 \right] + \frac{1}{0,59} \cdot \left[\frac{1}{0,75} - 1 \right] \cdot [66,66 + 1] \right]$$

$$\Delta T_{QT} = 3,55K$$

The total temperature rise between the suction and pressure side in accordance with (Formula 3-27) is:

$$\Delta T_{ges} = \frac{9,81 \cdot 46,6}{4,99 \cdot 10^3} \cdot \left[\left[\frac{1}{0,59} - 1 \right] + \frac{1}{0,59} \cdot \left[\frac{1}{0,75} - 1 \right] \right]$$

$$\Delta T_{ges} = 0,115K$$

HEAT BALANCE OF CANNED MOTOR PUMP

CNF 50/200

Impeller diameter	195.00	mm
Q-Optimum	50.00	m3/h

Performance curve for water:

Q m3/h	H m	P kW	NPSH m
0.0	52.0	4.50	1.20
25.0	50.0	7.00	0.50
50.0	43.0	9.50	1.20

AGX8.5

No load losses	P0	1.31	kW
Load parameter	A1	0.0138	
Motor cooling flow	Q9	0.60	m3/h

NH3

Temperature		50.00	grd	
Density		0.56	kg/1	
specific heat		4.99	kJ/kgK	
Vapor pressure		20.30	bar abs	
Vapor pressure acc. VDI-Atlas Da8				
calculated with 2 points:		50.00	grd	20.30 bar abs
		30.00	grd	11.70 bar abs

RESULTS

Q m3/h	H m	P kW	N1 kW	N2 kW	QT m3/h	DT grd	DHT m	Hmin m	DTges grd	DTS grd	NPSH* m
5.0	52.0	2.8	2.4	1.4	0.6	4.0	38.9	52.0	1.0	0.4	4.3
10.0	51.8	3.1	2.3	1.4	0.6	3.6	34.4	51.8	0.5	0.1	1.4
15.0	51.4	3.4	2.2	1.5	0.6	3.5	33.2	51.4	0.3	0.1	0.6
20.0	50.8	3.6	2.1	1.5	0.6	3.4	33.0	50.8	0.2	0.0	0.3
25.0	50.0	3.9	2.0	1.5	0.6	3.5	33.1	50.0	0.2	0.0	0.1
30.0	49.0	4.2	2.0	1.6	0.6	3.5	33.5	49.0	0.2	0.0	0.1
35.0	47.8	4.5	1.9	1.6	0.6	3.5	34.0	47.8	0.1	0.0	0.1
40.0	46.4	4.8	1.9	1.6	0.6	3.6	34.6	46.4	0.1	0.0	0.2
45.0	44.8	5.0	2.0	1.7	0.6	3.7	35.3	44.8	0.1	0.0	0.4
50.0	43.0	5.3	2.0	1.7	0.6	3.8	36.1	43.0	0.1	0.0	0.7

Notation:

N1	Pump power losses
N2	Motor power losses
QT	Motor cooling flow
DT	Temp. increase of motor cooling flow above suction temperature
DHt	Increase of vapor pressure head by DT
Hmin	Minimum head at reentrance of motor cooling flow
DTges	Total temperature increase between suction- and discharge side
DTS	Increase of suction temperature because of backflow
NPSH*	modified NPSH-required due to heat balance

Date:

06-03-1996

Example (3-D): Heat balance of canned motor pump

3.2.8 Friction losses on the rotor

Due to the friction moments which occur, a rotor rotating in a fluid is subject to an additional load P_r. This friction power loss affects the economy of the entire unit. It depends on the rotor diameter $D = 2 \cdot R$, its length L, the gap S between the rotor and casing, the distance between the rotor faces and the casing (refer to Fig. 3.24), the speed n and the kinematic viscosity of the surrounding fluid.

The rotor imparts a rotary motion to the adjacent fluid whose circumferential component v_u at the rotor corresponds to the particular peripheral velocity $u = \omega \cdot r$ on radius r and whose velocity at the casing wall is zero, so that the surrounding components of the velocity profile form a shear flow (refer to Fig. 3-24). This produces a linear distribution of the velocities conforming to a Couette flow only where there are narrow gaps and correspondingly small Reynolds numbers. The velocity components of a through flow, e.g. a Poiseulle or a secondary flow, e.g. Taylor vortex (refer to Fig. 3-25) can be overlaid vertical to the peripheral components of the shear flow.

The input power of the rotor is:

$$P_r = \omega \cdot [M + M_a] \qquad (3 - 30)$$

In this case is the angular velocity ω, M is the friction moment of the cylinder gap and M_a the friction moment of the outer front surfaces of the rotor at a distance a from the casing.

The friction moment of the outer front surfaces M_a can be calculated using the dimensionless coefficient of torque

$$C_{Ma} = \frac{M_a}{[0,5 \cdot \rho \cdot \omega^2 \cdot R^5]} \qquad (3 - 31)$$

The torque coefficient depends on the shape of the shear flow between the outer front surfaces of the rotor and the casing. The shape of the shear flow can be determined using the Re_u number formed by the peripheral speed U and radius R of the rotor. Depending on the value of this Re_u number

$$Re_u = \frac{U \cdot R}{\nu} = \omega \cdot \frac{R^2}{\nu} \qquad (3 - 32)$$

a laminar (L) or turbulent (T) shear flow is established with the boundary layers of the rotating front surfaces and stationary casing passing directly into each other (case S) only where there is a small relative clearance a/R. Larger relative clearances mean, however, that the boundary areas are distinctly separate (case G). The following empirical relationships are known (taken from [3-4]):

Re_u	$<10^3$	10^3 bis $5\cdot10^5$	$5\cdot10^5$ bis $7\cdot10^7$	$>5\cdot10^5$
a/R	$<0{,}015$	$>0{,}015$ $=1{,}0$	$<0{,}04$	$>0{,}04$ $=1{,}0$
Fall	LS	LG	TS	TG
C_{Ma}	$\dfrac{2\pi}{Re_u\cdot(a/R)}$	$\dfrac{3{,}7(a/R)^{0{,}1}}{Re_u^{0{,}5}}$ $3{,}7/Re_u^{0{,}5}$	$\dfrac{0{,}08(a/R)^{\bar{0}{,}25}}{Re_u^{0{,}25}}$	$\dfrac{0{,}0102(a/R)^{0{,}1}}{Re_u^{0{,}2}}$ $0{,}0102/Re_u^{0{,}2}$

Table 3-IV

The friction moment of the surface of the rotor cylinder jacket can be determined from a torque coefficient.

$$C_M = \frac{M}{[0,5 \cdot \rho \cdot \pi \cdot \omega^2 \cdot R^4 \cdot L]} \tag{3 - 33}$$

The shape of the flow in the annular gap between the rotor-jacket surface and the casing is also determinant for this torque coefficient. The following are used to identify the flow shapes and stability limits (Fig. 3.26 and 3.27):

Taylor number $\qquad Ta = \dfrac{\omega \cdot R \cdot s}{\upsilon} \cdot \sqrt{\dfrac{s}{R}} \tag{3 - 34}$

and where there is an axial gap flow, a

Reynolds number $\qquad Re = D_h \cdot \bar{v}_x/\upsilon = 2s \cdot \bar{v}_x/\upsilon \tag{3 - 35}$

with the hydraulic diameter D_h
and with the mean axial velocity of the gap flow $\bar{v}_x$

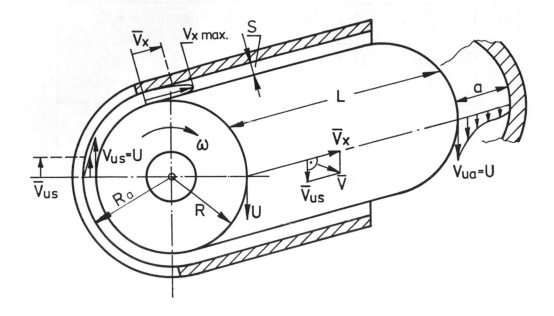

Fig. 3.24 Flows and velocities in the gaps and at the front faces of the rotor

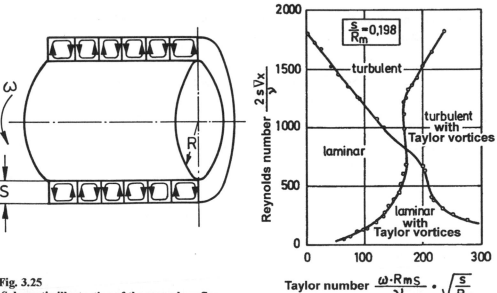

Fig. 3.25
Schematic illustration of the secondary flow in a concentric annular gap with a rotating inner cylinder (from [3-4])

Fig. 3.26
Forms of flow and stability limits for a concentric annular gap with axial through flow, with a rotating inner cylinder and a relative gap where s/Rm = 0.198 (from [3-4]).

Fig. 3.27 shows the dependence of the torque coefficient on the Taylor number for a Couette flow in an annular gap, i.e. a shear flow without an axial through flow. The range in which Taylor vortices are to be expected in laminar flow is approximately between Taylor numbers 40 and 400. A laminar Couette flow without vortices is found at Ta < 40. A turbulent shear flow can be expected at Taylor numbers Ta > 400.

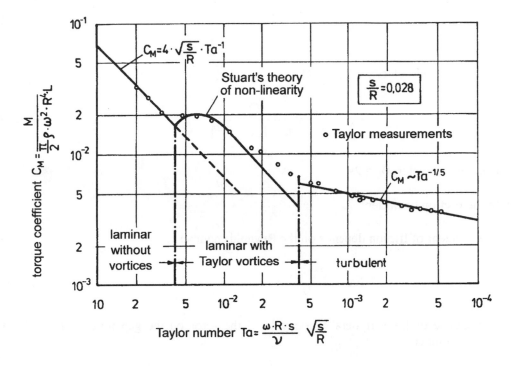

Fig. 3.27 Torque coefficient C_M for a pure Couette flow in a concentric annular gap with a relative gap s/R = 0.028 (from [3-7]).

From the torque coefficient for the face ends and cylinder jacket surface of the rotor we therefore get a load for the overall friction moment of the rotor of:

$$P_r = \frac{\rho}{2} \cdot \omega^3 \cdot R^4 \cdot [C_{Ma} \cdot R + C_M \cdot \pi \cdot L] \qquad (3 - 36)$$

where:

ρ	fluid density
ω	angular velocity of rotor
$R = D/2$	0,5 times pipe diameter
C_{Ma}	torque coefficient (rotor face ends)
C_M	torque coefficient (rotor cylinder jacket surface)

Where there is a pressure gradient, a through flow with an axial velocity component overlays the Couette flow in the annular gap. Where there is a stable laminar flow condition the velocity profile of the axial component corresponds to a Poiseulle flow. This superimposi-

135

tion of a pure Couette flow by a pure Poiseulle flow is known as a "mixed Poiseulle-Couette flow". Results of experiments show a clear influence of the rotation on the gap flow Q_s by the annular gap [3-4].

The gap flow Q_s causes a pressure drop in the gap over the length L in the axial direction

$$\Delta p = \lambda \cdot \frac{L}{D_h} \cdot \frac{\rho}{2} \cdot \bar{v}_x^2 \qquad (3 - 37)$$

with a mean axial velocity (Fig. 3-24)

$$\bar{v}_x = \frac{Q_s}{A_s} = \frac{Q_s}{\pi \cdot [R_a^2 - R^2]} \qquad (3 - 38)$$

a hydraulic diameter $D_h = 2s$

with the friction coefficient λ

The coefficient of friction depends on the Reynolds number

$$Re = \frac{2s \cdot \bar{v}_x}{\upsilon} \qquad (3 - 39)$$

The influence of the peripheral velocity U of the rotor on the gap flow is shown by a Reynolds number

$$Re_{sU} = \frac{2s \cdot U}{\upsilon} \qquad (3 - 40)$$

The influence of these Reynolds numbers on the coefficient of friction is shown in Fig. 3.28. The hatched area in Fig. 3.28 marks the increase in the coefficient of friction due to Taylor vortices in laminar flow. Experiments by Stampa [3-5] show that where turbulent flow forms the rotation does not influence the gap flow where $U/\bar{v}_x < 1$.

Regardless of the relative gap width s/R the coefficient of friction in the range

$$Re_{sU} > 10^4$$

The coefficient of friction can be approximately determined by the following equation

$$\lambda = 0,26 \cdot Re^{-0,24}\left[1 + \left[\frac{7}{16}\right]^2 \cdot \left[\frac{U}{\bar{v}_x}\right]^2\right]^{0,38}$$

(3 - 41)

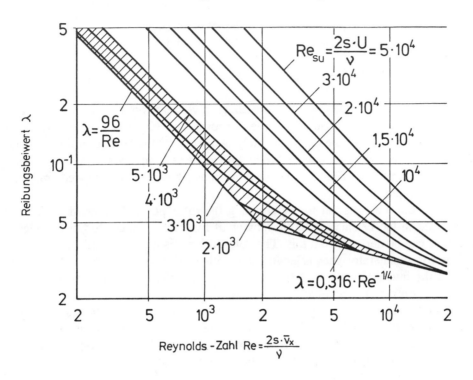

Reibungsbeiwert   = coefficient of friction

Reynolds-Zahl   = Reynolds number

Fig. 3.28 Results of measurements by Stampa [3-5] for the friction coefficient λ of an axial-flow, concentric annular gap with a rotating inner cylinder (from [3-4])

A rotor tested on the motor brake rig has, for example, the following data:

Example (3-E)

Length of rotor	L	= 322 mm		
Diameter of rotor D = 2R		= 191,5 mm		
Width of gap	s	= 1,35 mm		
Speed	n	= 2900 r.p.m		

With the measured
and the data for
and of the

gap current	Q_s	= 2,5 m³/h	
dinematic viscosity	ν	= 100 mm²/s	1 mm²/s
density	ρ	= 1 kg/dm³	1 kg/dm³

we obtain the following values:

$$\text{Reynolds number } Re = 23{,}1 \quad - \quad 2038$$
$$Re_{sU} = 393 \quad - \quad 39000$$

Rotor power loss due to the friction moment
$$Pr \text{ in kW} \quad = 23{,}6 \quad - \quad 3{,}3$$

These results agree well with the measured values.

It can be seen from the example given that the effect of viscosity on the rotor power loss is considerable. The higher power loss at higher viscosities is, however, found only in the start-up phase. After a short time (approximately two seconds) the heating of the relatively small gap flow Q_s produces a pronounced reduction in the viscosity thus reducing the rotor power loss Pr accordingly. The starting viscosity must, however, be allowed for when dimensioning the motor.

It can be seen from the equation for determining the rotor power loss Pr that the rotor diameter is used in the calculation to the power of five. The demand for the slenderest possible rotors can be derived from this. The rotor volume is a measure of the achievable power. If the rotor and stator pack is lengthened, with a reduction in the outer diameter of the rotor and the internal diameter of the stator, a higher efficiency can be achieved with the same power output. The amount to which the rotor diameter can be reduced is, however, limited because this is accompanied by an increase in the magnetic losses. To obtain efficient canned motors it is therefore necessary to optimize the length-diameter ratio of the rotor. Fig. 3.29 shows friction power loss of a rotor for a 100 mm length relative to rotor diameter and the kinematic viscosity of the pumped liquid.

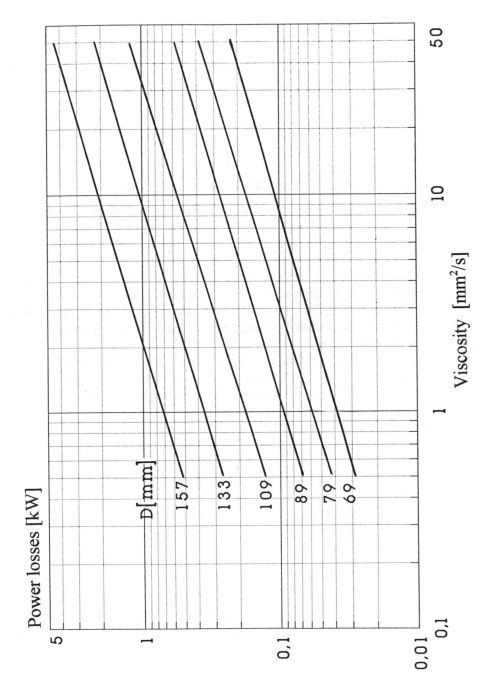

Fig. 3.29 Friction losses of rotors as a function of their diameter and the viscosity of the partial flow Q_T for a length of 100mm where n = 2900 r.p.m.

139

3.2.9 Radial forces thrust at the rotor

Due to the fact that their bearings lie in the product, canned motor centrifugal pumps are extremely sensitive to radial or axial forces, as the lubricating effect of most fluids is extremely small. These factors must taken into account in the design of these types of pumps and the decentralizing forces must be determined.

The radial forces acting on the rotor are shown in Fig. 3.30. Centralizing and decentralizing forces must be in equilibrium. The main radial forces are the hydraulic force F_r and the weight of the rotor F_G. F_r is however only relevant for pumps with blade-less volute or annular casings, whilst with multistage pumps with diffusers for pressure conversions or pumps with volute casings and an integrated diffuser device it is either absent or only present to a slight degree.

The force due to weight G of the rotor can be determined simply by weighing and the hydraulic force F_r must be determined by calculation.

3.2.9.1   Radial force F_r for volute casing pumps

To gain an impression of the radial forces which occur, we must look at the flow conditions at the circumference of the impeller and in the downstream flow equipment. The following applies to the duty point Q_{Opt} (point of best efficiency = design point) of a volute housing pump: the discharge from the impeller with a rotating angle of 360 ° has an adequate entry cross section into the downstream flow guide devices with the result that identical flow velocities are present at the impeller circumference and at the entry into the flow guide devices (Fig. 3.31 and 3.32). This is not valid without reservation, however, as these relationships do not apply in the area of the casing; even at the point of best efficiency, the radial force F_r is not absolutely zero.

F_r increases considerably in the partial flow rates and overload ranges, as the changes in cross-section of the volute guide (e.g. with circular cross-section) change quadratically via φ, whereas the increase or decrease of the transport flow is linear. Where Q is constant at any desired duty point, F_r changes linearly with the pumping head.

The magnitude of the radial force can be calculated on the basis of the following relationship:

$$F_r = K_r \cdot b_2 \cdot D_2 \cdot p_t \tag{3 - 42}$$

where: K_r Radial force coefficient
              b_2 Impeller width including the cover discs at the outlet
              D_2 Impeller diameter
              p_t Delivery pressure of impeller

F_r is thus the resulting force acting vertically relative to the shaft, arising as a consequence of the varying pressure conditions at the impeller circumference and the projected impeller discharge surface, including the cover discs.

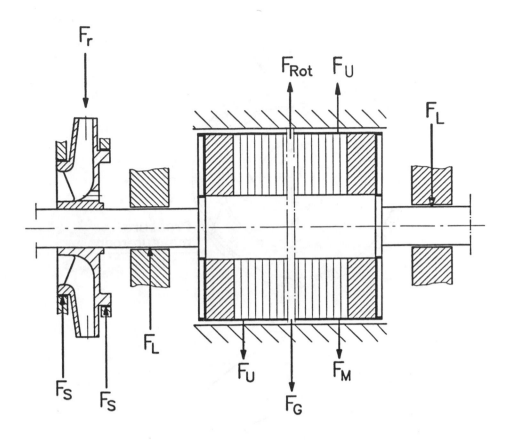

$F_s; F_{Rot}; F_L$ = Plain bearing and Lomakin force are effective if the bore and rotor are eccentric with respect to each other

F_r = Radial force. Acts on the impeller. This force occurs on pumps with volute casings due to the uneven pressure distribution on the circumference of the impeller. For multistage pumps with a diffuser this force is zero or only slight.

F_M = Force on the rotor due to a unilateral magnetic pull

F_G = Force due to the weight of the rotor

F_U = Centrifugal force due to imbalance

Fig. 3.30 Radial forces on the impeller

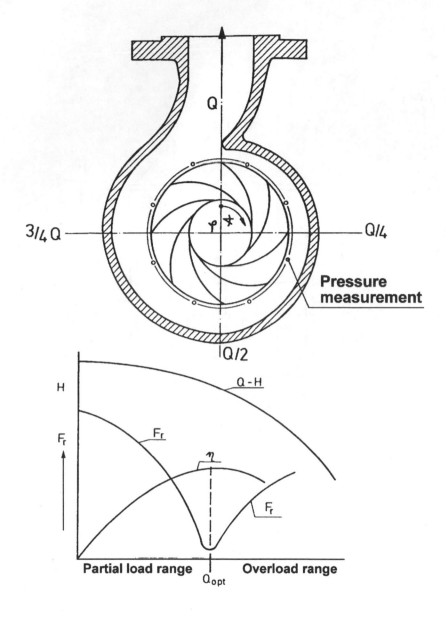

Fig. 3.31 Radial force F_r on the impeller

The radial force coefficient K_r depends on various factors, for example on the specific speed n_q [3-8]. K_r increases with nq (refer to Fig. 3.33) whereas there is also a relationship between n_q and the stagger angle λ of the volute guide. K_r is further dependent on the size of the space between the suction and pressure side of the impeller and casing. An increase in the impeller shroud space leads to an increase in K_r. The type of axial thrust relief device also influences the radial pressure (with a pressure side casing wear ring or with back shroud blades), whereby K_r depends on their diameter. A reduction in the diameter of the thrust balancing device means that the reduction in the effect on K_r is cubed.

142

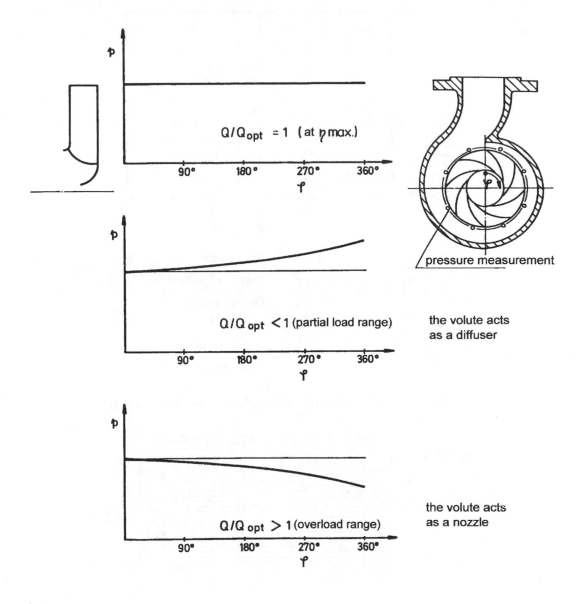

Fig. 3.32 Pressure distribution in the volute guide

Radial forces can reach considerable levels and they must be taken into account for conventional pumps and more so for hermetic centrifugal pumps. Fig. 3.34 shows the course of the radial force in a standard chemical pump, type 100-315, for two impeller diameters each at speed n = 2900 and 1400 r.p.m for pumping water at 20 °C.

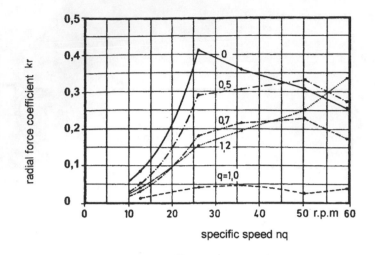

$$K_r = \frac{F_r}{\rho \cdot g \cdot H \cdot D_2 \cdot b_2}$$

$$q = \frac{Q}{Q_{Opt}}$$

F_r	Radial force
ρ	Density of pumped fluid
g	Gravity acceleration
H	Total head
D_2	Impeller diameter
b_2	Impeller width at outlet

Fig. 3.33 Radial force coefficient K_r relative to specific speed [3-8]

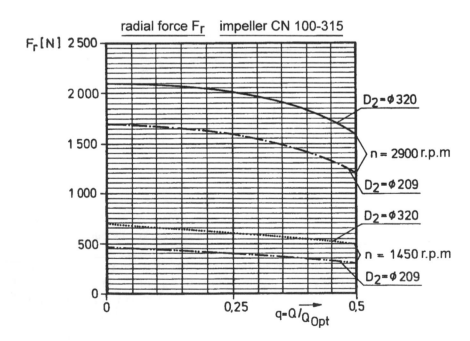

Fig. 3.34 Radial force F_r relative to rate of flow q, speed and impeller diameter D_2 for a standard chemical pump, type 100-315.

144

3.2.9.2 Balance-reduction of radial force F_r

The radial force on volute casing pumps can be considerably reduced by using a double volute as the guide device. Although the radial forces do not cancel each other out completely, reduction achieved is sufficient to provide satisfactory results in many cases. It may be assumed that the remaining forces are somewhere in the region of between 15 and 20 % of the respective radial force of the single volute casing pump (Fig. 3.35).

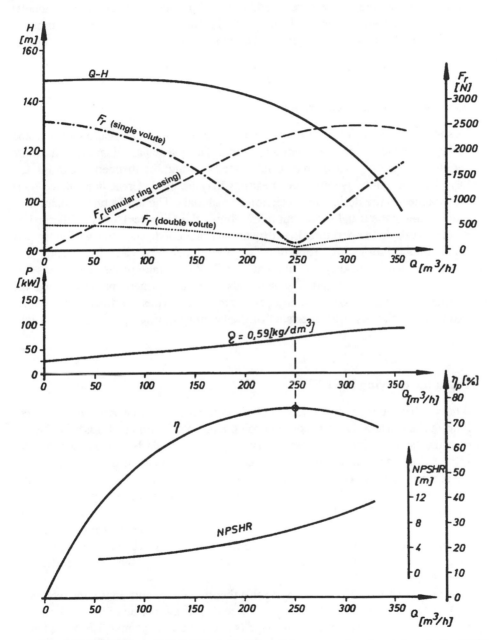

Fig. 3.35 Radial force F_r for single or double volutes or annular ring casings in the H(Q) diagram of a standard 100-315 centrifugal pump.

An almost complete balance of the static reaction forces can be achieved by connecting a diffuser as a velocity conversion mechanism downstream of the impeller. The effect of this would be to provide identical conditions for the approach flow to the guide blades under all operating conditions, thus ensuring that stationary reaction forces would also be identical. In the case of diffuser pumps, however, non-stationary radial forces can occur, especially where the impeller and the diffuser have the same number of blades. Radial forces can also be generated if the impeller is bedded eccentrically with regard to the diffuser. Particularly unfavorable radial pressure conditions are produced by concentric annular chambers. Whereas the total pressure balance at the throttling point $Q = 0$, F_r rises more or less linearly with Q. This type of speed energy conversion should be avoided for reasons of efficiency.

3.2.10 Bearings and radial thrust relief

The bearings in hermetically designed pumps must be located in the pressure chamber and they are therefore immersed in the operating fluid. The particular demands on these bearings require the use of hydrodynamic lubricated bearings or magnetic bearings too [3-9], with a few exceptions where roller bearings may be used. Great importance must therefore be attached to the design and selection of materials. The lubrication-coolant fluid is seldom a lubricant in the usual sense, but more often a fluid like acid, caustic, thermal oil (sometimes at very high temperatures), hydrocarbons, liquefied gas, etc. with extremely low-level or negligible lubricating properties; i.e. of extremely low viscosity. It cannot be assumed, therefore, that hydrodynamic lubrication will always take place: this means that mixed friction is also to be expected. The materials of which hermetic pump bearings are made must therefore possess good emergency running properties. Particular importance should be attached to the best possible relief of the bearings during operation in the radial direction.

3.2.10.1 Hydraulic carrying force F_L

It is well-known that the inner rotor acts like an oversized journal bearing when it is covered by fluid in the can and that it makes a considerable contribution towards relieving the bearings. It thus seems obvious that these carrying powers should be exclusively used for bedding of the rotor. The flow in the gap between rotor and can can be described as a flow in a cylindrical ring gap between a rotating inner cylinder with a radius Ri and a fixed outer cylinder with a radius Ra and the gap length L. The circumferential velocity of the inner cylinder is termed U (Fig. 3.36). If the inner cylinder is eccentrically positioned with regard to the outer cylinder, this results in centralising forces due to uneven pressure distribution along the circumference which act against the weight of the inner cylinder and any existing decentralising forces. The generation of these forces is described by the standard hydrodynamic bearing theory (Fig. 3.37).

Estimation of the carrying forces according to this theory gives results approximately equal to those in [3-10]. Fig. 3.38 shows the carrying force F_{turb} as a function of the smallest lubricant film thickness h_0 for a typical gap shape (see calculation example 3-F) for a liquid viscosity of $v = 1$ mm²/s and a speed of n = 2900 r.p.m. It can be seen that the carrying force increases considerably with increasing eccentricity of the inner cylinder.

146

The traditional hydrodynamic bearing theory (Fig. 3.37), however, proceeds from the assumption that the flow in the gap is of laminar nature and that forces of inertia are negligible compared to friction forces. Neither of these preconditions applies in the case of pumps with permanent-magnet or canned motor drive. Due to the large gap widths and low viscosities, the flow in the gap is extremely turbulent. In addition, due to the large radii and circumferential velocities, centrifugal forces are effective which cast fluid particles in the vicinity of the inner cylinder outwards due to the drag effect, resulting in instability of flow. Such flows were first investigated in detail by Taylor [3-11]. Above a critical Reynolds number, vortices with their axes along the circumferential direction and alternately rotating clockwise and counter-clockwise occur (Fig. 3.36).

The condition for the substitution of such vortices can be described, according to Taylor, as follows:

$$\text{Re}_{sU} = \frac{U \cdot S}{\upsilon} \geq \text{Re}_c = \frac{41,2}{\sqrt{s/Ri}} \qquad (3 - 43)$$

$$S = Ra - Ri$$

or else more simply using the Taylor number Ta:

$$Ta = \frac{U \cdot S}{\upsilon} \cdot \sqrt{S/Ri} \geq Ta_{krit} = 41,2 \qquad (3 - 44)$$

In the calculation example (3F) (Fig. 3.38) the numerical values Re_{su}, Re_c and Ta formed in accordance with (3-43) and (3-44) are given. The turbulent character of the flow with superimposed Taylor vortices can be clearly seen.

Calculation example (3-F):

Given:

Radius of rotating cylinder	Ri	53,1 mm
Radius of bore	Ra	53,6 mm
Radial gap width	S	0,5 mm
Gap length	L	85 mm
Viscosity	v	1 mm²/s
Speed	n	2900 r.p.m
Weight of impeller		108 N

Calculated:

Reynolds number	Re_{su}	8063
Critical Reynolds number	Re_c	424
Taylor number	Ta	782

Table 3-V:

Thickness of lubricating film h_o (µm)	Load capacity force, laminar as per [3-10] F_{lam} (N)	Load capacity, turbulent as per [3-12] F_{turb} (N)
10	687	9653
60	115	1554
100	69	905
200	34	418
300	22	245
400	11	111
500	0	0

Refer also to Fig. 3.38

The plain bearing forces where there is turbulent gap flow can be calculated using the following formula:

$$\frac{F_{turb}}{F_{lam}} = 1 + \frac{7}{600}[1+\varepsilon]^{0,725} \cdot Re_{su}^{0,725} \qquad (3-45)$$

where

$$\varepsilon = \frac{S-ho}{S}$$

as follows by the work carried out by Constantinescu [3-12]. Compared to the laminar gap flow the load capacity with turbulent flow in the example calculation is increased by a factor of 10 - 13 (Fig. 3.38). Lomakin forces at the rotor were investigated in addition to the pure friction forces of the plain bearing. Estimation of these forces (Fig. 3.39) according to Freese [3-13], however, shows that these are approximately a factor 100 lower. This can be explained by the strong turbulence in the rotor-stator gap. In addition to the carrying forces on the rotor there are further centralizing forces at the casing wear rings of the impeller and the radial bearing. Estimation of these bearing forces from Taylor vortices or Lomakin forces due to the axial flow over the gap shows that the Taylor forces dominate in the case of plain bearings but the Lomakin forces dominate for the casing wear ring. A stronger axial flow is obtained here because of the smaller gap length compared to the diameter of the annular gap.

The load capacity of the rotor can be explained clearly with the aid of calculation example (3-F):
In order to carry the rotor of 108 N, the rotor need be eccentrically bedded only within 1/10 mm, which is equivalent to a relative eccentricity of 20%. In this case, the smallest thickness of the lubricating film is still 0.4 mm which is extremely large for plain bearings.

The foregoing considerations show that the hydrodynamic plain bearings have a bearing function to perform mainly in the start-up and run-down phases and during shutdown and that the rotor performs the actual bearing function when running.

148

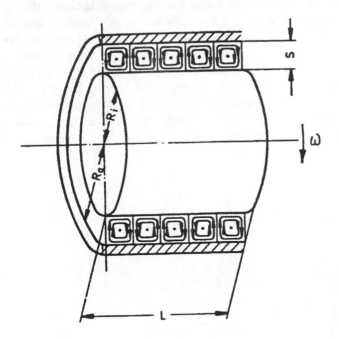

Fig. 3.36 Taylor vortexes at the rotor

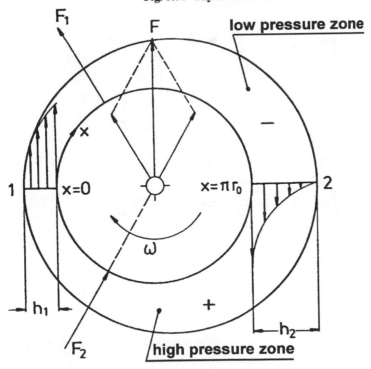

Flow direction radial

Fig. 3.37 Hydrostatic shaft lift according to Reynolds and Sommerfeld

It should be noted that in addition to the hydraulic forces the magnetic field also performs a function. For trouble-free running, therefore, the hydraulic center axis must be aligned with the magnetic and mechanical axis. If this is achieved then the life of the bearings depends, in addition to the cleanliness of the pumped material, particularly on the number of starts. There are individual cases where during continuous operation, e.g. in cooling plants, more than 250.000 running hours have been achieved without a bearing change. The start-up time to reaching the rated speed is between 0,6 and 1,6 seconds, depending on output, for the most common canned motors, whilst the run-down time takes approximately seven times as long. In the run-up and run down phases the actual bearing loading for mixed friction is present.

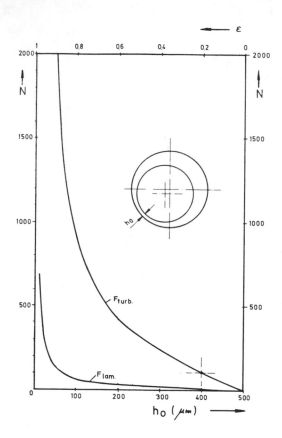

Fig. 3.38 Hydraulic carrying forces on a rotor (Example 3-F)

It is therefore understandable that the bearing materials must have the best possible running properties. The bearing materials used today are mainly technical ceramic, with a silicon carbide/silicon carbide (SiC/SiC) combination of material being preferred. In addition to its great hardness (Vickers hardness HV 2700), this bearing material has very good properties which makes it corrosion resistant with respect to almost all the fluids in use. Fig. 3.40 shows such a bearing constructed using the correct materials. Fig. 3.41 is a view of a plain bearing with a material pairing consisting of 1.4581 Tenifer-treated against carbon with a Teflon-carbon thrust disc.

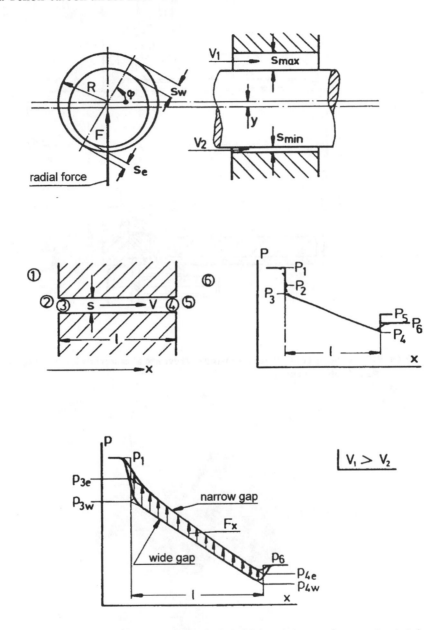

In an annular gap with an eccentric shaft (which rotates or does not rotate) the velocity of the gap flow and thus the pressure drop at the gap inlet is lower (and therefore the static pressure in the gap is higher) than on the side with the wide gap.

Fig. 3.39 Lomakin forces on the rotor, in the plain bearings and the annular gaps on the impeller

151

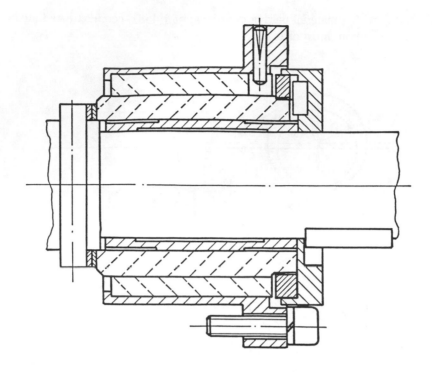

Fig. 3.40 Plain bearing construction showing correct use of materials for bearing shell pairing SiC/SiC (Hermetic)

Fig. 3.41 Plain bearing of a canned motor pump (Hermetic)

Table (3-VI) gives an overview of the current bearing material combinations.

Table (3-VI) Bearing materials

Bearing sleeve	Materials										
	VS	VC	V/0-32	V/0-32	V/LW5	H	Ox	Hartm.	Hartm.	Zi	SiC
Bearing bush	K	K	Ox	V/Si	V/Si	K	Ox	K	Ox	Zi	SiC

Key:

VS	1.4581 / Tenifer-treated
VC	1.4581 / Hard chromium
V/0-32	1.4581 / Chromium oxide coated
V/LW5	1.4581 / Tungsten coated
H	Hastelloy
Ox	Aluminum oxide-ceramic
Hartm.	Hard metal
Zi	Zirconium oxide
V/Si	1.4581 / silicized carbon
K	Carbon (in various qualities)
SiC	Silicon carbide

153

3.2.11 Axial thrust at the impeller and its balancing devices

The development of hermetic pump systems depended on the solution of a central problem, namely the elimination of axial thrust at the rotor equipment. The various fluid properties exclude the possibility of fitting mechanical relief bearings, particularly in the case of high thrusts. The only generally valid solution to this problem thus lay in hydraulic balance of the rotor. There are several ways of achieving this. A basic distinction must be made between balance in the motor section and balance in the pump section. Distribution of balance between pump and motor is also practised. The main cause of thrust is the impeller (in multistage pumps the set of impellers). It is therefore a good idea to try and effect balance in the places where the main thrust forces have their effect. The first possible methods involve traditional balance devices (pressure-side wear ring with holes in the impeller hub, rear blades on the impeller or fixed ridges in the pump casing). These types of balancing, however, can only be used with complete success in conventional pumps, as the roller bearings outside the medium are able to absorb the residual forces.

In the case of hermetic pumps, especially those with low specific speeds ($n_q = 10$ to $n_q = 30$ r.p.m), this type of balance is generally inadequate; moreover, additional thrust forces are created at the rotor. In the following sections, we describe two rotor balance methods with automatic thrust balance which have proven succesful in practical operations.

3.2.11.1 Hydraulic balancing using valve control at the impeller of single stage pumps Fig. 3.42 / [3-14]

Basic relief is provided by a considerably increased wear ring diameter on the impeller pressure-side (1). This method uses both the inside and outside of the wear ring collar as a throttle section on the way to the downstream pressure balance chamber (2). The impeller hub is fitted with holes which run into a valve axial clearance on the rear (3). This provides for control of the pressure in the balance chamber. Depending on the axial position of the rotor, this valve opens or closes; in other words, the balance holes are covered to a lesser or greater degree. In the case of excursion towards the suction side, the valve opens and the pressure in the balance chamber drops, whereas the pressure on the front cover disk of the impeller remains constant. This causes the impeller to react with a corresponding thrust in the direction of the motor. In the reverse case (excursion towards the motor), the valve closes, the pressure in the balance chamber (2) rises, and the pressure on the cover disk remains constant. The impeller reacts with a corresponding thrust towards the suction side, against the excursion. In a condition of equilibrium, the rotor automatically adjusts the required valve clearance (Sh). In contrast to other balance devices, this type of balance is extremely compact and, above all, it possesses a high degree of stability over the entire characteristic curve. Due to the large pressure-fed area within the twin wear ring, slight pressure changes i.e. extremely small excursions of the rotor, are sufficient to generate relatively large reaction forces. Fig. 3.42 shows measured force-clearance curves for a single-stage hermetic centrifugal pump with standard hydraulics, type 65-315. A condition of equilibrium is achieved with a valve clearance of 0,4 - 0,5 mm. The reaction forces are dependent on the volume flow.

154

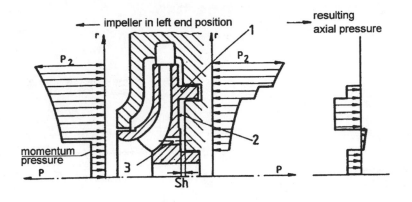

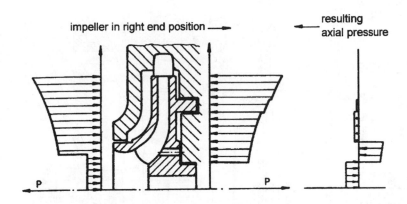

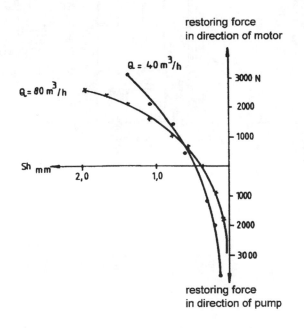

Fig. 3.42 Axial forces on impeller

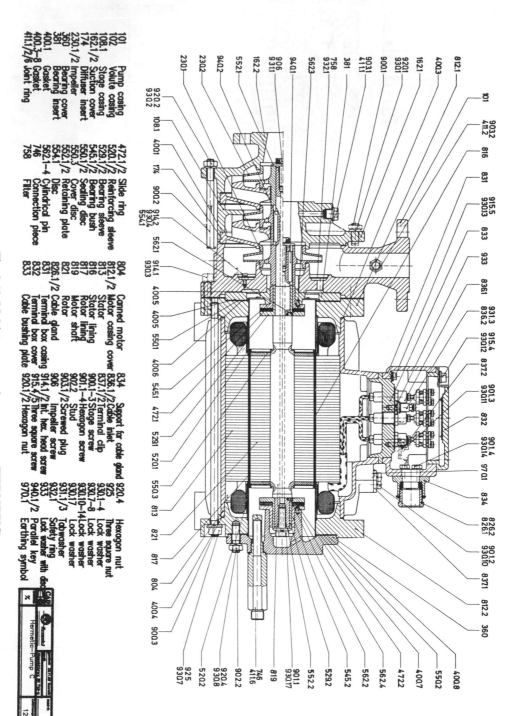

Fig. 3.43 Multistage canned motor pump (HERMETIC)

101	Pump casing	920.2		804	Canned motor	834	Support for cable gland	920.4	Hexagon nut
102	Volute casing	930.2		812.1/2	Motor casing cover	836.1	Cable inlet	925	Three square nut
108.1	Stage casing	108.1		813	Reinforcing sleeve	836.2		930.1-4	Lock washer
162.1/2	Suction cover	400.1		529.1/2	Bearing sleeve	837.1/2	Terminal clip	930.1-8	Lock washer
174	Diffuser insert	174		545.1/2	Bearing bush	900.1-3	Stage screw	930.7-8	Lock washer
230.1/2	Impeller	900.2		550.1/2	Sealing disc	901.1-4	Hexagon screw	930.10-14	Lock washer
360	Bearing cover	914.2		550.3	Cover disc	902.2	Stud	931.1/2	Lock washer
381	Bearing insert	554.1		552.1/2	Retaining plate	903.1/2	Screwed plug	931.1/3	Tab washer
400.1	Gasket	562.1		554.1	Cover disc	906	Impeller screw	932.1/	Safety ring
400.3-8	Gasket	746		562.1-4	Cylindrical pin	914.1	Int. hex. head screw	933	Lock ring
411.1/2/6	Joint ring	758				915.4/5	Three square screw	940.1/2	Parallel key
						920.1/2	Hexagon nut	970.1/2	Earthing symbol

3.2.11.2 Thrust balance due to valve control at the rotor of multistage pumps

A different method of thrust balance (DBP no. 12 57581) is practiced for multistage pumps (up to three stages), particularly those with an impeller/rotor diameter ratio of $L_D/R_D < 1.3$ (Fig. 3.43). In this method, the bearing collars are used as valves to control the thrust force. If we assume that the impellers always exhibit a thrust in a negative direction towards the suction side, then a thrust of the same force in the opposite direction must be realized in order to achieve a condition of equilibrium. This can be achieved by using the motor partial flow Q_T with its different pressure zones in front of and behind the rotor for this task. The partial flow entering the rotor chamber has almost the same pressure as the end pressure p_D of the pump, with the result that a corresponding thrust force is exercised on the rotor in a positive direction (Fig. 3.44).

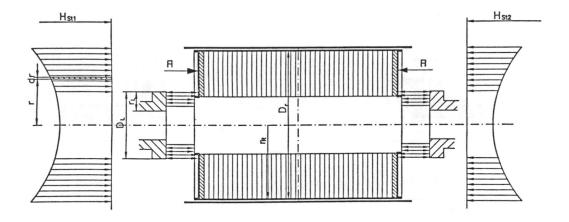

Fig. 3.44 Axial thrust forces on the rotor

This force is always greater than the thrust forces of the impellers towards the suction side. If the rotor is bypassed, the partial flow loses a large proportion of its inner pressure, and the thrust force exercised in a negative direction on the rear side of the rotor is correspondingly smaller. In order to achieve equilibrium, however, the negative thrust forces of the impeller and the rear-side rotor thrust must be of the same size as the positive thrust on the front face end of the rotor. If we call the positive thrust on the front face end of the rotor K, then K is proportional to the pressure difference Δp between rotor front side and rear side times the circle ring area A (outer diameter of the rotor bearing collar) (Fig. 3.44) or:

$$K = p \cdot A \qquad (3 - 46)$$

For its part, Δp changes approximately proportionally to Q_T.
Thus

$$\Delta p \sim Q_T \qquad (3 - 47)$$

However, this gives us a linear relationship between Q_T and K. To change K, we must simply change Q_T. In pumps in compliance with Fig. 3.43, a self-regulating valve is used, by utilizing the bearing collars of the slide bearings, to control the partial flow. If there is excursion of the rotor in a negative direction out of the central position, a greater Q_T will flow off via the rotor through the open valve (rear bearing), through the hollow shaft to the suction side. However, this means a larger Δp and thus a more powerful thrust force K, with the result that the rotor is affected by a restoring force towards the central position. In the reverse case, namely if there is excessive rotor excursion in a positive direction, Q_T is reduced by the closing of the rear-side valve, and Δp thus also reduced; as a consequence, K decreases and the rotor is set back to the central position. Fig. 3.45 shows a force-clearance diagram of such a thrust balance system. It should be noted, however, that this type of control mechanism does not possess the stability of the type of thrust balance described previously (Section 3.2.11.1 Hydraulic balancing using valve control at the impeller of single stage pumps).

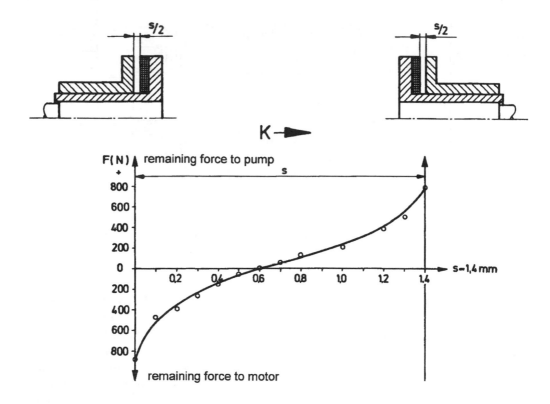

Fig. 3.45 Path of hydraulic restoring force at the impeller

Balance discs as shown in Fig. 3.46 have proved useful for pumps with more than three stages which deliver very high pressures. The same applies to certain parameters of the pumped material, e.g. fluids with very high viscosities.

An axial thrust measuring device as shown in Fig. 3.47 is used to measure the axial thrust forces on the impeller and determine the force-path diagram.

158

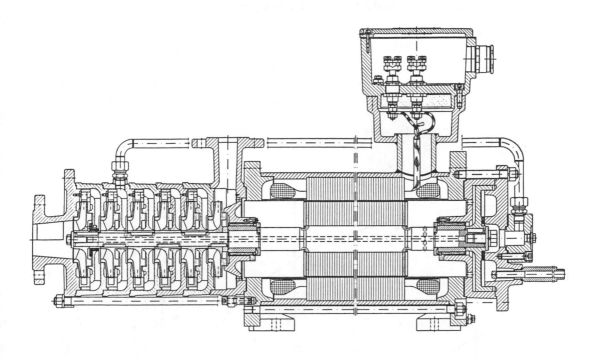

Fig. 3.46 Six-stage canned pump motor with balance disc to compensate for axial thrust (Hermetic)

Fig. 3.47 Axial thrust measuring device

3.3 Operating range of canned motor centrifugal pumps

Centrifugal pumps are generally limited to their operational range as determined by the particular characteristic data, which if exceeded, or undershot can lead to operating malfunctions or failure of the pump. It is therefore particularly important to know and comply with the maximum and minimum flow rates Q_{max} and Q_{min}. This applies to conventional centrifugal pumps and it is even more important for canned motor centrifugal pumps.

Fig. 3.48 shows the operational range of a canned motor centrifugal pump.

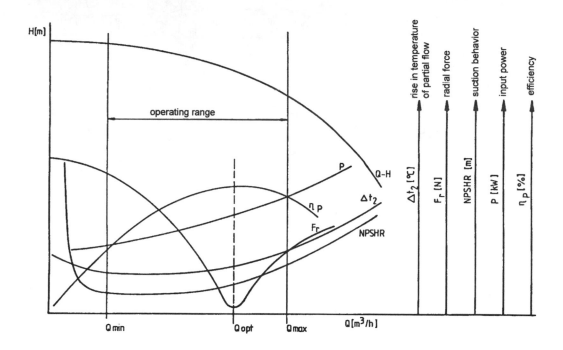

Fig. 3.48 Operating range of canned motor pumps

3.3.1 Q_{min} limitation

Q_{min} limitations are determined by one or more of the following requirements:

I	adequate cooling of canned motor
II	minimization of radial and axial forces
III	non-vaporization of pumped liquid on the suction side (avoidance of cavitation)
IV	non-vaporization of the motor cooling/lubricating flow

160

Requirements III and IV play a particularly important role when pumping fluids near or at boiling point. Requirement III is a very important factor which unfortunately is always overlooked. This is the effect of the thermal balance and the physical characteristics of the fluid on the NPSHR (refer to Section 3.2.7.2 "Heating of partial flow Q_T").

3.3.1.1 Q_{min} limitation due to NPSHR

It is normally assumed that the path of the NPSHR curve is independent on temperature and the physical properties of the fluid.
This is, however, only true up to a point, although the influences which lead to the determination of the NPSHR do not enable this to be detected because the thermodynamic effects are not taken into account.
Section 2.9.9 ("Increase in NPSHR values in the partial load range due to heating up of the fluid on the inlet side; effect of pre-rotation and recirculation") dealt with these relationships.

3.3.1.2 Q_{min} limitation due to motor cooling/lubrication flow

In order to fulfill requirement IV (3.3.1), it must be ensured that the static pressure at the hottest point of the motor partial flow is greater than the vapor pressure at this temperature.

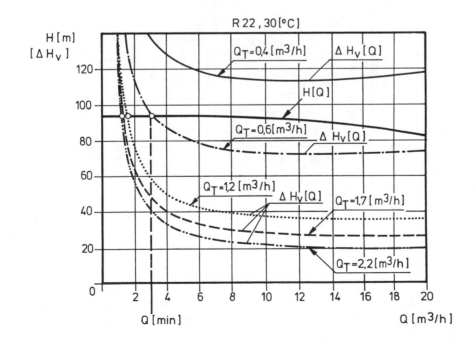

Fig. 3.49 **Influence of the partial flow Q_T on the vapor pressure head ΔH_v for Freon R 22 at 30 °C**

In this case, it is also expedient to refer all vapor pressure heads to the vapor pressure head H_v at the inlet of the pump, as was the case with the NPSHR. Fig. 3.49 shows, from example (2-A) (2.9.9.1 "Dependency on physical properties of the fluid and temperature particularly for canned motor and magnetically-coupled pumps") the rise in the vapor pressure head at the hottest point of the motor relative to the partial flow volume Q_T for R22 at 30 °C as a function of the capacity Q, and compares this rise with the static capacity H(Q) at this point.

It can be seen from this that, for example, at $Q_T = 0{,}4$ m³/h the absorbed heat loss would cause the partial flow to vaporize. This would result in the motor running hot and failure of bearing lubrication and thus failure of the pump itself. The condition for non-vaporization of the liquid is:

$$H \geq H_v \qquad\qquad (3 - 48)$$

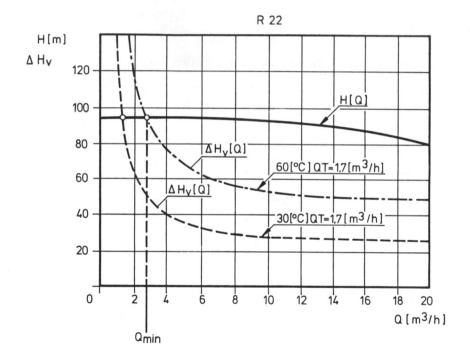

Fig. 3.50 **Influence of the temperature of the motor partial flow Q_T on Q_{min} for Freon R22**

With a partial flow of 0,6 m³/h, a Q_{min} of 3 m³/h must be present to prevent vaporization in the rotor chamber. For reasons of safety and the dependence of the properties of the various fluids, however, the pump used in our example is operated with a Q_T of 1,7 m³/h as standard value.

This would require a Q_{min} of 1,3 m³/h, far below the minimum volume of 6 m³/h necessary due to the NPSHR, so that in this respect the pump is operated with a high degree of safety. The increase in vapor pressure of the partial flow above the rotor is also a function of the flow temperature, as shown in Fig. 3.50.

162

This, in turn, is clearly shown by Fig. 2.69 (Section 2.9.9.1); the steep rise in vapor pressure in the range of higher temperatures. Whereas from the point of view of temperature influence a Q_{min} of 2,8 m³/h is required where Q_T = 1,7 m³/h for R 22 and 60 °C, vaporization can be prevented at 30 °C with a Q_{min} as low as 1,1 m³/h. Because in our example the pump is operated with a Q_{min} of 6 m³/h, there are ample reserves.

3.3.2 Maximum discharge flow Q_{max}

The maximum flow that can be taken off at the pressure nozzle of the pump is limited by the following factors:

- the requirement of non-vaporization of the partial flow Q_T in the motor section. This applies above all to pumps which are not equipped with a pressure increase impeller in the rotor chamber, as an increase in the flow is countered by a fall in the pressure difference between the front and rear ends of the rotor. This means that from a certain Q_{max} onwards the partial flow Q_T falls below the required minimum volume:

- limitation due to NPSHR. With increasing flow rate Q, an increase in NPSHR with a progressive curves recorded, particularly in the overload range,

- the power requirement P. This rises more or less linearly together with the flow Q, so that the pump power must not exceed the output power of the drive motor. This applies particularly to motors which have to conform to the regulations governing explosion-protected electrical equipment to EN 50014 to 50020/VDE 0171,

- the radial force F_r. As described in section 3.2.9 ("Radial forces on impeller") the radial force increases considerably in the partial load and overload ranges in the case of single-stage pumps with single-volute. If such pumps are used, the radial force occurring must be taken into account when fixing the maximum flow.

3.3.3 Efficiency by limitation of Q_{min} - Q_{max}

Compliance with the Q_{max} - Q_{min} limit is obviously desirable for economic reasons, i.e. the pump should operate within an acceptable efficiency range. Today, energy saving is assuming ever increasing importance.

3.4 Noise emissions from canned motor pumps

Mechanical energy is converted to flow energy in the impeller, thus leading to a high pressure rise in the flow through the impeller blade channels causing boundary layer separation and the formation of vortices. Furthermore, the finite number of blades causes periodic pressure fluctuations. All these influences cause the flow in the pump to fluctuate, resulting in vibration of the pump casing and connected pipelines which is transmitted to the surrounding air. Such vibration is perceived as noise or airborne sound. Further sources of noise are present on conventional centrifugal pumps, at roller bearings and couplings.

DIN 45 635 "Noise measurement on machines" was drawn up to provide pre-conditions which would enable the noise output from machines to the air "noise emissions" to be determined using a standard method and thus enable results which could be compared. According to this standard the sound emissions are measured using the enveloping surface method with the measuring points of the machine surface being spaced 1 m apart, thus providing the simplest geometric shape, a cuboid. The acoustic measuring variable used is the A-weighted sound pressure level $\overline{L}_{pA}$, the dimension dB.

The letter A represents the filter characteristic curve in accordance with DIN 45 633 which corresponds approximately to the sensitivity of the human ear to moderate technical noises. The A sound pressure level averaged over the measured area "S" is represented by $\overline{L}_{pA}$.

A relationship exists between the sound pressure level $\overline{L}_{pA}$ and the input power of the pump, which can be determined by formula (3 - 49):

$$\overline{L}_{pA} = 61,1 + 11,1 \cdot \lg P_2 \qquad (3 - 49)$$

$$\text{where } n = 1450 \text{ and } 2900 \text{ r.p.m } [3 - 15]$$

According to measurements by Gordzielik [3 - 16], the average difference between both speeds is only 3 dB.

The A-weighted sound pressure level L_{WA} is expressed in dB as a measure of the total noise radiated from the pump to the surrounding air. Its empirical validity can be expressed as:

$$L_{WA} = 72,7 + 12,7 \cdot \lg P_2 \qquad (3 - 50)$$

Centrifugal pumps are in most cases driven by three-phase motors. The sound emissions from these are always higher than those from the centrifugal pumps themselves, which means that when determining the sound pressure level of the complete unit the motor must be regarded as the main source of emissions.

This is due mainly to the noise caused by the fan. For canned motor pumps the sound emissions are substantially better relative to the complete unit. Canned motor pump units have no roller bearings and no couplings. The motor is liquid cooled so that fan and rotor

164

noise is absent. The rotor is immersed in liquid and this provides particularly good sound insulation. The following formulae can be used for determining the sound characteristic values of canned motor pumps:

Sound pressure level $\bar{L}_{pA}$ in dB

$$\bar{L}_{pA} = 47,5 + 10 \cdot \lg P_2 \qquad (3 - 51)$$

where n = 2900 r.p.m

in this case P_2 is the rated pump output in kW.

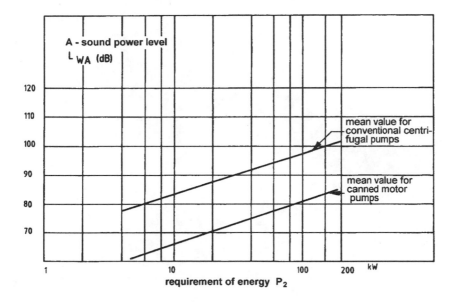

Fig. 3.51 Comparison between the sound power level relative to motor power output for conventional and canned motor centrifugal pumps

Sound power level L_{WA} in dB

$$L_{WA} = \bar{L}_{pA} + L_s \qquad (3 - 52)$$

Level of measuring surface L_S in dB

Table 3 - VII

Size	L_S
25 to 80	12
100 to 200	13
250 to 400	14

165

Fig. 3.51 shows the measuring surface sound-pressure levels relative to motor power output of canned motor pump units and conventional centrifugal pumps without a drive motor. The graph shows that canned motor pump units run on average 18 dB (A) quieter than similar pumps without a motor.

This is explained by the type of shaft bearings (hydrodynamic plain bearings instead of roller bearings).

3.5 Examples of hermetic centrifugal pumps with a canned motor drive

The wide variety of tasks which industry requires to be performed by pumps demands great flexibility and adaptation to particular operating circumstances on the part of manufacturers of leak-free-pumps. Examples of these are the pumping of fluids with high or low temperatures, pumping liquids from a vacuum or against high pressure, or recirculation in the high pressure range, circulating highly toxic substances which form highly volatile explosive mixtures, have a tendency to coagulate or contain solid matter, and also pumps to perform charging and transfer operations. Such tasks also frequently occur in combination.

The planning engineer clearly requires a great deal of specialist knowledge. Decisions of wider significance frequently have to be made, which are required not only to guarantee the operating sequence but also the safety of the plant and therefore of the life and limb of the operating personnel and furthermore that of the population in the immediate vicinity. These tasks are no less problematic to the engineer engaged in tendering. On one hand he is obliged to guarantee the safety-relevant requirements and on the other hand he has to offer the operator pumps which work economically, are completely adequate for the task, are as reliable as possible and still value for money. He can only solve these tasks successfully if he receives the relevant data from the planning engineer, if possible also with a sketch of the plant.

The manufacturers have produced suitable data sheets for this purpose, which can be obtained on request. They guarantee that all the data necessary for the particular offer is taken into account. The more precisely and completely these forms are filled in the easier it is to choose the most satisfactory pump for the requirement. The purpose of this book is not only to provide information on the state of the art for hermetic centrifugal and rotating displacement pumps but also to provide help in deciding which type of pump is best suited to a particular task.

3.5.1 Single-stage canned motor pumps

The large number of single-stage standard chemical pumps to DIN 24256/ISO 2858 operating in chemical plants cannot fully meet today's demands for zero emission of hazardous substances given in the TLV and BAT lists. Added to this is the reduced availability due to the unavoidable wear of the axial or radial sealing faces of the shaft gland, which results in the increased escape of hazardous substances and interruptions in operation.

These disadvantages can be avoided with canned motor pumps at the same output using the hydraulics of standard chemical pumps (Fig. 3.52). Their normal operating range for drive speeds n = 1450 r.p.m and 2900 r.p.m is given in Fig. 3.53. Larger pump units such as those shown in Fig. 3.54 are also manufactured.

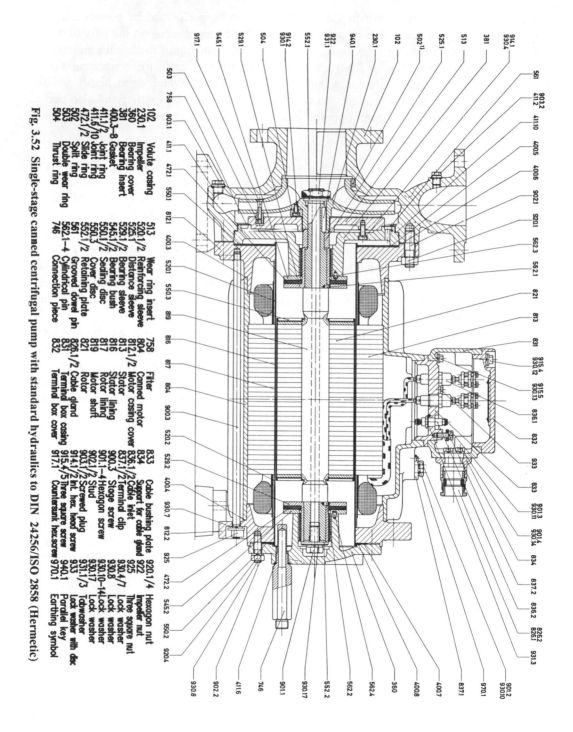

Fig. 3.52 Single-stage canned centrifugal pump with standard hydraulics to DIN 24256/ISO 2858 (Hermetic)

102	Volute casing	513	Wear ring insert	758	Filter	833	Cable bushing plate	920.1/4	Hexagon nut
230.1	Impeller	520.1/2	Reinforcing sleeve	804	Support for cable gland	834		922	Impeller nut
360	Bearing cover	525.1	Distance sleeve	812.1/2	Motor casing cover	836.1/2	Cable inlet	925	Three square nut
381	Bearing insert	529.1	Bearing sleeve	813	Stator	837.1/2	Terminal clip	930.4/7	Lock washer
400.3-8	Gasket	545.1/2	Bearing bush	816	Stator lining	900.3	Stage screw	930.8	Lock washer
411.6/10	Joint ring	550.1/2	Sealing disc	817	Rotor lining	901.1-4	Hexagon screw	930.10-14	Lock washer
411.1/2	Joint ring	550.3	Cover disc	819	Motor shaft	902.1/2	Stud	930.17	Lock washer
472.1/2	Slide ring	552.1/2	Retaining plate	821	Rotor	903.1/2	Screwed plug	931.1/3	Tabwasher
502	Split ring	561	Grooved dowel pin	826.1/2	Cable gland	914.1/2	Int. hex. head screw	933	Lock washer with disc
503	Double wear ring	562.1-4	Cylindrical pin	831	Terminal box casing	915.4/5	Three square screw	940.1	Parallel key
504	Thrust ring	746	Connection piece	832	Terminal box cover	917.1	Countersunk hex.screw	970.1	Earthing symbol

168

Allgemeine Kennlinienfelder
n = 1450 U/min

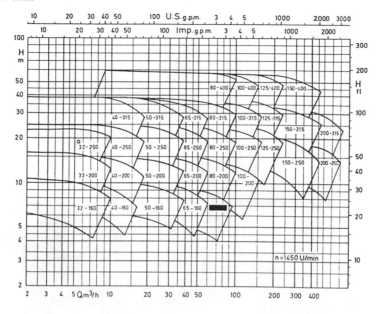

Allgemeine Kennlinienfelder
n = 2900 U/min

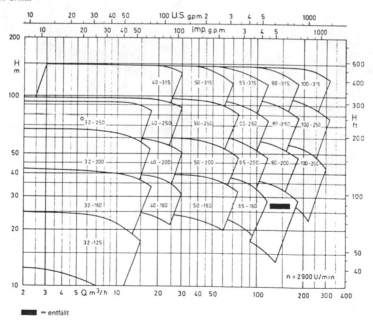

▬ = entfällt

Allgemeine Kennlinienfelder = general families of charakteristik ; U/min = r.p.m. ; entfällt = omitted

Fig. 3.53 Families of characteristics for a series of single-stage canned motor chemical pumps with standard hydraulics to DIN 24256/ISO 2828 (Hermetic)

169

This pump has the following delivery parameters:

Pumped fluid = 12 dichloroethane
Q = 700 - 1050 - 1100 m³/h
H = 30 - 25 - 24 m
ρ = 1250 kg/m³
P_2 = 150 kW

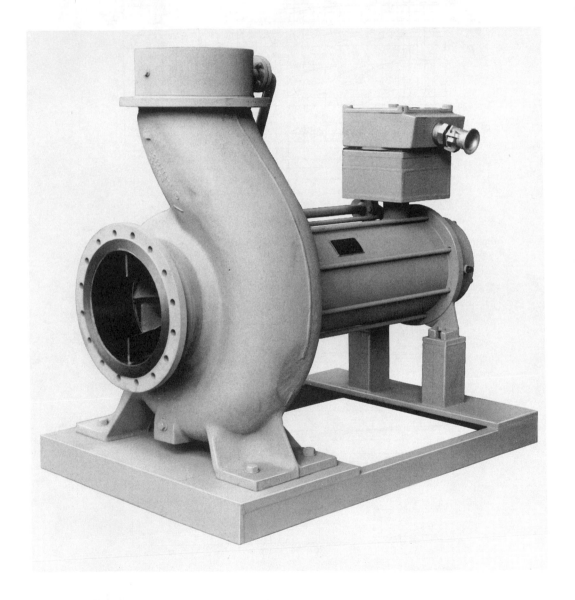

Fig. 3.54 Single-stage canned motor centrifugal pump with standard hydraulics for large performance parameters. Input drive power 150 kW (Hermetic)

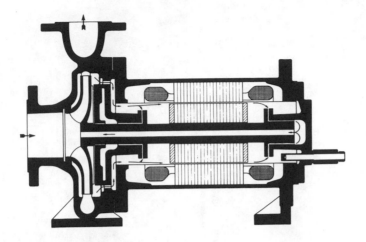

Fig. 3.55 Functional diagram of a single-stage canned motor centrifugal pump with standard hydraulics

Fig. 3.56 Single-stage canned centrifugal pumps in a technical process plant (Hermetic).

Fig. 3.57 Singe-stage canned motor centrifugal pumps in a chemical plant (Hermetic)

In this case a canned motor with a substantially reduced sound power level is used in place of the normal three-phase motor with its relatively high noise emissions. An internal lubricating-cooling circuit passing from the pressure area of the pump through the rotor stator gap to the pump inlet on the suction side dissipates the electrical and mechanical losses to the pumped medium (Fig. 3.55). Fig. 3.56 and 3.57 show a single stage canned motor pump with standard hydraulics in a technical process plant. Fig. 3.56 illustrates the size comparison between a canned motor pump and a conventional centrifugal pump with the same output. An inline arrangement as shown in Fig. 3.58 can be particularly space saving.

172

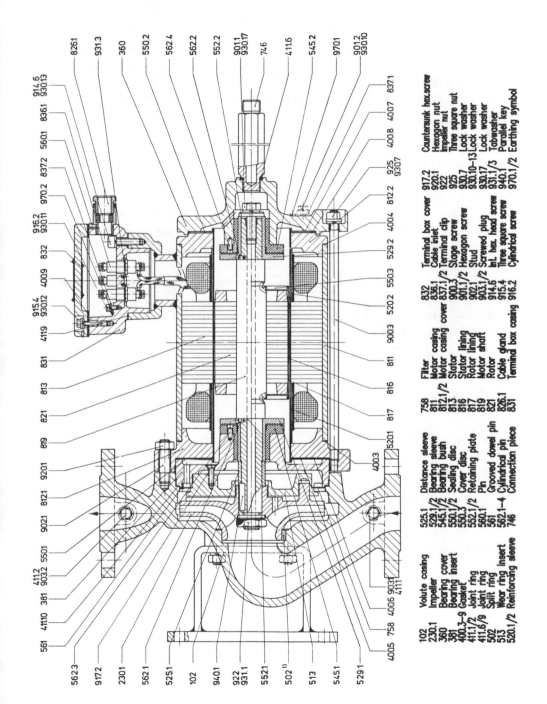

Fig. 3.58 Single-stage canned motor centrifugal pump in inline arrangement (Hermetic)

102	Volute casing	525.1	Distance sleeve
230.1	Impeller	529.1/2	Bearing sleeve
360	Bearing cover	545.1/2	Bearing bush
381	Bearing insert	550.1/2	Sealing disc
400.3–9	Gasket	550.3	Cover disc
411.1/2	Joint ring	552.1/2	Retaining plate
411.6/9	Joint ring	560.1	Pin
502	Split ring	561	Grooved dowel pin
513	Wear ring insert	562.1–4	Cylindrical pin
520.1/2	Reinforcing sleeve	746	Connection piece

758	Filter	832	Terminal box cover
811	Motor casing	836.1	Cable inlet
812.1/2	Motor casing cover	837.1/2	Terminal clip
813	Stator	900.3	Stage screw
816	Stator lining	901.1/2	Hexagon screw
817	Rotor lining	902.1	Stud
819	Motor shaft	903.1/2	Screwed plug
821	Rotor	914.6	Int. hex. head screw
826.1	Cable gland	915.4	Three square screw
831	Terminal box casing	916.2	Cylindrical screw

917.2	Countersunk hex.screw
920.1	Hexagon nut
922	Impeller nut
925	Three square nut
930.7	Lock washer
930.10–13	Lock washer
930.17	Lock washer
931.1/3	Tabwasher
940.1	Parallel key
970.1/2	Earthing symbol

3.5.2 Multi-stage canned motor pumps

The understandable desire on the part of the operator, to have single-stage pumps where possible, can no longer be met where good efficiencies still have to be achieved at small flow rates and large heads. In the specific speed range of $n_q = 1 - 40$ r.p.m the speed n_q can be increased by distributing the total head of the pump over several stages at the same speed and flow rate thus leading to substantial improvement in efficiency (refer to Section 2.8 "Efficiency and specific speed" Table 2-I and Fig. 2.29). Values above $n_q = 40$ r.p.m. result in only slight increases in efficiency which are simply not reasonable compared with the necessary structural expenditure.

The multistage construction shown in Fig. 3.59 enables a flexible matching to the particular operating conditions in the plant. Fig. 3.60 shows a section view of a model of a five-stage canned motor pump and Fig. 3.61 shows the pattern of the partial flow for heat dissipation and lubrication of bearings. Fig. 3.62 shows such pumps installed in a large chemical plant. The drive motors of such pumps are mainly of the two-pole type which provides good bearing stability. Fig. 3.63 is an overview of the normal service range of multistage canned motor centrifugal pumps.

Pumps are also manufactured with performance parameters going beyond the standard production program shown in Fig. 3.63.
An example (3 - G) from practice is used to show the characteristic values of a four-stage canned motor centrifugal pump in accordance with Fig. 3.64 and 3.65 measured and recorded on a test bench and represented with the aid of data processing.
In Table 3-VIII shows the measured actual values against the specified data.

Fig. 3.66 shows the printout of the following values:

> H (Q) curve
> Motor input power P_1
> Motor output power P_2
> Current consumption I
> Hydraulic power P_{hydr}

Table 3.VIII.1 particularly contains the measured values of the NPSHR which are entered in the diagram, Fig. 3.67.

174

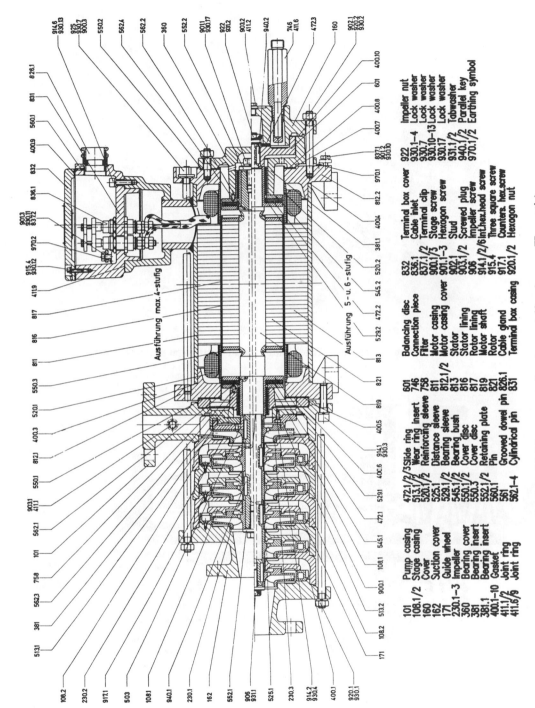

101	Pump casing	472.1/2/3	Slide ring	601
108.1/2	Stage casing	513.1	Wear ring insert	746
160	Cover	520.1/2	Reinforcing sleeve	758
162	Suction cover	525.1	Distance sleeve	811
171	Guide wheel	529.1/2	Bearing sleeve	812.1/2
230.1–3	Impeller	545.1/2	Bearing bush	813
360	Bearing cover	550.1/2	Cover disc	816
381	Bearing insert	550.3	Cover disc	817
381.1	Bearing insert	552.1/2	Retaining plate	819
400.1–10	Gasket	560.1	Pin	821
411.1/2	Joint ring	561	Grooved dowel pin	826.1
411.6/9	Joint ring	562.1–4	Cylindrical pin	831

601	Balancing disc	832	Terminal box cover	922	Impeller nut
746	Connection piece	836.1	Cable inlet	930.1–4	Lock washer
758	Filter	837.1/2	Terminal clip	930.7	Lock washer
811	Motor casing	901.1/3	Stage screw	930.10–13	Lock washer
812.1/2	Motor casing cover	901.1–3	Hexagon screw	930.17	Lock washer
813	Stator	902.1	Stud	931.1/2	Tabwasher
816	Stator lining	903.1/2	Screwed plug	940.1/2	Parallel key
817	Rotor lining	906	Impeller screw	970.1/2	Earthing symbol
819	Motor shaft	914.1/2/6	Int.hex.head screw		
821	Rotor	915.4	Three square screw		
826.1	Cable gland	917.1	Counters. hex.screw		
831	Terminal box casing	920.1/2	Hexagon nut		

Fig. 3.59 Multistage canned motor centrifugal pump (Hermetic)

175

Fig. 3.60 Section of a model of a five-stage canned motor centrifugal pump (Hermetic)

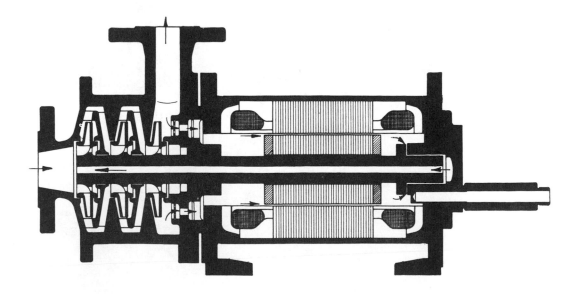

Fig. 3.61 Partial flow path in a multistage canned motor centrifugal pump

Fig. 3.62 Canned motor centrifugal pumps in a large chemical plant (Hermetic)

n = 2900 U/min

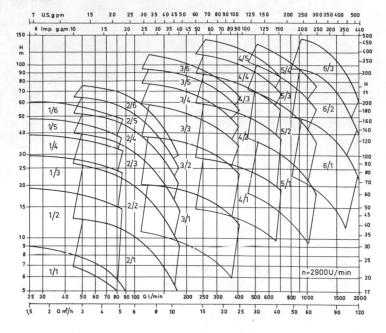

Fig. 3.63 Family of characteristics of multistage canned motor centrifugal pumps (Hermetic)

Fig. 3.65 View of the canned motor centrifugal pump shown in Fig. 3.64 (Hermetic)

178

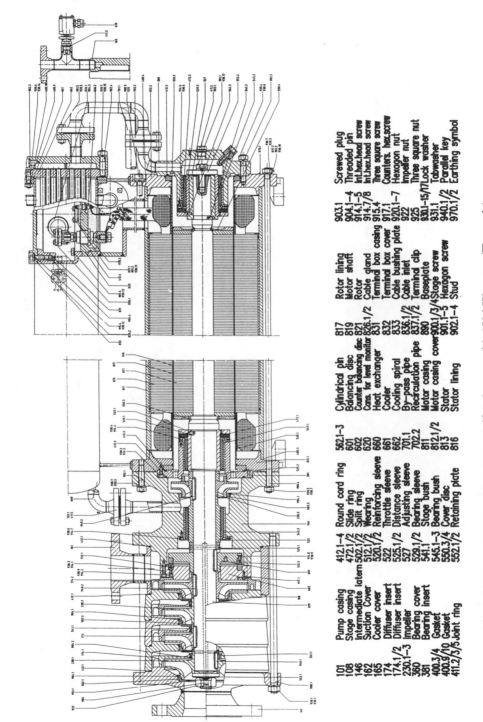

101	Pump casing	412.1–4	Round cord ring	562.1–3	Cylindrical pin	817	Rotor lining	903.1	Screwed plug
108	Stage casing	472.1/2	Slide ring	601	Balancing disc	819	Motor shaft	904.1–4	Threaded pin
146	Intermediate latern	502.1/2	Split ring	602	Counter balancing disc	821	Rotor	914.1–5	Int.hex.head screw
162	Suction Cover	512.1/2	Wearing	620	Cons. for level monitor	826.1/2	Cable gland	914.7/8	Int.hex.head screw
165	Cooler cover	520.1/2	Reinforcing sleeve	660	Heat exchanger	831	Terminal box casing	915.4	Three square screw
174	Diffuser insert	522	Throttle sleeve	661	Cooler	832	Terminal box cover	917.1	Counters. hex.screw
174.1/2	Diffuser insert	525.1/2	Distance sleeve	662	Cooling spiral	833	Cable bushing plate	920.1–7	Hexagon nut
230.1–3	Impeller	527	Adjusting sleeve	701.1	By-pass pipe	836.1/2	Cable inlet	922	Impeller nut
360	Bearing cover	529.1/2	Bearing sleeve	702.2	Recirculation pipe	837.1/2	Terminal clip	925	Three square nut
381	Bearing insert	541.1	Stage bush	811	Motor casing	890	Baseplate	931.1–15/17	Lock washer
400.3/4	Gasket	545.1–3	Bearing bush	812.1/2	Motor casing cover	900.1/3/4	Stage screw	931.1	Tabwasher
400.9/10	Gasket	550.3/4	Cover disc	813	Stator	901.1–5	Hexagon screw	940.1/2	Parallel key
411.2/3/5	Joint ring	552.1/2	Retaining plate	816	Stator lining	902.1–4	Stud	970.1/2	Earthing symbol

Fig. 3.64 Four-stage canned motor centrifugal pump with 150 kW motor (Hermetic)

Test report

HERMETIC - Pumpen G.m.b.H.
79194 Gundelfingen b. Freiburg

Order-No. :	CH 63529/93	
Commission-No:	CH 63529	
Repair-No :		
Customer :	605000 BASF AG	
Pump type :	CAMK 52/4	
Material :	1.4571, 1.4408	
Date :	2. 7.93	
Name :	Scherer	

Pump datas

Q : 20.0 95.0 110.0 m^3/h   H : 390.0 335.0 310.0 m
τ :
n : 2960 1/min   Impeller φ D2

Back vane φ DR : 255.0 mm
Balancing hole φ d2 : 0.0 mm
d2 Impeller φ d2 : *φ mm
Partial Flow Exit : *φ mm
Shaft boring D11 : 1./2.φ8   0.0 mm

Running test
Fluid : Wasser
Temp. : 20.0 °C
Test Time : 30.0 min

Pressure test

	Water	Air / N2
Pump	64.0 bar	6.0 bar
Heating Jacket :	bar	bar
Stator :	bar	6.0 bar

Remarks
Impeller φD2: 255.0 mm
Impeller φR2: 0.0 mm

Motor datas

Typ : CKPK85Z-2h
Motor No :
Motor No : 285
Motor kW : 150.0 kW   cosPhi : 0.92
U : 380.0 V   F : 50 Hz   I : 325.0 A
Iso-Klasse : H   Schaltung : Dreieck
Insul.Resistance : 1000.0 MΩ
Winding-Resistance :
Temperature :

n	Q	H	I	P1 Motor	P2 Motor	ETA Motor	cosPhi	P2 Pump	ETA Pump	SH	Axialthrust MS	Axialthrust PS	Clea-rence	U	NPSH Pump	H rel	H duty
1/min	m^3/h	m	A	kW	kW	%	-	kW	%	mm	kg	kg	mm	V	m	%	m
2982	403	0.00	193	115	77.1	67.0	0.88	0.00	0.00	0.08	132	1089	2.70	394			
2980	390	20.0	208	126	87.8	69.6	0.89	21.2	24.2	0.19	29.9	1101	2.70	394			
2978	381	40.4	226	139	100	72.2	0.90	42.0	41.8	0.16	291	1226	2.70	394			
2973	365	70.2	257	161	121	75.5	0.92	69.8	57.5	0.16	165	1221	2.70	393			
2970	335	95.2	282	178	138	77.5	0.93	86.8	63.1	0.14	164	1025	2.70	393			
2969	310	111	294	186	146	78.3	0.93	93.5	64.2	0.12	215	1248	2.70	393			

Table 3-VIII Operating parameters for canned motor centrifugal pump shown in Fig. 3.64 and 3.65

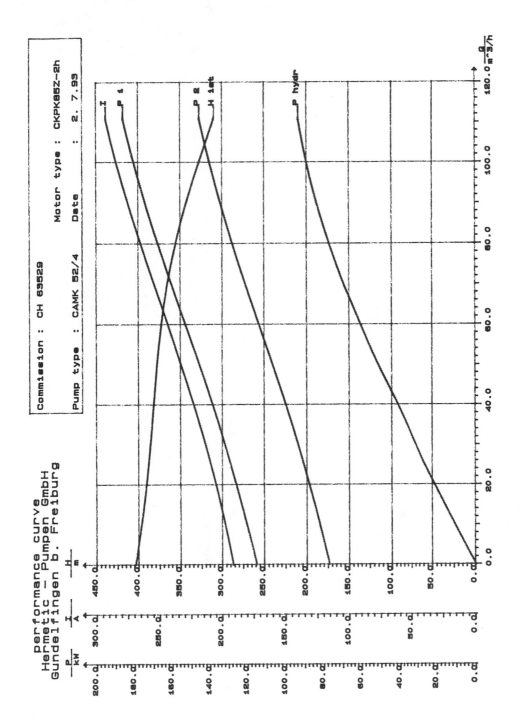

Fig. 3.66 NPSHR (pump) curve for canned motor centrifugal pump as shown in Fig. 3.64 and 3.65 (Hermetic)

181

Order-No.	:	CH 6329/93
Commission-No:		CH 63529
Repair-No	:	
Customer	:	605000 BASF AG
Pump type	:	
Material	:	CAMK 52/4
Date	:	1.7.93
Name	:	Scherer
Material	:	1.4571, 1.4408

Pump datas

Q :	20.0	95.0	110.0	m^3/h	H :	390.0 335.0 310.0 m
τ :						
n :	2960	1/min		Impeller φ D2		

Back vane φ DR	:	255.0 mm
Balancing hole φ d2	:	0.0 mm
d2 Impeller φ d2	:	*φ mm
Partial Flow Exit	:	1./2.68
Shaft boring D11	:	0.0 mm

Impeller φD2: 255.0
Impeller φR2: 0.0 mm

Running test

Fluid	:	Wasser	
Temp.	:	20.0	°C
Test Time	:	30.0	min

Pressure test

	Water	Air / N2
Pump :	64.0 bar	6.0 bar
Heating Jacket :	bar	bar
Stator :	bar	6.0 bar

Motor datas

Typ	:	CKPK85Z-2h
Motor No	:	285
Motor KW	:	150.0 kW
U : 380.0 V	F : 50 Hz	cosPhi : 0.92	I : 325.0 A	Schaltung : Dreieck
Iso.-Klasse	:	H
Insul.Resistance	:	
Winding-Resistance	:	1000.0 MΩ
Temperature	:	

Remarks

n	Q	H	I	P1	P2	ETA	cosPhi	P2	ETA	SH	Axialthrust		Clea-rence	NPSH	H rel	H duty
				Motor	Motor	Motor		Pump	Pump		MS	PS		Pump		
1/min	m^3/h	m	A	kW	kW	%	-	kW	%	mm	kg	kg	mm	m	%	m
2968	111		300	188			0.93							3.10		
2970	94.9		289	180			0.93							2.86		
2973	71.3		265	164			0.92							2.50		
2977	40.7		232	142			0.91							2.19		
2979	19.9		215	130			0.90							2.10		

Table 3-VIII.1 Operating parameters of canned motor centrifugal pump as shown in Fig. 3.64 and 3.65

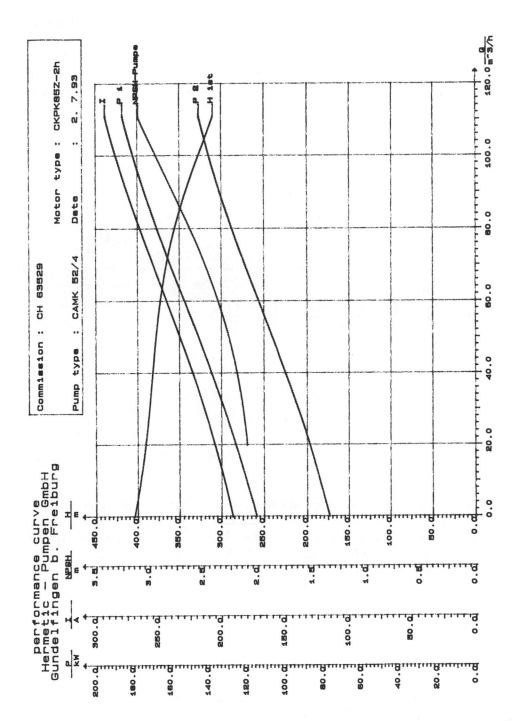

Fig. 3.67 NPSHR (pump) curve for canned motor centrifugal pump as shown in Fig. 3.64 and 3.65 (Hermetic)

183

3.5.3 Multi-stage canned motor pumps of tandem design

If due to demands of economical operation and the operating parameters, the number of stages (five to six) in canned motor pumps is exceeded, this leads to deflection problems at the gap seals of the impellers. The required impeller wear ring clearance depends on the configuration of the plain bearings and the tangent of the angle α, which is formed by the triangle created when the rotor shaft deflects, due to the pressure of its own weight and possible radial loads, towards the theoretical center of the pump (Fig. 3.68).

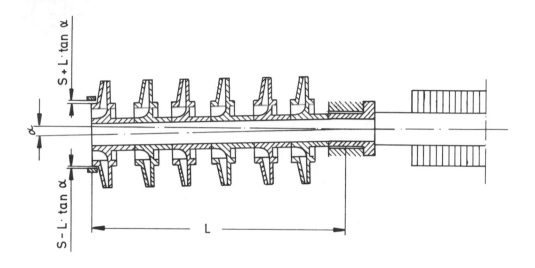

Fig. 3.68 Deflection of a multi-stage rotor

Due to the special nature of the applications for which hermetic centrifugal pumps are used, one almost general requirement is that austenitic materials are used for construction. The well-known poor friction properties of these materials requires an additional safety spacing between rotating and fixed wear rings in order to totally exclude the possibility of contact. The constructional tolerances required by this factor result in certain gap widths with correspondingly large gap flows; this either qualifies or totally cancels out the advantages of multi-stage design.

3.5.3.1 Multi-stage design and axial thrust balance

In the centrifugal pump sector there are multi-stage assemblies in which the pumping head is divided up between two multi-stage centrifugal pumps in back-to-back design with reversible flow direction in order to eliminate axial thrust. The precondition for such constructions is a flow supply line from pump 1 to pump 2 and an identical number of stages per pump. Due to the dangerous nature of fluids in the chemical and nuclear sector, it seems desirable, apart from using dynamic sealings, to keep the number of static gaskets to a minimum and to avoid the use of pipelines.

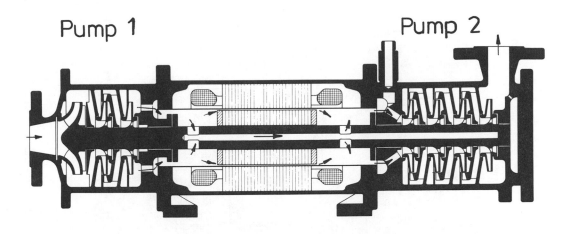

Fig. 3.6.9 Functional diagram of a tandem canned motor pump (Hermetic)

The pump shown in fig. 3.69 is based on a single-suction design with one-sided direction of flow, in which the total volume flow, after leaving pump 1, flows through the motor inside the rotor chamber, and from there to pump 2. This is possible either if clearance is created between motor shaft and rotor packet or if the major part of the volume flow is transported through the hollow shaft-design motor shaft from the front to the rear rotor chamber (Fig. 3.70). In both designs, flow over the rotor-stator gap for dissipation of motor losses is ensured, the gap flow rejoining the main flow in the rear rotor chamber. This method ensures excellent cooling of the motor without additional partial flow. In canned motor pumps of one-sided design, the cooling flow is booked as a loss factor in the economy calculation. This advantage of tandem-design is, however, compensated by the loss of pumping head which affects the fluid when it flows over the rotor. With the one-sided pump arrangement, the partial (loss) flow increases linearly with increasing pumping head.

In the tandem design, on the other hand, the losses in pumping head also increase linearly with increasing volume flow, as measurements have shown. In this type of arrangement, the number of stages of the individual pumps can either be the same or different. The required axial thrust balance is effected by a hydraulic balance device mounted in pump 2. The one-sided flow direction system also allows operation of the pump in in-line mode. When used in this mode, however, care should be taken to ensure the pump is adequately vented. Tandem pumps are manufactured with up to 12 stages and heads in the standard speed range (n = 2900 r.p.m) up to 500 m. This range can be further extended by converting the rotational frequency of the input motor power.

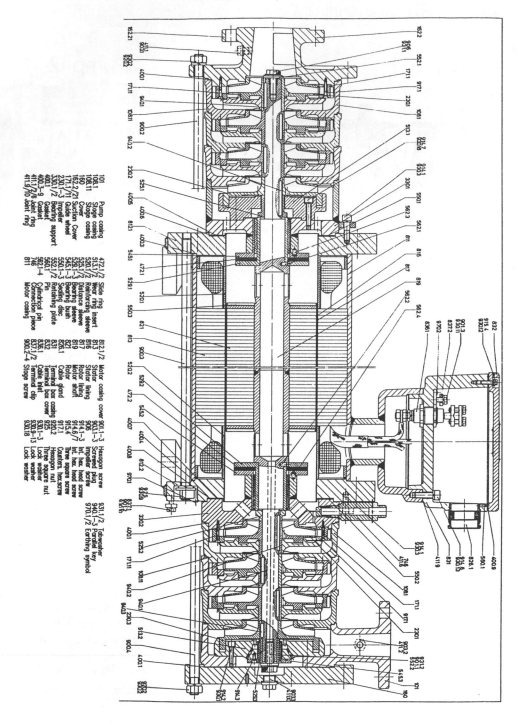

101	Pump casing	472.1/2	Slide ring	812.1/2	Motor casing cover	901.1-3	Hexagon screw	931.1/2	Tidewasher
108.1	Stage casing	513.1/2	Wear ring insert	813	Stator	903.1-3	Screwed plug	940.1-3	Parallel key
108.11	Stage casing	520.1/2	Reinforcing sleeve	816	Stator lining	906	Impeller screw	970.1/2	Earthing symbol
160	Cover	525.1/2	Distance sleeve	817	Motor lining	914.1-3	Int. hex. head screw		
162.2/21	Suction Cover	529.1-3	Bearing sleeve	819	Motor shaft	914.6/7	Int. hex. head screw		
171.1/11	Guide wheel	545.1-3	Bearing bush	821	Rotor	915.4	Three square screw		
230.1-3	Impeller	560.1-3	Sealing disc	826.1	Cable gland	917.1	Counters. hex.screw		
330.1/2	Bearing support	550.1/2	Retaining plate	831	Terminal box casing	920.2	Hexagon nut		
400.1	Gasket	552.1/2	Gasket	832	Terminal box cover	925	Three square nut		
400.3-9	Gasket	562.1-4	Pin	836.1	Cable inlet	930.1-3	Lock washer		
411.1/2/6	Joint ring	746	Connection piece	837.1/2	Terminal clip	930.8-13	Lock washer		
411.9/10	Joint ring	811	Motor casing	900.2-4	Stage screw	930.18	Lock washer		

Fig. 3.70 Cross-section of a canned motor centrifugal pump of tandem design (Hermetic)

186

Fig. 3.71 shows a cross-section of an eight-stage tandem pump. Fig. 3.72 shows the practical application of tandem pumps.

Fig. 3.71 Cross-section of a canned motor pump of tandem design (Hermetic)

3.5.3.2   Improved NPSHR characteristic in pumps of tandem design

In standard canned motor pumps the partial flow necessary to cool the motor passes through the rotor-stator gap and then continues through the hollow bore of the shaft to the suction side, or in pumps in accordance with standard hydraulics to DIN 24256. It passes into the area in the pump housing at the level of the impeller outer diameter which is subject to the pressure of the pumped fluid. In the first case a mixed temperature is created by the mixture of the pumped flow travelling in the direction of the pump with the heated partial flow. This mixed temperature adversely affects the NPSHR of the pump. In the second case, mixing takes place in the pressure chamber and here there is also an increase in temperature at the impeller intake due to the gap flow which is affected by the mixed temperature. This in turn increases the suction pressure of the pump. This is not the case in tandem design so that these pumps provide a better NPSHR response than the standard canned motor pumps, particularly in the partial load range.

3.5.4 Hermetic centrifugal pumps in barrel design for highly toxic fluids

The number of static gaskets required, and therefore the risk of leakage, increases with the number of stages of a pump. The barrel design shown in Fig. 3.73 and 3.74 can substantially increase the safety of such machines and reduce the number of static gasket to two. These seals can then be designed as twin or safety seals consisting of two O-rings with a

Fig. 3.72 Canned motor centrifugal pumps of tandem design in a technical process plant (Hermetic)

188

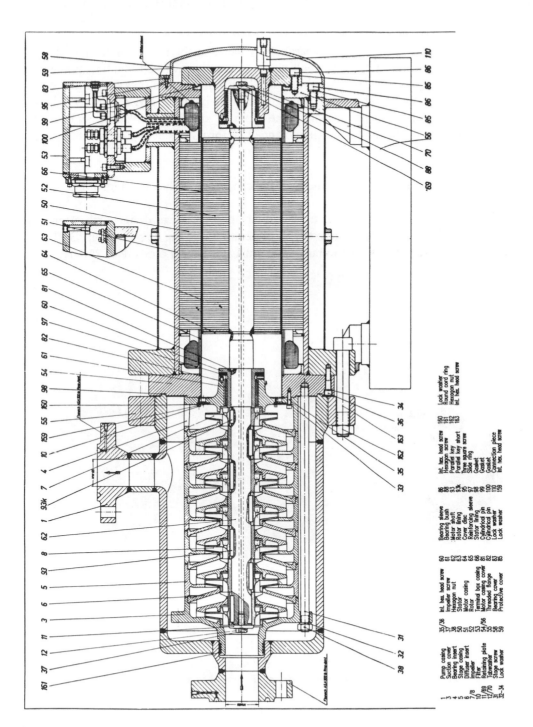

Fig. 3.73 Multistage centrifugal pump of barrel design (Hermetic)

189

control bore between them, which terminates in a trap enabling leakage of the first seal to be detected. With this arrangement the outer casing of the pump should preferably be made of forged material to satisfy the increased tightness requirements. Pumps of this kind are given a helium leak test.

This type of construction has proved particularly valuable for pumps of tandem design (Fig. 3.75, 3.76 and 3.77), because multiple stages are always available and only two static seals are necessary even for 12-stage pumps. The high delivery pressures frequently required for liquefied gas can be achieved by arranging tandem pumps in series. Fig. 3.79 shows a pump set of this kind for pumping phosgene at 95 °C at a head of 900 m, an example which illustrates the particular advantage of canned motor pumps over pumps with mechanical shaft seals.

A particular variant, the inline arrangement, of the can design is shown in Fig. 3.80.

Where high delivery and system pressures are demanded, the can design with a tandem arrangement provides a simple solution to this difficult task as can be seen in Fig. 3.81 and 3.81.1. This 12-stage pump is used to pump ammonia at a system pressure of 260 bar.

Fig. 3.74 View of the canned motor centrifugal pump shown in Fig. 3.73 (Hermetic)

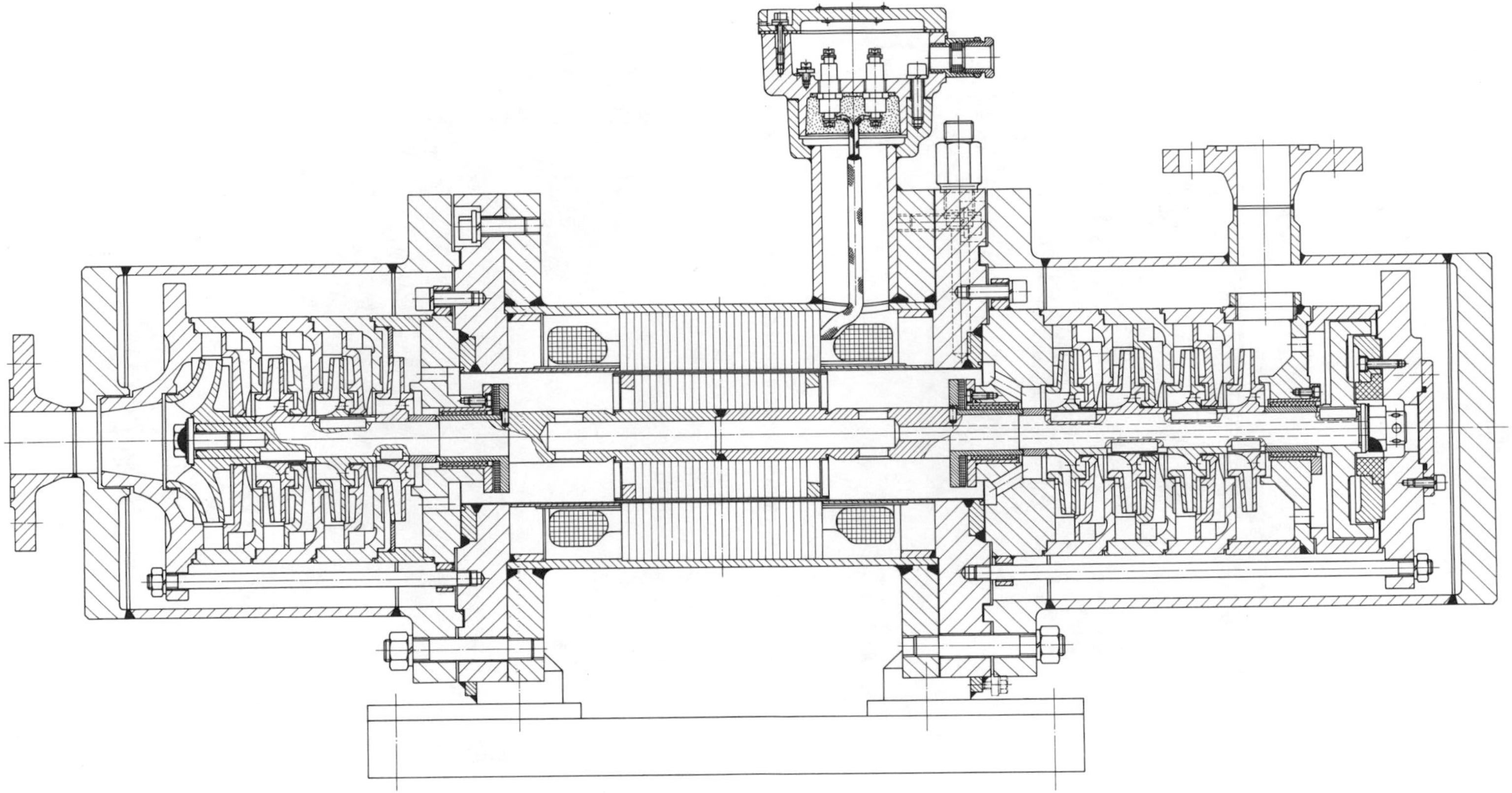

Fig. 3.75 Section view of a tandem centrifugal pump of barrel design (Hermetic)

Fig. 3.76 View of the hermetic tandem centrifugal pump shown in Fig. 3.75

Fig. 3.77 Hermetic tandem centrifugal pump for high-temperature fluids

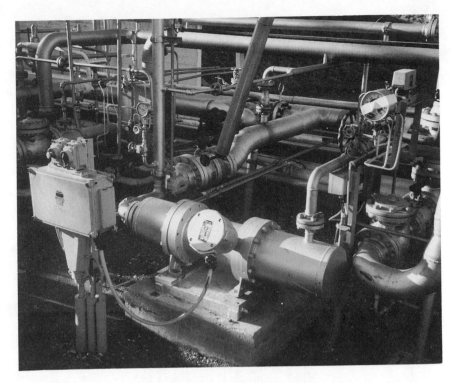

Fig. 3.78 Tandem canned centrifugal pump, as shown in Fig. 3.75, in a chemical plant (Hermetic)

Fig. 3.79 Tandem canned motor centrifugal pumps arranged in series (Hermetic)

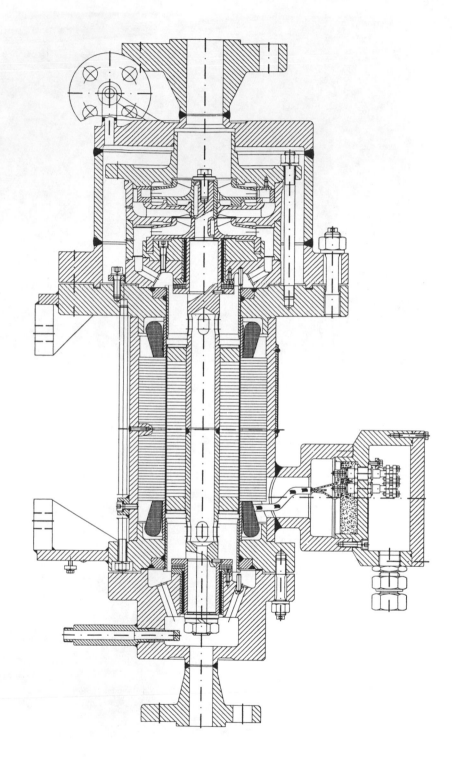

Fig. 3.80 Hermetic centrifugal pump with the hydraulic part mounted in a barrel in inline design (Hermetic)

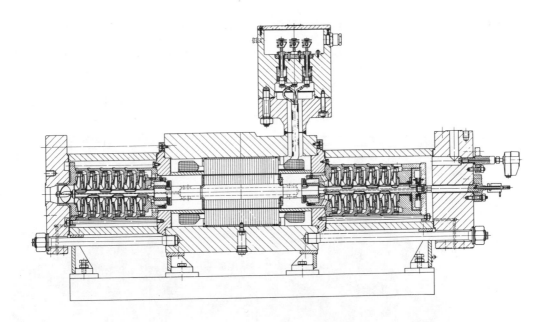

Fig. 3.81 12-stage, high-pressure, hermetic centrifugal pump of barrel design (Hermetic)

Fig. 3.81.1 View of high-pressure tandem centrifugal pump of barrel design, as shown in
Fig. 3.81 (Hermetic)

3.5.4.1 Canned motor pumps in nuclear auxiliary system

Nuclear auxiliary plants are designed for supply, cleaning and processing of the cooling water required by the reactor plant. Pumps used in nuclear process systems are divided into requirement categories (as well as nuclear safety classes). The requirement category to which a pump belongs depends on the safety class of the system in which it is operated. Each main component is then of the same safety class.

Fig. 3.8.2 Centrifugal canned motor pump in a nuclear reactor plant (Hermetic)

Safety class 1 includes all pumps which are characterized by a high degree of reliability and availability according to technical safety criteria. Canned motor pumps of special design meet these requirements, not least due to the fact that they allow fully hermetic flow circuits as they do not require dynamic shaft sealing. The remaining static gaskets mostly present no problems, but can nevertheless be a latent source of danger, and their number should thus be kept to a minimum. Pumps of this kind are subjected to a helium leak test.

They guarantee a leak rate of $< 10^{-6}$ mbar·l/s. Fig. 3.82 shows this kind of pump in use in a nuclear reactor plant.

3.5.5 Canned motor pumps used at high temperatures

Heat transport using thermal oils in the 150 - 400 °C range and the pumping of molten liquids up to 500 °C, e.g. sodium chloride, aluminum sulphate, sodium silicate, sodium sulphate, lithium chloride, ammonium sulphate etc. poses sealing problems for conventional centrifugal pumps which are often difficult to solve. Canned motor pumps solve these problems in a relatively simple manner. Heat transfer oils can be pumped with canned motor pumps both with a cooling medium and without a cooling medium, whilst the design without a cooling medium is used for molten substances.

3.5.5.1 Externally-cooled canned motor centrifugal pumps

With this design, the pump and the motor are physically separated from each other by a spacer (Fig. 3.83, 3.84 and 3.85) to prevent heat being transferred from the pump to the motor.

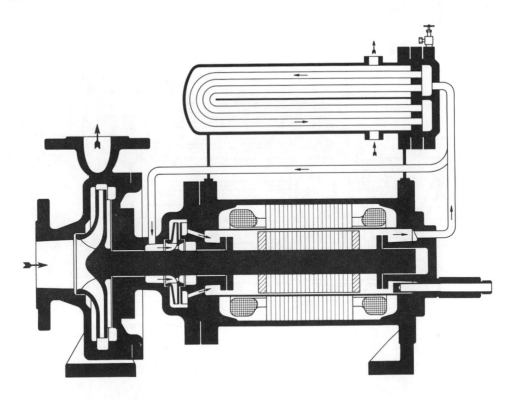

Fig. 3.83 Centrifugal pump for thermal oil with externally cooled canned motor (functional diagram)

Pressure between the pump and rotor chamber is equalized through a relatively narrow, long annular gap. There is an impeller in the motor itself and the same liquid in the rotor chamber is circulated through a external cooler which is either arranged around the motor or mounted separately. Heat due to energy losses in the motor is absorbed by a cooling liquid. In this way, two pumping circuits are formed with different temperature levels. The operating circuit can withstand temperatures up to 400 °C, whilst the secondary cooling circuit remains at substantially lower temperatures ranging from 60 to 80 °C. The motor winding is designed to insulation class H. An exchange of liquids between the two temperature zones is practically eliminated because of the pressure equalization in the annular gap. By the use of a separate cooling circuit it is not necessary to cool the motor partial flow from the high operating temperature level to a level acceptable for normal canned motors, in order to then add it to the delivery flow. This would cause an unacceptably high level of energy loss. A further application for this type of pump is in the conveyance of liquids for which a temperature increase, such as for example occurs in the flow through the stator-rotor gap, would tend towards crystallization or polymerization. The externally cooled design has the advantage that the motor meets the explosion protection requirements of protection type EEx de II CT1-T4.

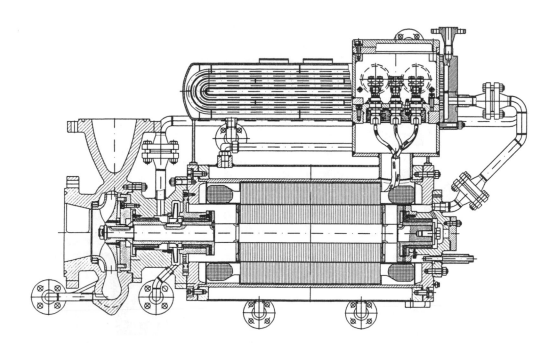

Fig. 3.84 Cross section of a single staged canned motor centrifugal pump acc. to DIN 24256/ ISO 2858 with externally cooled motor

Fig. 3.85 Cross-section of a canned motor centrifugal pump with external cooling (Hermetic)

The disadvantage in this case is the need to use a cooling liquid, mainly water, which poses the danger of freezing during shutdown of a system installed outdoors. For large rates of delivery, the use of pumps with API hydraulics in accordance with Fig. 3.86 with the casing bearings in the center axis is recommended, to allow for the expansion due to thermal stress. Fig. 3.87 and 3.88 show plants in which canned motor pumps with externally-cooled motors are installed.

3.5.5.2 Canned motor pumps with self-cooled motors

There is not always sufficient cooling water of adequate quantity, or an alternative cooling medium, available for motor cooling. The drainage channels are frequently at the limit of their thermal stress, so that a further rise in temperature would be unacceptable for reasons of environmental protection. With molten materials of various kinds there is always a requirement to heat the pumped material before use both in the pump and in the motor. The temperatures necessary for this are usually in a range which exceeds the permissible maximum temperature of the aforementioned insulation class H.

Fig. 3.86 Canned motor centrifugal pump with externally-cooled motor and API hydraulics (Hermetic)

Tasks in the high temperature range can be easily solved using the so-called "hot motor", the motor with a special winding to insulation class C, shown in Fig. 3.89 and 3.90. Silicon ceramic is used as the insulating material, with oxidation of the copper being avoided by suitable measures. The wire-slot insulation and impregnation medium are resistant to prolonged high temperature and have a high insulation strength whilst at the same time providing adequate flexibility to compensate for the thermal expansion of the wire. Windings of this type can withstand continuous temperatures of 420 °C in the winding. This enables economical motor loading up to an inlet liquid temperature of 360 °C (Fig. 3.91).

Fig. 3.87 Canned motor centrifugal pumps used in a chemical plant for heat transport
(Hermetic)

Fig. 3.88 Canned motor centrifugal pumps used in a technical process requiring high fluid
temperatures and large delivery rates

The testing authorities of Federal Republic of Germany still issue explosion protection certificates for Temperature Class T1. These motors differ structurally from canned motors only in that the terminal box is spatially separate from the stator casing, to keep the temperature low at that point because of the incoming normal feed cable. Fins located radially on the motor casing improve the heat dissipation to the environment due to natural convection, for models which are not cooled (Fig. 3.89 upper half and 3.90).

The equilibrium temperatures on the surface in this case are between 270 and 340 °C depending on the radiation and size.

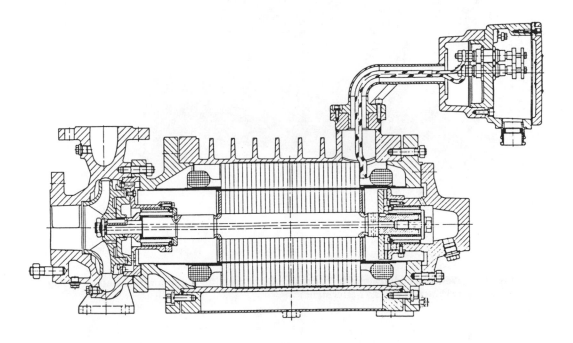

Fig. 3.89 Single-stage canned motor centrifugal pump with standard hydraulics to DIN 24256/ISO 2858 and a self-cooling motor, for high fluid temperatures (Hermetic)

Pumping of molten substances requires the aggregate to be heated before the pump is started and therefore pumps and motor casings are provided with a heating jacket (Fig. 3.93 and 3.94). The heating medium, thermal oil or steam, required for heating before startup is also used during running as external cooling for the canned motor.

Particularly high liquid temperatures require devices in addition to the measures already outlined, to protect the motor from excessive thermal stress. Fig. 3.95 is a functional diagram of a canned motor pump provided with positive circulation and a "hot" motor, which is fitted with an air cooler to reduce the winding temperature. In this case the inlet temperature is 400 °C with a circulation equilibrium temperature in the motor of approximately 200 °C. Fig. 3.96 is a view of a pump of this kind.

Fig. 3.90 View of a high temperature, canned motor centrifugal pump with a stator winding as shown in Fig. 3.91 (Hermetic)

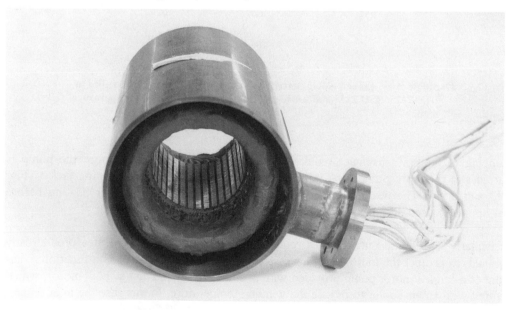

Fig. 3.91 Stator of a canned motor with a silicon-ceramic winding

Fig. 3.92 Self-cooled canned motor centrifugal pumps in a production plant

The lower circuit temperature means that the motor can take a higher electrical stress. The lubricating-cooling circuit cooled to half the temperature of the operating liquid has a higher viscosity, which has a favorable effect on the hydrodynamic lubrication of bearings. The partial flow Q_T can also be cooled by means of a honeycomb radiator with a fan connected to it as shown in Fig. 3.97.

The canned motor pumps shown so far for fluids heated to high temperature are fitted with single-stage hydraulics to DIN or API. Obviously multi-stage pumps can also be provided with "hot motors" (Fig. 3.98) in the normal and also high pressure range.

The unavoidable heat loss from fluids heated to high temperature to the surrounding atmosphere due to radiation can be substantially reduced if the pump set is insulated as shown in Fig. 3.99. It should, however, be noted that motors provided with insulation to class C are only permitted up to 200 °C with this arrangement.

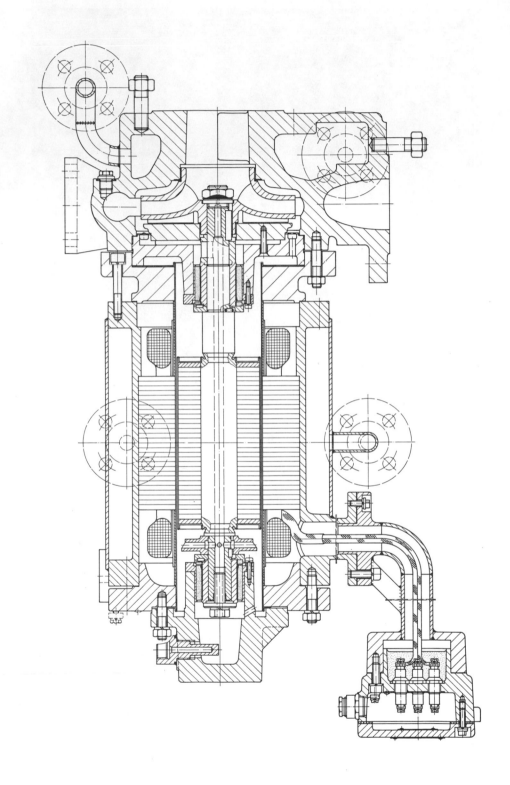

Fig. 3.93 Canned motor centrifugal pumps fitted with heating jacket for pumping molten liquids (section view) (Hermetic)

206

Fig. 3.94 View of canned motor centrifugal pump shown in Fig. 3.93

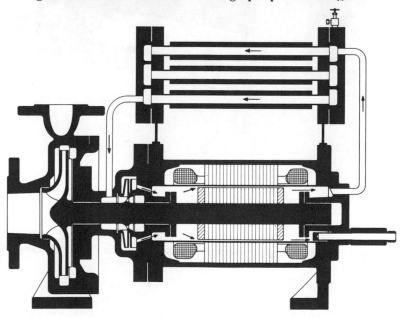

Fig. 3.95 Functional diagram of a canned motor centrifugal pump with air-cooled motor for pumping fluids heated to high temperatures

Fig. 3.96 View of high temperature pump as shown in Fig. 3.95 (Hermetic)

Fig. 3.97 Canned motor centrifugal pump for high liquid temperatures with honeycomb radiator and fan (Hermetic)

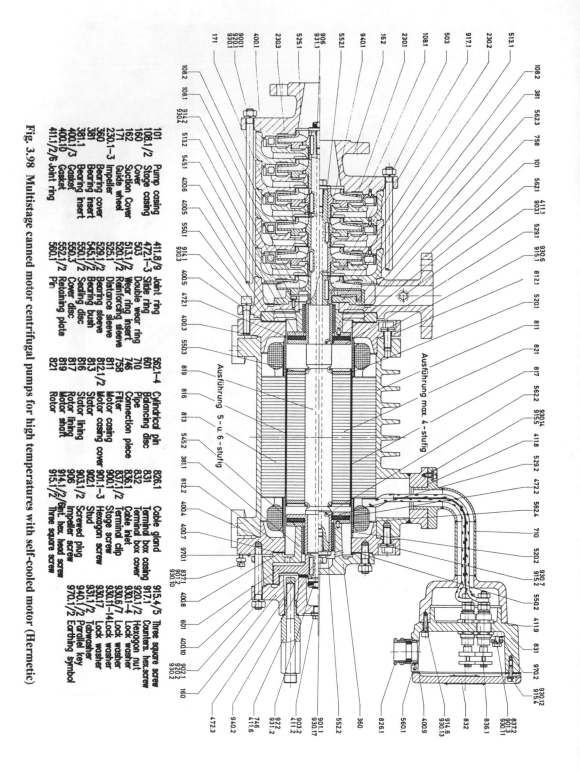

Fig. 3.98 Multistage canned motor centrifugal pumps for high temperatures with self-cooled motor (Hermetic)

101	Pump casing	411.8/9	Joint ring	560.1	Pin	826.1	Cable gland	915.4/5	Three square screw
108.1/2	Stage casing	472.-3	Slide ring	562.1-4	Cylindrical pin	831	Terminal box casing	917.1	Counters. hex.screw
160	Cover	503	Double wear ring	601	Balancing disc	832	Terminal box cover	920.1-4	Hexagon nut
162	Suction Cover	513.1/2	Wear ring insert	710	Pipe	836.1	Cable inlet	930.1-4	Lock washer
171	Guide wheel	520.1/2	Reinforcing sleeve	746	Connection piece	837.1/2	Terminal clip	930.6/7	Lock washer
230.1-3	Impeller	525.1	Distance sleeve	758	Filter	900.1	Stage screw	930.11-14	Lock washer
360	Bearing cover	529.1/2	Bearing sleeve	811	Motor casing	901.1-3	Hexagon screw	930.17	Lock washer
381	Bearing insert	545.1/2	Bearing bush	812.1/2	Motor casing cover	902.1	Stud	931.1/2	Lock washer
381.1	Bearing insert	550.1/2	Sealing disc	813	Stator	903.1/2	Screwed plug	940.1/2	Parallel key
400.1/3	Gasket	550.3	Cover disc	816	Stator lining	906	Impeller screw	970.1/2	Earthing symbol
411.1/2/6	Joint ring	552.1/2	Retaining plate	817	Rotor lining	914.1/2	Cyl. hex. head screw		
				819	Rotor	915.1/2/6	Three square screw		
				821	Motor shaft				

Ausführung 5- u. 6-stufig

Ausführung max. 4-stufig

Fig. 3.99 Insulated canned motor pump unit in a chemical plant (Hermetic)

3.5.5.3 Canned motors with at-rest heating

The different temperature and viscosity behavior of the fluids to be pumped means that the operational readiness of the pump cannot always be guaranteed. Some fluids become solid at normal ambient temperature or become so viscous that they are not suitable for operation with canned motor pumps. If steam or heat transfer oil heating cannot be used, at rest heating by means of the motor winding can be used to keep the fluid to the necessary minimum temperature to guarantee constant operational readiness. To do this, a low voltage is applied to the motor winding via a transformer to keep the winding at a suitable temperature. A control mechanism using a thermostat (PT 100) is used to ensure that the motor is kept warm without overheating. Fig. 3.100 and 3.100.1 are diagrams of an at-rest heating system with a star-delta motor circuit.

3.5.6 Canned motor pumps in the low temperature and liquid gas range

The liquid gas sector has a very important application area for hermetic centrifugal pumps. It is known that difficulties with shaft sealing occur when conventional pumps are used to pump liquid gases. Because of the high system pressures to which the liquid gases are subjected and the associated unavoidable leak rates, the pumped liquid vaporizes and ice crystals form in the area of the mechanical sealing. This can lead to damage to the sliding faces. Very often the liquid gases to be pumped are highly explosive or toxic (vinyl chloride, chlorine, phosgene, ethylene oxide), so that the use of a mechanical sealing in those cases represents an unacceptable safety risk. Canned motor pumps therefore have been used prefered for pumping liquid gases. Fig. 3.101 shows a photograph of a refrigeration plant with NH_3 circulation. The moisture content of the atmosphere causes an ice coating to form due to condensation on the surfaces of the pump and the motor after a short running time and this means that the use of "open" centrifugal pumps for such applications is questionable. This applies in particular to plants where in a dangerous situation, e.g. where there is a leakage of large amounts of NH_3 from a damaged shaft seal there is no means of escape, such as in ships' refrigeration plants. Some special features must be taken into account when pumping liquids which are in a state of vapor-pressure equilibrium:

With liquid gases almost all of the motor heat losses are dissipated to the flow, thus posing the danger of vaporization of the fluid. The formation of gas must, however, be avoided under all circumstances in order to maintain the functional efficiency of the pump unit. Several design solutions are available to prevent vaporization of the heated partial flow in the motor.

212

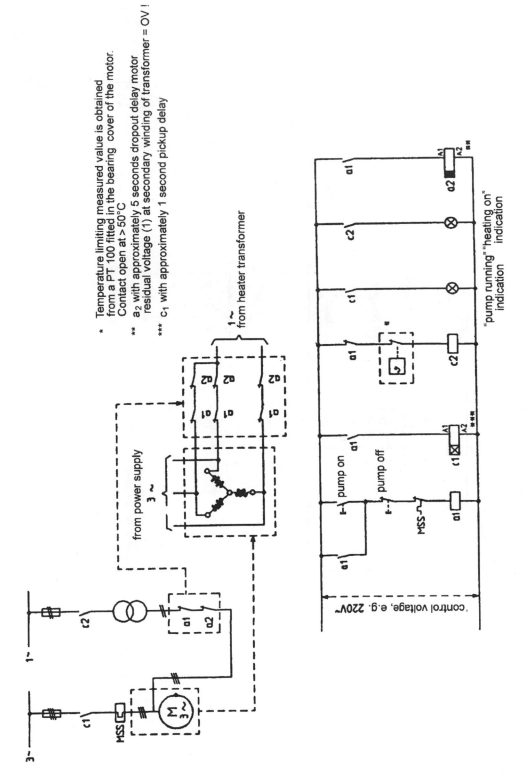

* Temperature limiting measured value is obtained
  from a PT 100 fitted in the bearing cover of the motor.
  Contact open at >50°C

** a_2 with approximately 5 seconds dropout delay motor
   residual voltage (1) at secondary winding of transformer = OV !

*** c_1 with approximately 1 second pickup delay

1~
from heater transformer

from power supply
3~

"pump running" "heating on"
indication   indication

pump on

pump off

control voltage, e.g. 220V~

Fig. 3.100 At-rest heating for canned motors with star connection

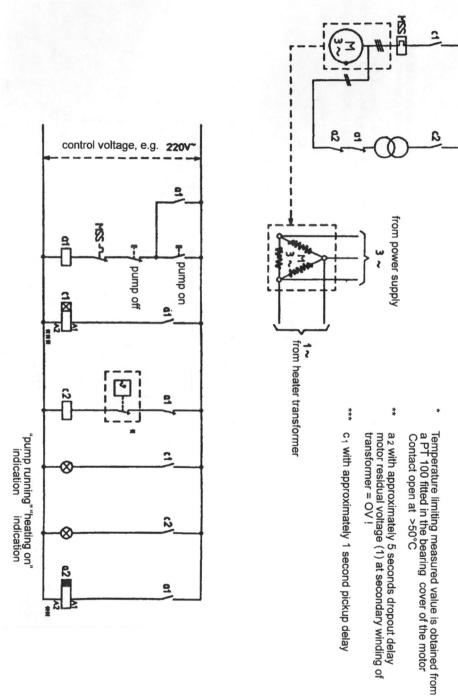

Fig. 3.100.1 At-rest heating for canned motors with delta connection

control voltage, e.g. 220V~

pump on
pump off

"pump running" "heating on"
indication   indication

from power supply
3 ~

from heater transformer
1 ~

* Temperature limiting measured value is obtained from
a PT 100 fitted in the bearing cover of the motor
Contact open at >50°C

** a 2 with approximately 5 seconds dropout delay
motor residual voltage (1) at secondary winding of
transformer = OV !

*** c 1 with approximately 1 second pickup delay

214

Fig. 3.101 Canned motor centrifugal pumps in a refrigeration plant (Hermetic)

3.5.6.1 Guidance of partial flow

A precondition for the use of plain bearings is that the cooling lubricating flow is always in a liquid state even after absorption of the motor heat losses. In principle, vaporization within the pump is always precluded provided the corresponding static pressure is still above that of the vapor pressure of the heated liquid. Liquid gas pumps with canned motors are essentially used according to three functional principles (Fig. 3.102)

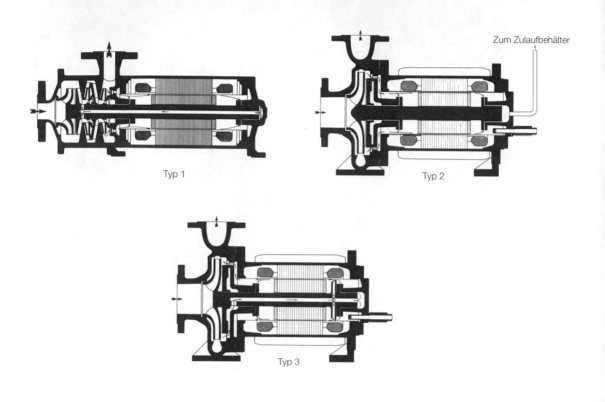

Zum Zulaufbehälter = to supply tank

Typ = type

Fig. 3.102 Functional principle of canned motor centrifugal pumps for pumping liquid gases

- *Type 1 multistage liquid gas pumps (Fig. 3.102)*

The partial flow branched off at the discharge side passes through the motor and is then fed not to the suction side but instead passes between two impellers into an area which has a high pressure potential with respect to the inlet pressure of the pump. Fig. 3.103 shows the construction of a pump of this kind and Fig. 3.104 shows one with a sectioned housing.

216

- Type 2 single-stage liquid gas pumps

The partial flow raised to a higher temperature by the heat losses from the motor are returned to the inlet tank via a separate return line.

- Type 3 single-stage liquid gas pumps

The return of the heated partial flow to the discharge side of the pump is aided by a pressure-increasing impeller in the rotor space, thus ensuring a constant cooling-lubricating flow independent of the particular duty point of the pump. This results in constant cooling conditions over the complete range of the characteristic curve.

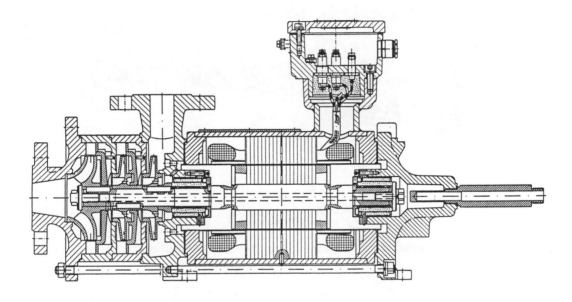

Fig. 3.103 Section view of a multistage canned motor centrifugal pump for liquefied gases (Hermetic)

With types 1 and 3 the heat losses from the motor are absorbed by the main delivery flow on the discharge side. The arrangement of type 2 shows that the heat losses from the motor are absorbed by the liquid in the inlet tank. Liquid gas pumps of this type have a disadvantage in that additional pipelines are needed from the pump to the inlet tank. Type 3 is particularly advantageous for pumping liquid gas as its partial flow which is constant independent of the head enables precise heat balance and Q_{min} - Q_{max} calculations (refer to Section 3.2.7 "Heat balance of a canned motor pump"). A schematic diagram (Fig. 3.105) is used to further explain its functioning.

Fig. 3.104 Sectioned model of a canned motor centrifugal pump as shown in Fig. 3.103

A self-cleaning filter (1) is fitted in the pump section. A portion of the partial flow taken from the main flow for cooling the motor and lubricating the plain bearings is passed through this filter and bores (2) through a chamber (3). This "partial flow" enters the radial bores (4) in the hollow shaft (5) of the impeller and passes via radial bores (6) into an auxiliary impeller (7) within the rotor space, which returns the flow through the clearance (8) between the stator and rotor and through bore (9) to the pump discharge side. The pressure level at the take-off point corresponds approximately to the delivery pressure of the pump. This pressure is therefore superimposed on the complete rotor space. Vaporization is precluded because the static pressure is always greater than the vapor pressure of the heated cooling-lubricating flow. The auxiliary impeller (7) which provides the necessary pressure increase for the gap flow in the motor is designed such that it can take the partial flow directly from the bores in the shaft. The advantage of this is a compact construction, i.e. it can be fitted inside the rotor space and an additional sealing gap is completely unnecessary, thus improving the efficiency of the motor.

This type of partial flow guidance also enables the same construction and dimensions to be used for pumping both liquid gas and non-liquid gas.

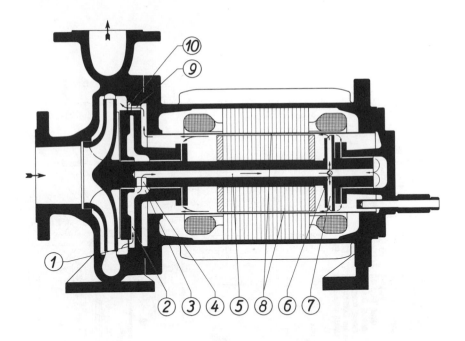

Fig. 3.105 Schematic view of the partial flow in a single-stage canned motor centrifugal pump for near-boiling fluids

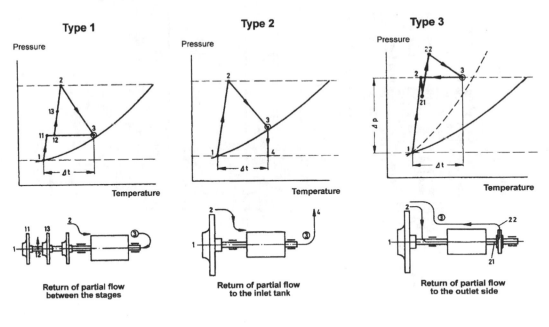

Fig. 3.106 Temperature-pressure behavior of partial flow Q_T in canned motor centrifugal pumps

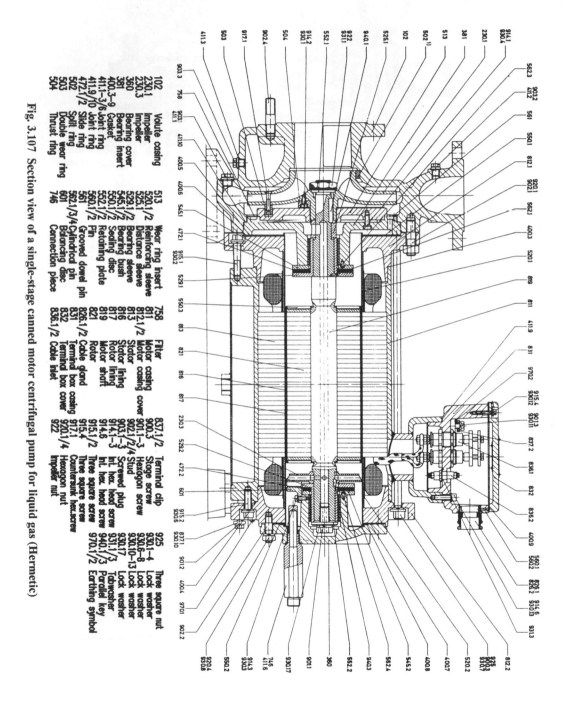

102	Volute casing	513	Wear ring insert	758	Filter	837.1/2	Terminal clip
230.1	Impeller	520.1/2	Reinforcing ring	811	Motor casing	900.3	Stage screw
230.3	Impeller	525.1	Distance sleeve	812.1/2	Motor casing cover	901.1–3	Hexagon screw
360	Bearing cover	529.1/2	Bearing sleeve	813	Stator	902.1/2/4	Stud
381	Bearing insert	545.1/2	Bearing bush	816	Rotor lining	903.1–3	Screwed plug
400.3–9	Gasket	550.1/2	Sealing disc	817	Motor shaft	914.1–3	Int. hex. head screw
411.1–3/6	Joint ring	552.1/2	Retaining plate	819	Motor lining	914.6	Int. hex. head screw
411.9/10	Joint ring	560.1/2	Pin	821	Rotor	915.1	Three square screw
472.1/2	Slide ring	561	Grooved dowel pin	826.1/2	Cable gland	915.4	Three square screw
502	Split ring	582.1/3/4	Cylindrical pin	831	Terminal box casing	917.1	Three square screw
503	Double wear ring	601	Balancing disc	832	Terminal box cover	920.1/4	Hexagon nut
504	Thrust ring	746	Connection piece	836.1/2	Cable inlet	922	Impeller nut

925	Terminal clip	930.1–4	Lock washer
900.3	Stage screw	930.6–8	Lock washer
		930.10–13	Lock washer
		930.17	Lock washer
		931.1/3	Tabwasher
		940.1/3	Parallel key
		970.1/2	Earthing symbol

Fig. 3.107 Section view of a single-stage canned motor centrifugal pump for liquid gas (Hermetic)

Fig. 3.108 Sectioned model of a canned motor centrifugal pump as shown in Fig. 3.107

The pressure-temperature relationships within the pumps for the three different hermetic liquid gas pumps operating according to the three different principles is shown in Fig. 3.106.

From this it can be seen that fluids with an extremely steep temperature-vapor pressure characteristic can also be pumped by a type 3 liquid gas pump. The pump pressure is supermposed on the liquid in the rotor space, thus preventing an excessive temperature rise of the fluid and/or enabling liquids with a steep vapor pressure rise to be handled.

Fig. 3.107 and 3.108 show the construction of such pumps.

3.5.6.2 NPSHR and NPSHA for liquid gas plants (refer also to Section 2.9.7 "The behavior of the NPSHR of centrifugal pumps when pumping near boiling fluids" (liquid gas pumping))

Liquid gases are mainly pumped in a near boiling state. For economic reasons (overall heights of plants) the lowest possible NPSHR values for pumps are desirable. This can be achieved by design measures, examples of which follow:

- Impellers with double curvature blading

- A change in the impeller geometry of the first impeller on multistage pumps in that an impeller with a specific speed which is higher than the following impellers is fitted (Fig.3.109 and 3.110). This means that the first impeller runs in the partial load area, i.e. at lower NPSHR values. This improvement in the NPSHR is, however, obtained at the expense of the overall efficiency η_{ges}. In the example in Fig. 3.110 the NPSHR is reduced by 1 m at the cost of a reduction in the efficiency of the first stage by 4%. Because it is a four-stage pump, the overall efficiency of the pump is reduced by only 1%. This is completely acceptable in view of the reduced construction exrenditure building height of 1 m.

- A further possible way of reducing the NPSHR is by fitting an inducer in the suction connection of the pump (refer to Section 2.9.8 "Inducers, purposes and operating behavior").

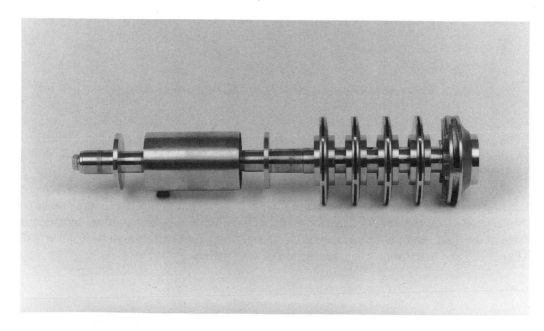

Fig. 3.109 Impeller set of a multi-stage canned motor centrifugal pump with a first stage suction impeller (Hermetic)

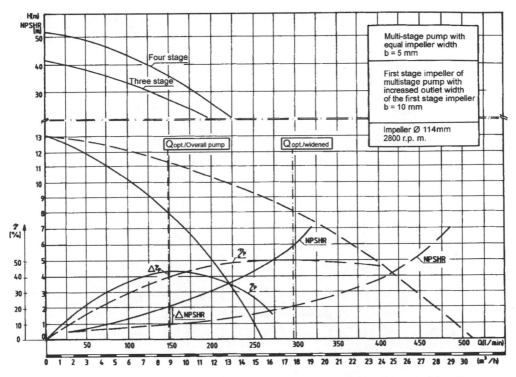

Fig. 3.110 Family of characteristics of a canned motor centrifugal pump with a first stage suction impeller (Hermetic)

3.5.7 Pumping suspensions

Impurities entrained in the pumped flow, such as welding beads, residues from tank wagons or other solids are prevented from entering the rotor space by the filtration of the partial flow as described in the introduction. Fluids with suspended solids of various kinds, particularly sludge and fibrous materials do, nevertheless, still require special measures for problem-free fluid pumping. A pump with a separate cooling-lubricating circuit as shown in Fig. 3.83 is used for this purpose. In addition, a clean liquid must in this case be fed into the rotor space (Fig. 3.111 and 3.112). This can be either the filtered vehicle of the suspension, a neutral liquid compatible with the pumped liquid or else a liquid which must be added to the circuit as part of the process. A piston diaphragm pump is suitable for this purpose (as this is also a hermetic pump), to generate the necessary higher pressure than that predominating in the radial clearance whilst at the same time maintaining precise control of the amount of flushing liquid injected. The amount of flushing liquid flow necessary is relatively small and varies between 2 and 10 l/h, depending on the size of the motor. If it is necessary to minimize the flushing flow entering the primary circuit, a mechanical seal is used instead of the radial clearance and the pressure conditions in the secondary circuit are appropriate. The injected amount is then reduced to the leak rate of the radial shaft seal. This is called a semi-hermetic centrifugal pump. The use of a canned motor pump with an integrated metering pump is shown in Fig. 3.113.

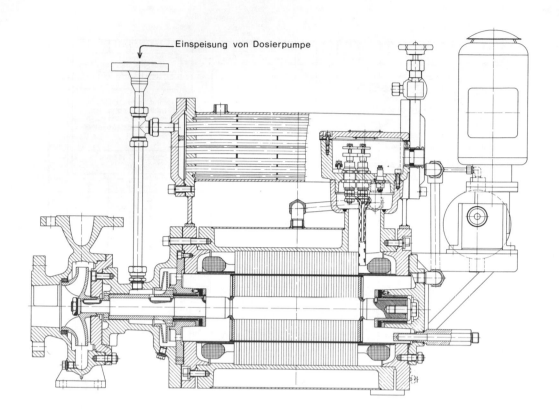

Einspeisung von Dosierpumpe

Einspeisung von Dosierpumpe = supply of metering pump

Fig. 3.111 Canned motor centrifugal pump with top-mounted piston diaphragm metering pump for suspensions (Hermetic)

3.5.8 Vertically-mounted, submerged canned motor pumps

There are many pumping processes in chemical plants and in nuclear plants where highly toxic substances must not be drained from vessels via ground drains. Submerged canned motor pumps as shown in Fig. 3.114 and 3.115 are used for such applications. The hydraulic balance in this case must be increased by the amount of the weight produced by the impeller. The connecting cable to the motor is brought out of the vessel through a separate conduit pipe welded to the stator casing, without contact with the product.

3.5.9 Canned motor pumps in high pressure systems

Applications which are mainly reserved for hermetic canned motor pumps are those tasks in which liquids must be continuously circulated under high system pressures (100 to 1200 bar). Conventional centrifugal pumps are mainly unsuitable for this task because it is either impossible to seal the pump shaft or to do so is unacceptably expensive. Canned motor pumps provide a technically simple solution to this problem. To be able to work under high pressure conditions, the complete pump set must be mounted in a high pressure body which can withstand the particular operating pressures and temperatures. This consists mainly of three main parts, the pump and motor sections and the intermediate section. The stresses produced by high internal pressures in these components are contained without danger if the enclosing body is machined from a forged blank. To ensure a low stress pattern which is easy to determine, the most satisfactory basic shape is a thick-walled hollow cylinder. The individual components are bolted together using high tensile expansion bolts with inserted comb-profiled high pressure seals (Fig. 3.116). Pure metal seals are also used for many types [3-17]. The critical component of the pump set is the thin-walled can (wall thickness 1 mm) which is subject to system and delivery pressure and must therefore be appropriately supported. The stator plates

Fig. 3.112 View of canned motor centrifugal pump for
suspensions as shown in Fig. 3.111

can be used to provide mechanical support to the can in the area of the stator pack, provided the stator pack has been specifically designed for this purpose. This requires the stator plate to be single-chambered on the outer circumference and also an axial support in the area of the stator, to avoid buckling or deflection. In modifying the stator cut of a standard canned motor for high pressure application the copper-carrying slots have to be displaced radially in order to get space for ceramic bars in order to close the slots for protection of the winding (Fig. 3.117 and 3.118). These are also necessary for force transmission and they must be capable of taking the high mechanical loads and also be made of nonmagnetic material to ensure that additional magnetic losses are obviated. The displacement of the winding slots outwards also has a beneficial effect on the constructive design on the canned motor from a static point of view. Suitable support rings are provided to support the ends of the can outside the stator pack. The additional space gained by shifting the winding

225

outwards can be used for this purpose. The previously-described arrangement of the punched cut-outs can be successfully used for internal pressures for up to 900 bar. At pressures ≥ 900 bar it is necessary for the punched cutouts to be completely closed up to the internal diameter (Fig. 3.119).

Fig. 3.113 Canned motor centrifugal pump for pumping suspensions in a technical process plant

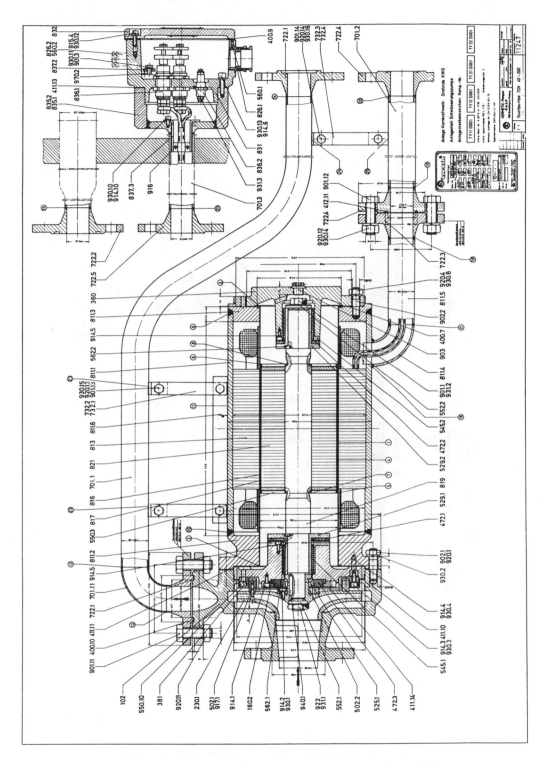

Fig. 3.114 Section view of a submerged canned motor centrifugal pump (Hermetic)

227

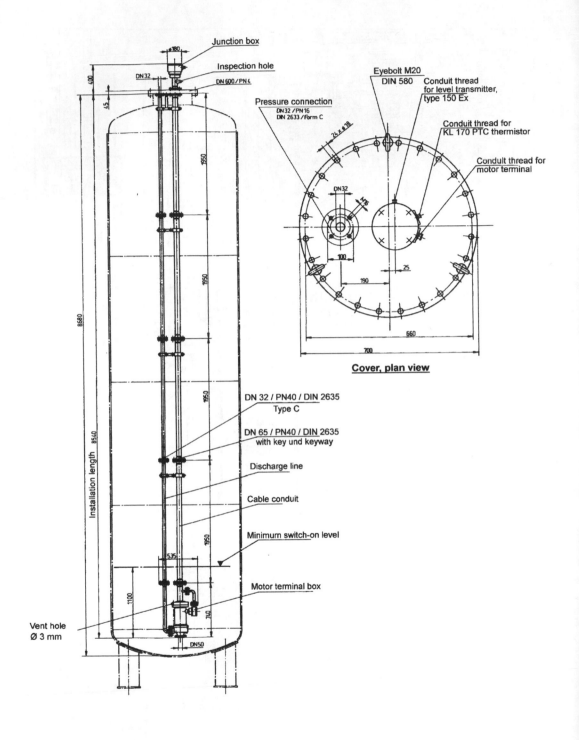

Fig. 3.115 Installation diagram of a vertical canned motor centrifugal pump in a vessel (Hermetic)

228

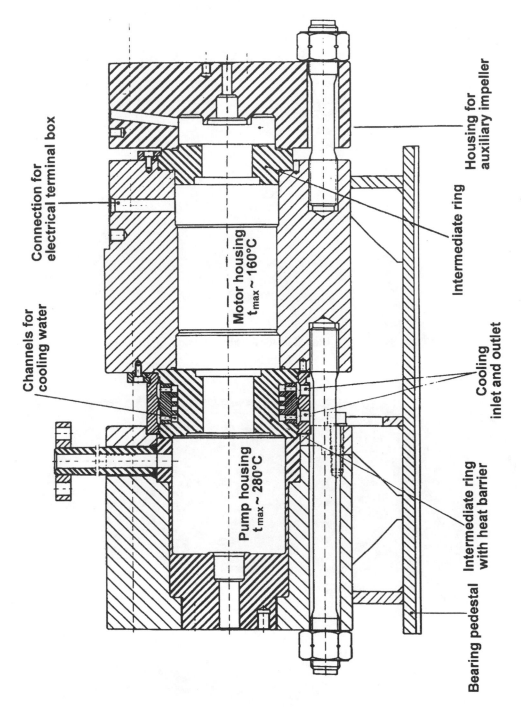

Connection for electrical terminal box

Housing for auxiliary impeller

Channels for cooling water

Motor housing $t_{max} \sim 160°C$

Intermediate ring

Cooling inlet and outlet

Pump housing $t_{max} \sim 280°C$

Intermediate ring with heat barrier

Bearing pedestal

Fig. 3.116 High pressure housing for centrifugal pump and canned motor

229

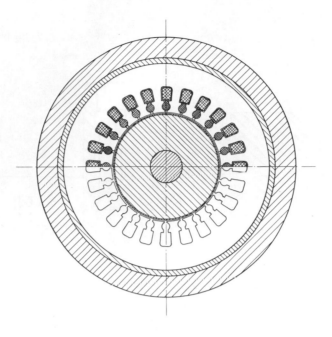

Fig. 3.117 Punched cut of a canned motor for high pressure applications

Fig. 3.118 Canned motor for high system pressures (stator) (Hermetic)

The can is sealed from the pressurized part by welding the can into the front and rear motor end covers [3-18].

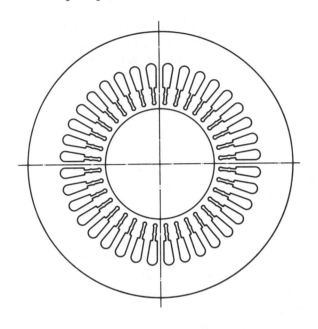

Fig. 3.119 Punched cut of a canned motor for system pressures > 900 bar

Thin walled cans can be used with the aforementioned construction, thus minimizing the electrical and magnetic losses and achieving good efficiency. The terminal box for the power supply deserves particular attention. This has to be suitably designed to withstand the rated pressure, so as to prevent the leakage of the pumped fluid from the terminal box in the event of an internal leak from the rotor space to the stator space. The regulations on the use of motors in explosion-protected rooms (VDE 0171) (EN 50014 to 50020) must be complied where necessary.

3.5.9.1 Typical uses of hermetic centrifugal pumps in high pressure applications

The advantages of centrifugal pumps - almost continuous delivery, matching of delivery rate by throttling, low power/weight ratio, energy saving, coupled with a hermetic drive - solves today's difficult pumping tasks with surprising simplicity and above all safely, economically and without environmental pollution.
A few examples:

- Recirculation pump in a recirculatory reactor [3-17]

Canned motor centrifugal pumps have been shown to be particularly successful in maintaining a driving jet in a recirculatory reactor under high pressure conditions (Fig. 3.120). The task shown in the example is to generate a recirculatory pressure head of 45 m with an aggressive suspension at an inlet system pressure of 700 bar and a temperature of 250 °C. This problem cannot be solved with conventional centrifugal pumps but the solution is relatively simple using a canned motor pump as shown in Fig. 3.121 and 3.122. The canned motor centrifugal pump is mounted in a high pressure cylinder and all the components in contact with the liquid are made of a material which can resist the aggressiveness of the pumped fluid. An auxiliary impeller in the end cover furthest from the motor keeps the liquid in the rotor space to a standard temperature of 60 to 80 °C by means of an external high pressure cooler.

A mechanically-pure flushing liquid is injected into the cooling circuit at a rate of 2 l/h by means of a piston diaphragm pump. This flows through an intermediate piece and a helical shaft seal, positioned between pump and motor front bearing, in the direction of the pump so that no solid particles can enter the rotor space.

This meets the requirements for:

- high internal pressure
- high temperature
- pumping of suspension
- resistance of materials

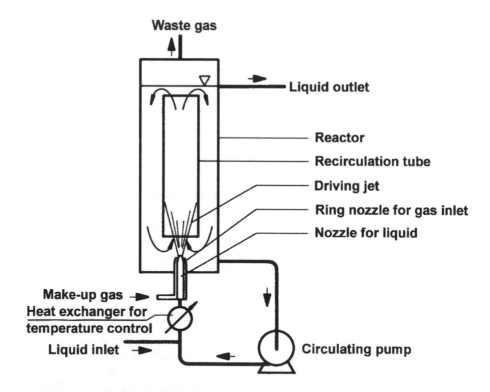

Fig. 3.120 Recirculatory reactor with circulating pump

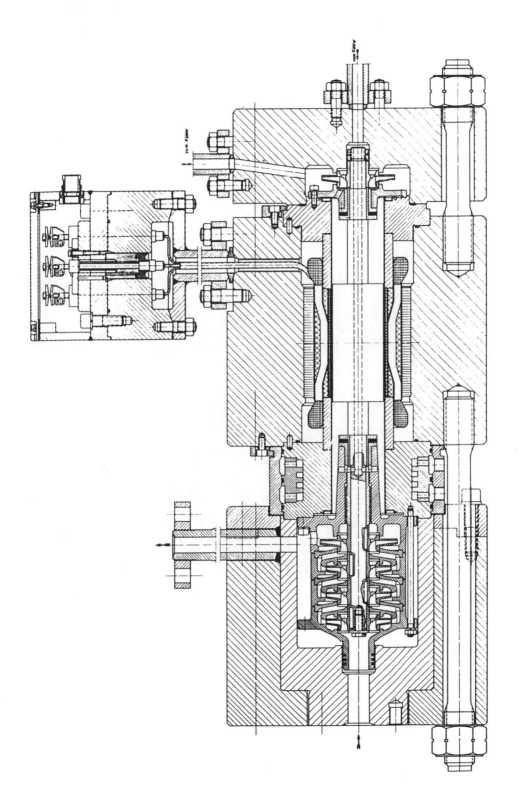

Fig. 3.121 Section view of a multistage high pressure canned motor centrifugal pump for pumping an aggressive suspension at 700 bar input pressure (Hermetic)

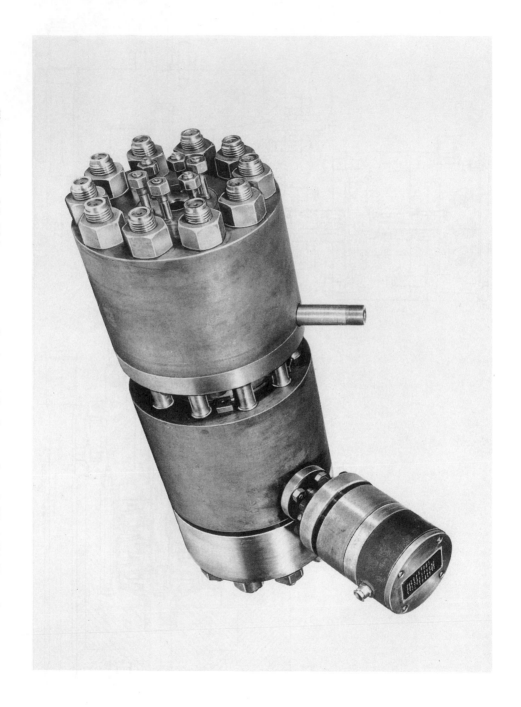

Fig. 3.122 View of the high pressure canned motor centrifugal pump shown in Fig. 3.121 (Hermetic)

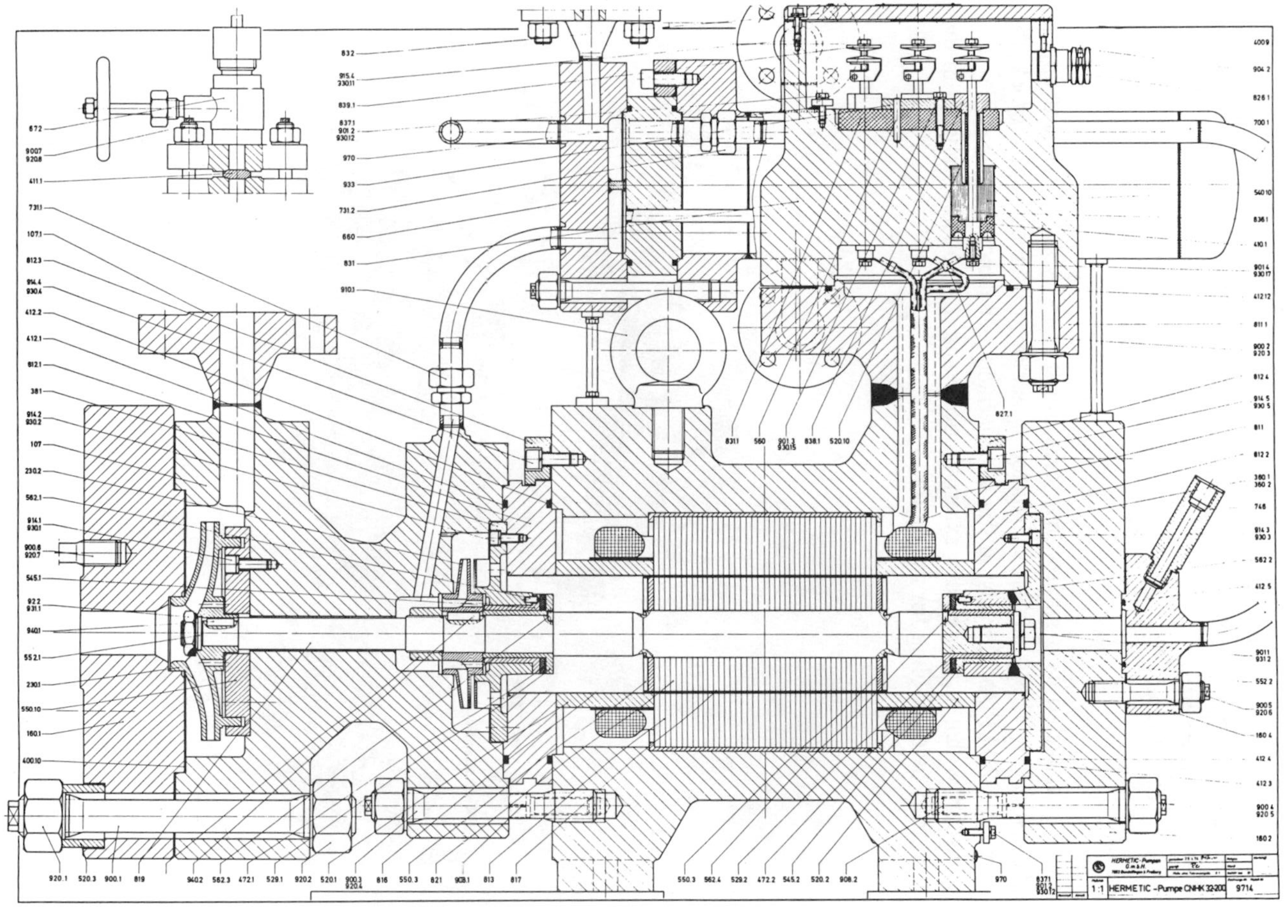

Fig. 3.123 High pressure canned motor centrifugal pump for pumping hot water (Hermetic)

235

- **Hot water delivery (H₂O)** (Fig. 3.123 and 3.124)

Operating parameters:

Operating temperature	300 °C
Density	712.4 kg/m²
Viscosity	0.1 mm²/s
Vapor pressure at operating temperature	87,6 bar
Design pressure	145 bar
Test pressure	250 bar
Motor output	6.5 kW

The pump is constructed as a single-stage centrifugal pump with a standard impeller to DIN 24256/ISO2858. The axial thrust balance is achieved by means of a block control, refer to Section 3.2.11 ("Axial thrust at the impeller and its balancing devices").
A cooling circuit with a high pressure external cooler and a secondary impeller in the motor are used to provide cooling for the motor.

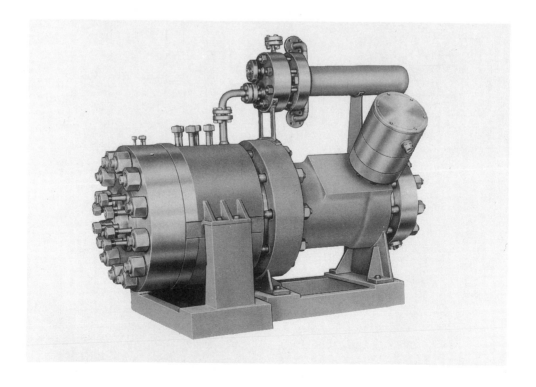

Fig. 3.124 View of a high pressure canned motor centrifugal pump for high temperature fluids (Hermetic)

- Pumping demineralized water

To test fuel and control rods in nuclear reactors, special slide-in pumps were developed which are integrated into the rod to be tested, thus enabling a space saving pump-fuel and control rod unit which required no pipelines to be constructed (Fig. 3.125) with the following operating data:

Fluid: Demineralised water (pH 5-10)

Q	= 500	600	650	m^3/h
H	= 43	35	30	m
T_e	=	330		°C
ρ	=	640		kg/m^3
System pressure	=	160		bar
Motor output	=	125		kW

- Integrated cooling water pump in uranium fuel rods

To provide intense cooling of uranium fuel rods for reactors by circulating fully demineralized water, high pressure hermetic pumps were installed in the rods, proof of the operating safety of such machines. A particularly slim construction was necessary for this application (Fig. 3.126 and 3.126.1).
To improve the NPSHR, the first impeller has a higher specific speed. The pump draws from its central pipe within the fuel rod and pumps the water back in the opposite direction at the tube wall. The technical data of this pump is as follows:

- Operating fluid	=	fully demineralized water
- Rate of flow	=	4 m^3/h
- Head	=	14 m
- Operating temperature	=	350 °C
- Viscosity	=	$0,1 \cdot 10^{-6}$ m^2/s
- Operating pressure	=	160 bar
- Design pressure	=	190 bar
-Radiation field of reactor		
-Helium leak rate	=	$< 10^{-6}$ mbar $\cdot$ dm^3 $\cdot$ s^{-1}
-Minimum life	=	3000 Stunden

The pump is integrated into the fuel rod so that it is activated in the radiation field of the reactor, or heavily contaminated by activated abrasion or activated corrosion products. This means that replacement or repair of the pump in the event of failure is impossible and means that pumps of this kind then must have a minimum life of 3000 running hours [3-19].

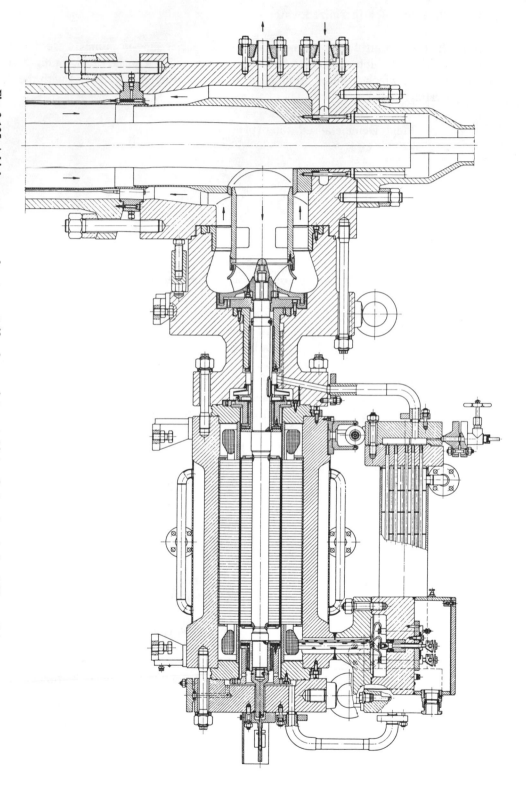

Fig. 3.125 A high pressure canned motor centrifugal pump integrated into a fuel and control rod (Hermetic)

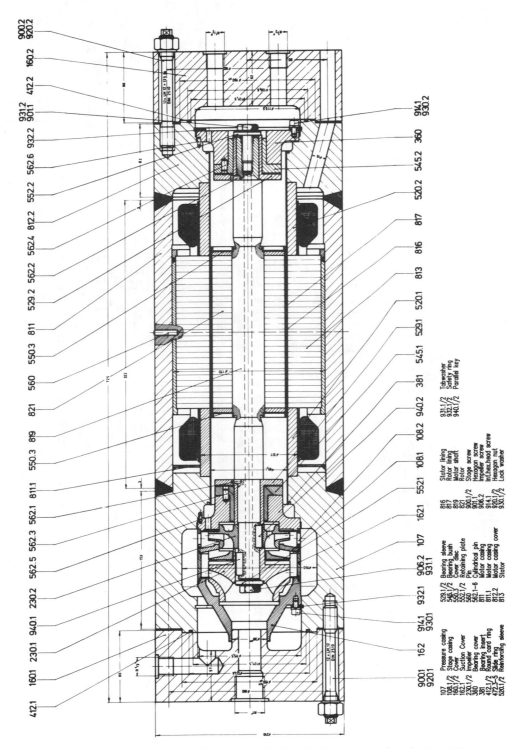

Fig. 3.126 Canned motor centrifugal pump for installation in a uranium fuel rod (Hermetic)

107 Pressure casing
108.1/2 Stage casing
160.1 Cover
162.1 Suction Cover
230.1/2 Impeller
360 Bearing cover
381 Bearing insert
412.3-5 Round cord ring
472.3-5 Slide ring
520.1/2 Reinforcing sleeve

523.1/2 Bearing sleeve
545.1/2 Bearing bush
550.1/3 Cover disc
560 Retaining plate
562.1-6 Pin
811 Cylindrical pin
811.1 Motor casing
812.2 Motor casing
813 Stator

816 Stator lining
817 Rotor lining
819 Motor shaft
821 Rotor
901.1/2 Stage screw
901.1 Hexagon screw
906.2 Impeller screw
914.1 Int.hex.head screw
920.1/2 Hexagon nut
930.1/2 Lock washer

931.1/2 Tabwasher
932.1/2 Safety ring
940.1/2 Parallel key

239

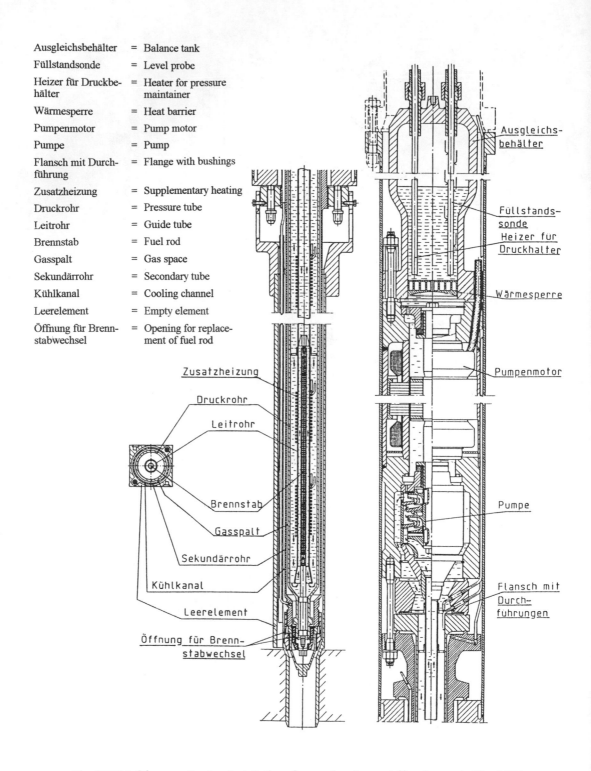

Ausgleichsbehälter	= Balance tank
Füllstandsonde	= Level probe
Heizer für Druckbe-hälter	= Heater for pressure maintainer
Wärmesperre	= Heat barrier
Pumpenmotor	= Pump motor
Pumpe	= Pump
Flansch mit Durch-führung	= Flange with bushings
Zusatzheizung	= Supplementary heating
Druckrohr	= Pressure tube
Leitrohr	= Guide tube
Brennstab	= Fuel rod
Gasspalt	= Gas space
Sekundärrohr	= Secondary tube
Kühlkanal	= Cooling channel
Leerelement	= Empty element
Öffnung für Brenn-stabwechsel	= Opening for replace-ment of fuel rod

Zusatzheizung

Druckrohr

Leitrohr

Brennstab

Gasspalt

Sekundärrohr

Kühlkanal

Leerelement

Öffnung für Brenn-stabwechsel

Ausgleichs-behälter

Füllstands-sonde

Heizer für Druckhalter

Wärmesperre

Pumpenmotor

Pumpe

Flansch mit Durch-fuhrungen

Fig. 3.126.1 Diagram showing installation of canned motor centrifugal pump in Fig. 3.126

240

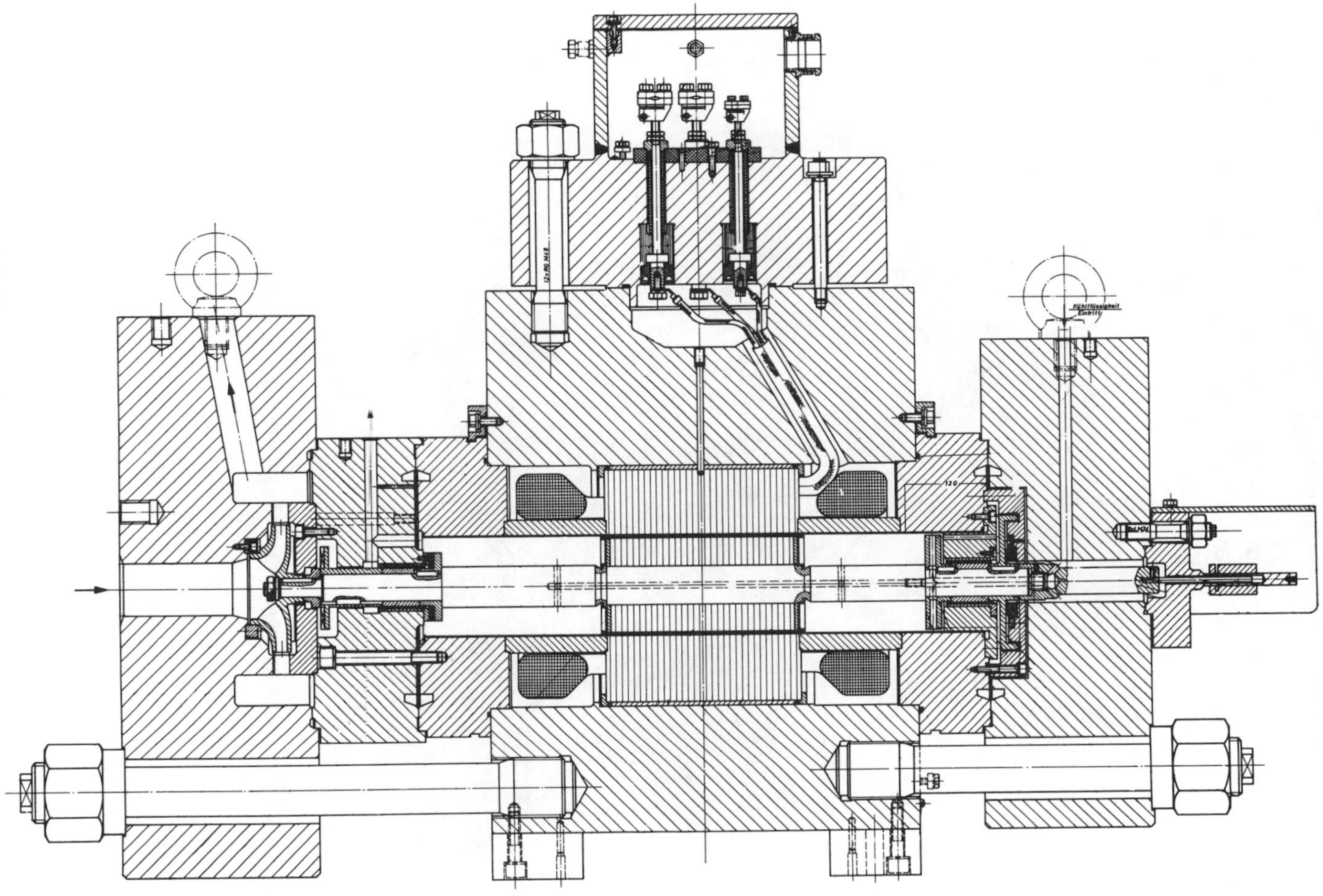

Fig. 3.127 Hermetic high pressure canned motor centrifugal pump for 1200 bar design pressure (Hermetic)

241

- Canned motor pump for pressures above 1000 bar

The problem of circulating fluids in extremely high pressure systems can be solved with centrifugal pumps by appropriate dimensioning of the wall components exposed to pressure. The pump shown in Fig. 3.127 and 3.128 is designed for an operating pressure of 1200 bar.

Fig. 3.128 View of a high pressure canned motor centrifugal pump as shown in Fig. 3.127 (Hermetic)

3.5.9.2 Canned motor pumps for supercritical gases

The transport of fluids in gaseous, supercritical states, for example used for extraction from vegetable matters, requires systems with internal pressures of several 100 bar. Under such operating conditions, gases assume high, liquid-like densities due to the high levels of compression. The <u>delivery head</u> of centrifugal pumps is always expressed in meters of the pumped fluid. Thus, the <u>delivery pressure</u> of such pumps increases with increasing density. Correspondingly high delivery pressures can therefore be achieved at normal rotational speeds although the fluid is in a gaseous state. Pumps designed for applications in this field are similar to those illustrated in Fig. 3.123 in their basic design except that a hydrodynamic mechanical seal is fitted in the spacer between the pump and motor. The rotor cham-

Fig. 3.129 und 3.130 High pressure canned motor centrifugal pumps in a chemical process plant (Hermetic)

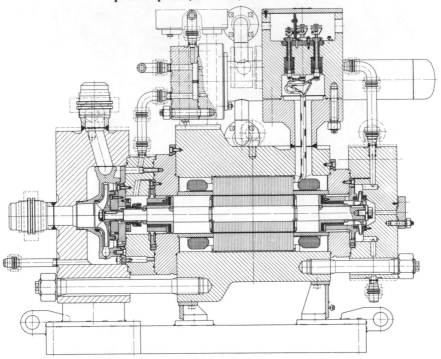

Fig. 3.131 Single-stage semi-hermetic high pressure canned motor centrifugal pump (Hermetic)

ber and the cooling system are filled with water or other liquid which is circulated via a high pressure cooler. This provides normal conditions for heat transfer, thrust compensation and bearing lubrication. The mechanical seals located between the primary and secondary circuits prevent uncontrolled escape of liquid into the pumped fluid. A balancing line provides pressure compensation between the rotor chamber and the pumped fluid (gas) to relieve the mechanical seals. A supply vessel, separately mounted on the pump, is installed in this connecting line in which the liquid level is monitored by means of maximum and minimum level sensors. When the minimum level is reached, liquid is automatically added by a metering pump. This provides a simple means of monitoring the leakage of the mechanical seal the pressure on each side of which is almost equal.

Fig. 3.132 View of the semi-hermetic high pressure canned motor centrifugal pump shown in Fig. 3.131 (Hermetic)

This type of pump is known as a semi-hermetic centrifugal pump as it is hermetically sealed externally but requires a mechanical seal between the internal and external circuits (Fig. 3.131 and 3.132). Fig. 3.133 shows a multistage type with an externally mounted balancing tank and metering pump.

The operating parameters of this pump are as follows:

Fluid	CO_2 (supercritical)
Operating temperature	55 °C
Density	820 kg/m³
Rate of flow	68 m³/h
Differential pressure at operating temperature	7,22 bar
Design pressure	327 bar
Test pressure	490 bar
Motor output	44 kW

Fig. 3.133 Multistage, semi-hermetic high pressure canned motor centrifugal pump with externally-mounted balance tank and metering pump

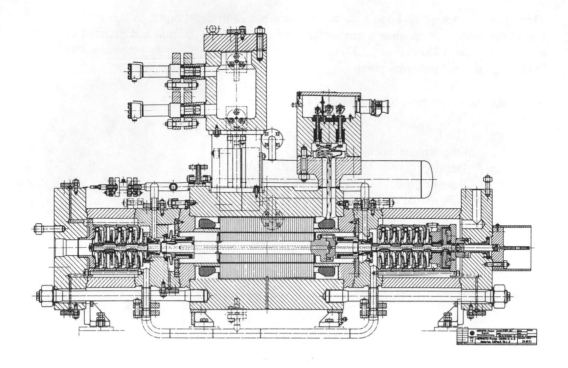

Fig. 3.134 Tandem high pressure canned motor centrifugal pump of semi-hermetic type with balance tank (Hermetic)

Tandem high pressure centrifugal pumps can be used with good success where there are high delivery and system pressures (Fig. 3.134 and 3.135). Fig. 3.135 shows the pressure balance tank with the optoelectronic level indicator, the cooler for the rotor chamber circulation fluid and the metering pump for re-supply of the rotor chamber fluid. The operating data of this pump is as follows:

Pumped fluid	$CO_2 + H_2O$		
Rate of flow	6	30	35 m³/h
Head	300	242	214 m
NPSHR	0,5	1,4	1,8 m
Temperature		40 - 80	°C
Density		950 - 830	kg/m³
System pressure		325	bar

Fig. 3.135 View of canned motor centrifugal tandem pump as shown in Fig. 3.134 (Hermetic)

3.5.9.3 Pumping two-phase mixtures

With mixtures of liquid gases with different vapor pressures and for gases which are in solubility equilibrium there is the danger of gas formation due to the increase in velocity at convergences in the pipe, e.g. at the pump inlet. The liquid part with the higher vapor pressure begins to vaporize, giving rise to a two-phase mixture. For reasons of motor cooling and to maintain the hydraulic axial thrust balance the gas bubbles must be prevented from entering the rotor chamber. A hydrodynamic shaft seal, shown in Fig. 3.136 and 3.137, in the form of an open impeller with straight blades is positioned between the pump and the motor to prevent an exchange of fluid and ensure that the gas bubbles do not enter the rotor chamber. Pumping a liquid-gas mixture requires an inducer to be fitted in front at the pump inlet. Only in this way the conventional impeller can perform its pumping function, as the amount of gas can be up to 40% by volume. The advantage of this arrangement is the omission of a pressure equalization line between the pump and motor sections and a greater resistance to wear because a mechanical seal is no longer required, and consequent increase in service life.

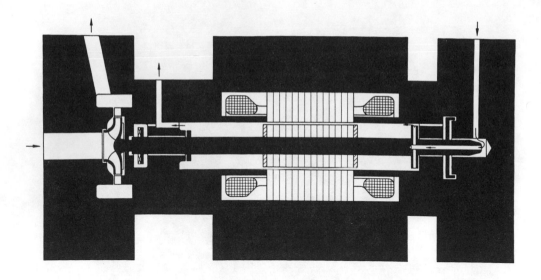

Fig. 3.136 Operating principle of a hermetic high pressure centrifugal pump with hydrodynamic shaft sealing

3.5.9.4 Canned motor pumps as the main circulating pumps in the primary circuit of ships' nuclear reactor

The main circulating pumps used in the primary circuit are required to pump radioactive coolant and therefore an absolute technical seal against fluid leakage must be guaranteed. This problem can only be solved using a canned motor drive. The three examples show canned motor pumps of nuclear energy ship propulsion units of an American, Russian and German ship (Fig. 3.138, 3.139 and 3.140) [3-20].

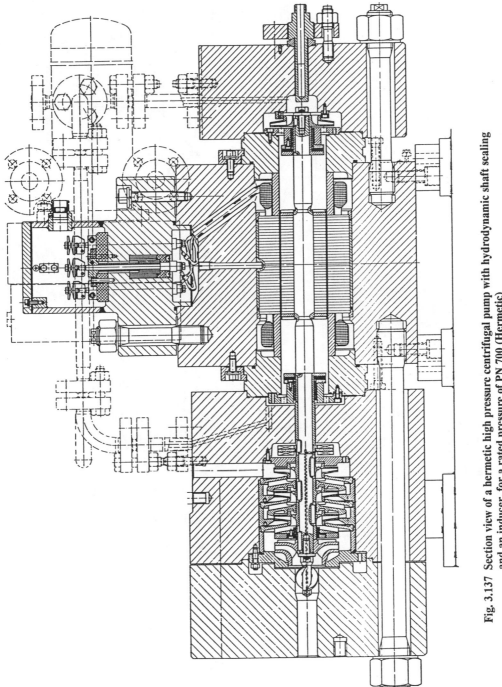

Fig. 3.137 Section view of a hermetic high pressure centrifugal pump with hydrodynamic shaft sealing and an inducer, for a rated pressure of PN 700 (Hermetic)

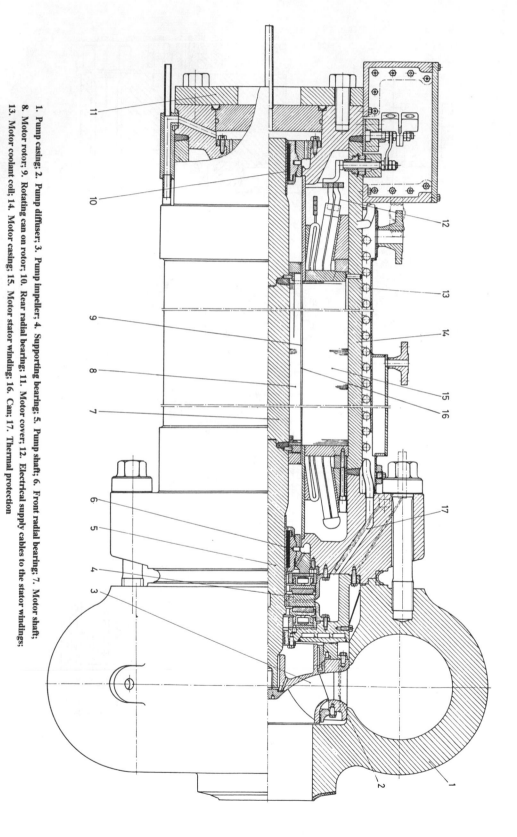

Fig. 3.138 Main primary circulating pump in the primary circuit of a conventional reactor of the NS "Savannah"

1. Pump casing; 2. Pump diffuser; 3. Pump impeller; 4. Supporting bearing; 5. Pump shaft; 6. Front radial bearing; 7. Motor shaft;
8. Motor rotor; 9. Rotating can on rotor; 10. Rear radial bearing; 11. Motor cover; 12. Electrical supply cables to the stator windings;
13. Motor coolant coil; 14. Motor casing; 15. Motor stator winding; 16. Can; 17. Thermal protection

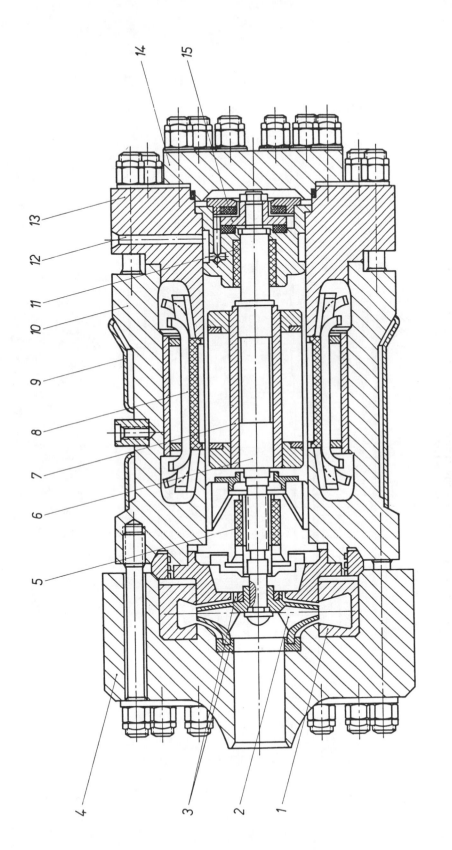

1. Pump collector channel; 2. Pump impeller; 3. Impeller seal; 4. Pump casing; 5. Front radial bearing;
6. Pump and motor shaft; 7. Motor rotor; 8. Motor stator; 9. Motor cooling jacket; 10. Motor casing; 11. Rear radial bearing;
12. Cooling water supply pipe to the support bearing; 13. Housing of radial and support bearing; 14. Cover; 15. Rear support bearing

Fig. 3.139 Auxiliary primary circulating pump in the primary circuit of the conventional reactor of the "Lenin"

251

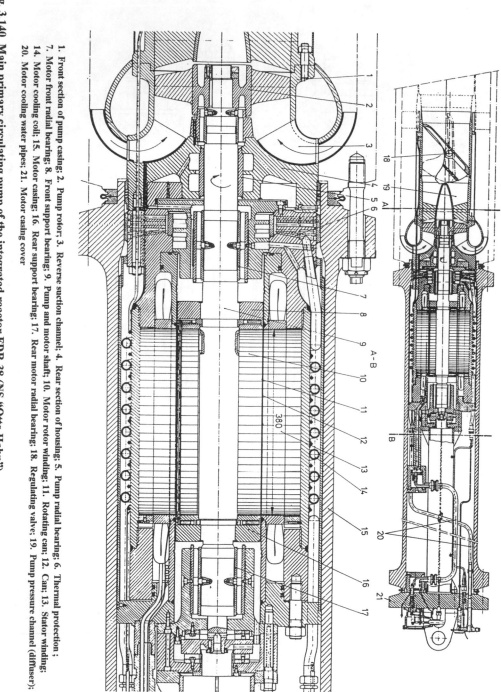

Fig. 3.140 Main primary circulating pump of the integrated reactor FDR-38 (NS "Otto Hahn")

1. Front section of pump casing; 2. Pump rotor; 3. Reverse suction channel; 4. Rear section of housing; 5. Pump radial bearing; 6. Thermal protection ;
7. Motor front radial bearing; 8. Front support bearing; 9. Pump and motor shaft; 10. Motor rotor winding; 11. Rotating can; 12. Can; 13. Stator winding;
14. Motor cooling coil; 15. Rear support bearing; 16. Rear support bearing; 17. Rear motor radial bearing; 18. Regulating valve; 19. Pump pressure channel (diffuser);
20. Motor cooling water pipes; 21. Motor casing cover

252

3.5.10 Canned motor pumps in the foodstuffs industry and bioengineering

The hermetic delivery system of the canned motor pump is suitable for the hygiene sector, whether for foodstuffs or in pharmaceutical or biotechnology plants. In addition to the absolute technical seal, the ability to maintain cleanliness and the surface condition of the parts in contact with the product are of great importance. Fig. 3.141 and 3.142 show a pump which meets these requirements. It is used in the foodstuffs industry particularly for dairy products. The same applies to plants where sterility is a pre-condition. In the production of foodstuffs for example, this enables uniform quality to be achieved and maintained on the basis of physiological, hygienic and nutritional values.

Fig. 3.141 View of a canned motor centrifugal pump for the hygiene sector as shown in Fig. 3.142 (Lederle)

These pumps do not cause a change in color, taste, aroma, texture or stability. In addition to these advantages the practically-noiseless pumping of fluids is a further bonus in research and production plants using canned motor pumps. The hermetic construction also allows pumps to be used under high vacuum. A pump constructed to meet hygiene requirements must be light, easily dismantleable, consisting of the minimum number of individual parts, the rotating unit must be removable as a single assembly, the pump materials must be resistant to corrosion and wear and also have a smooth pore-free surface. It must also

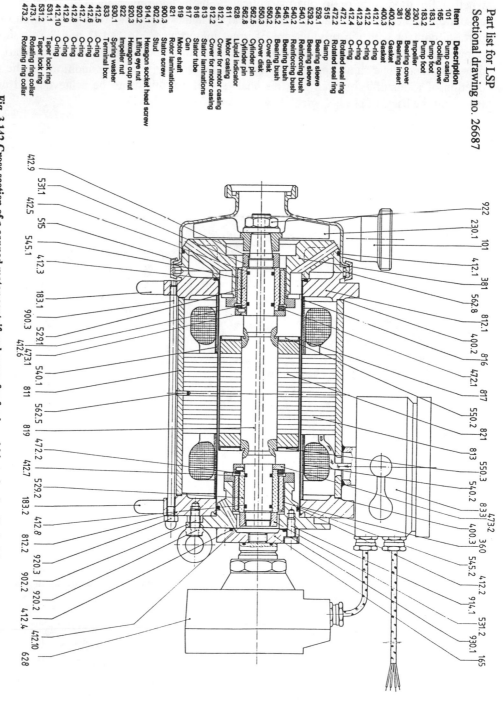

Fig. 3.142 Cross section of a canned motor centrifugal pump for food- and bio technolgy (Lederle)

254

be possible to dismantle the pump without the use of tools. Rolled Cr-Ni steel has been shown to be a particularly suitable material. A multistage rolling process gives these steels a high compaction thus providing the necessary smooth, pore-free surface. The complete unit must be spraywater proof.

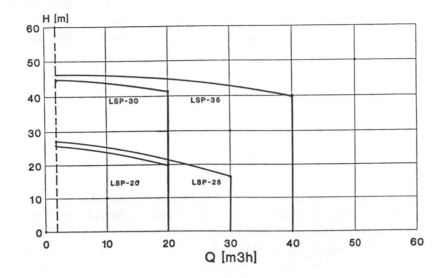

Fig. 3.143 Family of characteristics of a canned motor centrifugal pump for foodstuffs, as shown in Fig. 3.141 and 3.142 (Lederle)

The flow cleaning (CIP) specified by the milk industry is guaranteed with the pump shown in Fig. 3.142. The inlet and outlet connections are provided with round threads to DIN 11851. Tests carried out at the Dr. Oskar Farny Institute of the National Milk Industry Teaching and Research Establishment Wangen/Allgäu have shown that such pumps meet the strictest test criteria in accordance with the ATP measuring methods and therefore can be completely clean and when pumping milk with a fat content of 4.3% have a lower oxygen saturation compared with pumps with a mechanical seal. Fluids up to 120 °C can be pumped with these pumps. Fig. 3.143 shows the family of characteristics for a series of these pumps.

To avoid the danger of the pump running dry if the tank is empty and the danger of damage to bearings and motor windings, these canned motor pumps are fitted with a movement-free run-dry safety device and also a motor protection and safety start-up switch. The activation of a motor protection switch alone is not sufficient, because if the pump runs dry not just the bearing friction losses have to be allowed for but also there is no hydraulic output and the cooling afforded by radiation is not adequate. The pump is monitored by a pump protection probe mounted in the rear of the motor terminal cover. The probe has no moving parts and operates on the principle of induction. It has automatic deposit compensation with an electronic protective ring so that the liquid level of the pump can be reliably monitored including for products which tend to stick.

3.5.11 Self-priming centrifugal pumps with a canned motor drive

Operating states frequently occur in technical processes in which the suction line has to be automatically evacuated by the pump for energetic power transfer.

Let us assume for instance that the liquid level on the suction side is lower than the pump inlet cross-section or the liquid tank has to be emptied "over a hill" because there is no bottom outlet, or one is not permitted for safety reasons. The ability to evacuate the suction pipe by itself with the aid of centrifugal pumps can be achieved by three different systems.

- With side channel flow pumps
- With centrifugal pumps using impeller cell flushing
- With vacuum tank fitted before the pump

3.5.11.1 Side channel pumps

Side channel pumps are flow machines characterized by the fact that their delivery flow has a high energy density with a small structural volume. Their operating principle is based on an impulse exchange between a faster flow in the star-shaped impeller and a slower flow in the side channel of the casing (Fig. 3.144). A large energy transfer ψ coefficient with a small capacity coefficient φ is achieved by multiple impulse exchanges during the rotary motion of the impeller. It is 5 to 15 times that of normal radial pumps.

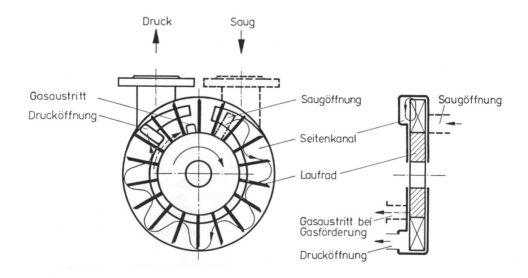

Druck = Pressure	Saugöffnung = Suction port	
Saug = Suction	Seitenkanal = Side channel	
Gasaustritt = Gas outlet	Laufrad = Impeller	Drucköffnung = Discharge port
Drucköffnung = Pressure port	Gasaustritt bei Gasförderung = Gas outlet for gas delivery	Saugöffnung = Suction port

Fig. 3.144 Operating principle of a side channel pump

Because the volume rate of flow delivered in this way is relatively small compared to the achieved heads, this results in low specific speeds which are in the $n_q = 2$ to $n_q = 12$ r.p.m range. Despite these facts, relatively satisfactory efficiencies are still achieved compared with radial pumps with the same n_q.

3.5.11.2 Self-priming centrifugal pumps using the side channel principle with a horizontal type canned motor drive [3-21]

The short, space-saving, maintenance-free construction of these pumps has proved useful in chemical industry plants and in mechanical engineering. Figures 3.145 and 3.146 show the design of a pump of this type. It can be seen that the first impeller does not operate on the side channel principle but is instead designed as a normally-aspirating impeller. This is the result of the requirement for minimum NPSHR values. Despite their ability to pump gases (air) and two phase mixtures, side channel pumps are less satisfactory with regard to their NPSHR behavior than pumps with radial impellers. Placing a normally-aspirating impeller (Fig. 3.147) in front enables the favorable flow conditions of this type of impeller to be exploited, particularly if the impeller is designed such that it operates in the partial load range, i.e. to the left of Q_{opt} in the co-ordinates system. The family of characteristics of a multi-stage self priming side channel pump, represented in Fig. 3.148, shows the NPSHR values of a type SRZ pump without a normal aspirating initial stage alternating with an SRZS with such an initial stage. The large NPSHR difference between these two systems speaks for itself. For a volume rate of flow of, for example, 16 m³/h the NPSHR improves from 2,6 to 0,5 m!

The suction connection on side channel pumps is normally designed such that the inlet flange is either directed outwards at the same level of the outlet flange of the discharge connection or the suction flange is offset 90° but with its inlet level above the casing diameter (Fig. 3.149 and 3.150). This is necessary to ensure that after shutdown the pump does not empty due to siphoning and to guarantee that a liquid ring can be formed on restarting, thus providing the required sealing of the individual vane cells. Placing a normally aspirating impeller in front does away with this arrangement in favor of a better flow and a horizontal, liquid entry which is centered on the axis. To prevent the operating liquid being completely siphoned out when the pump is shutdown and thus posing a hazard for restarting, an adequate volume of liquid is always guaranteed in the pump by inserting a suitably-designed spacer. In addition, the fluid level in the pump and in the rotor chamber of the motor is also monitored with the aid of an optoelectronic sensor (Fig. 3.145 and 3.151).

All these pump sets are fitted with an explosion-proof drive motor of flame-proof enclosure.

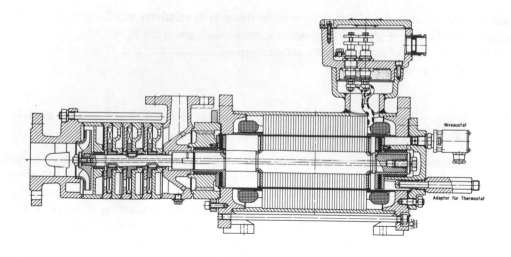

Niveaustat

Adaptor für Thermostat

Fig. 3.145 Self-priming canned motor side channel pump with a first stage radial impeller (Hermetic)

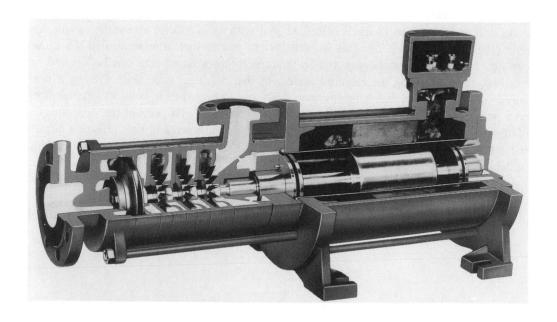

Fig. 3.146 Sectioned model of a self-priming canned motor side channel pump as shown in Fig. 3.145 (Hermetic)

Fig. 3.147 Normal suction impeller and side channel impeller (Sero)

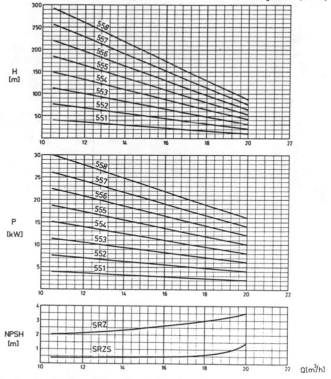

Fig. 3.148 Family of characteristics of a multi-stage side channel pump showing the NPSHR
with and without a normal suction impeller (Sero)

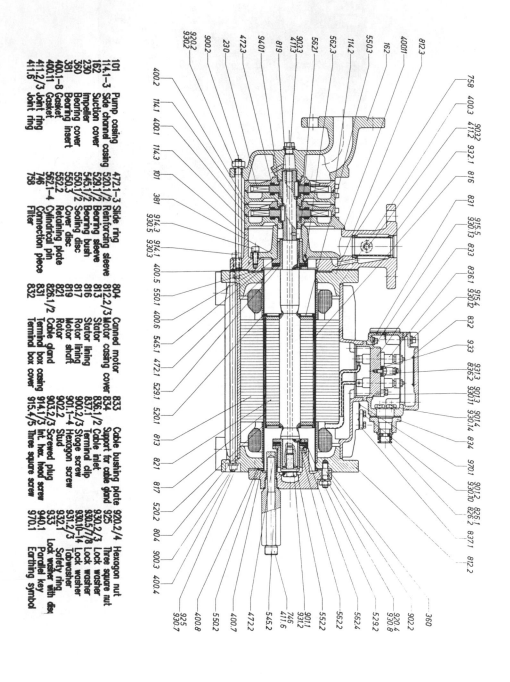

101	Pump casing	472.1-3	Slide ring	804	Canned motor	833	Cable bushing plate	920.2/4	Hexagon nut
114.1-3	Side channel casing	520.1/2	Reinforcing sleeve	812.2/3	Motor casing cover	834	Support for cable gland	925	Three square nut
162	Suction cover	529.1/2	Bearing sleeve	813	Stator	836.1/2	Cable inlet	930.2/3	Lock washer
230	Impeller	545.1/2	Bearing bush	816	Stator lining	837.1	Terminal clip	930.5/7/8	Lock washer
360	Bearing cover	550.1/2	Sealing disc	817	Rotor lining	900.2/3	Stage screw	930.10-14	Lock washer
381	Bearing insert	550.3	Cover disc	819	Motor shaft	901.1-4	Hexagon screw	931.2/3	Lock washer
400.1-8	Gasket	552.2	Retaining plate	821	Rotor	902.2	Stud	931.2/3	Tabwasher
400.11	Gasket	562.1-4	Cover plate	826.1/2	Cable gland	903.2/3	Screwed plug	932.1	Safety ring
411.2/3	Joint ring	746	Cylindrical pin	831	Termind box casing	914.1/3	Int. hex. head screw	933	Lock washer with disc
411.5	Joint ring	758	Filter	832	Termind box cover	915.4/5	Three square screw	940.1	Parallel key
								970.1	Earthing symbol

Fig. 3.149 Canned motor side channel pump without an inducer (Hermetic)

260

Fig. 3.150 Canned motor side channel pump as shown in Fig. 3.149, in a technical process plant (Hermetic)

Fig. 3.151 Canned motor side channel pumps in an operating plant, protected by opto-electronic level monitoring and thermostatic motor control (Hermetic)

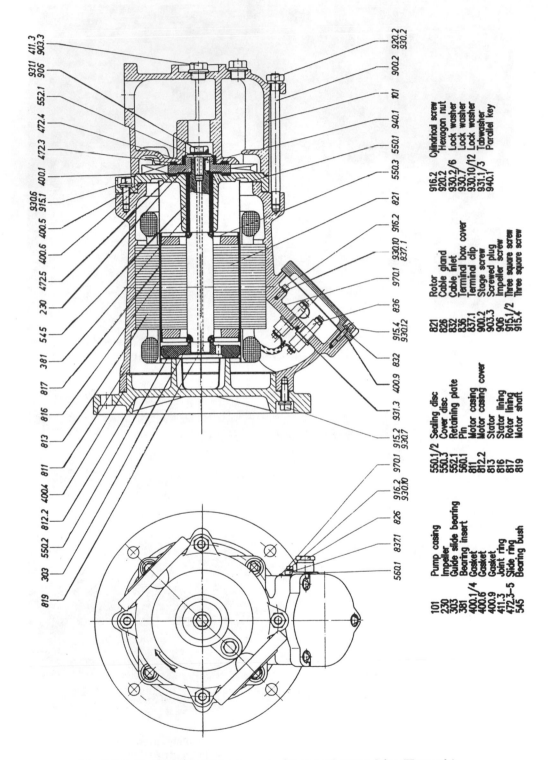

Fig. 3.152 Vertical side channel pump with canned motor drive (Hermetic)

819	
303	
550.2	
812.2	
400.4	
811	
813	
816	
817	
381	
545	
230	
472.5	
400.6 400.5	
400.1	
915.1 930.6	
472.3 472.4 552.1	
906 9.31.1 411.3 903.3	

101   Pump casing
230   Impeller
303   Guide slide bearing
381   Bearing insert
400.1/4   Gasket
400.6   Gasket
400.9   Gasket
411.3   Joint ring
472.3–5   Slide ring
545   Bearing bush

550.1/2   Sealing disc
550.3   Cover disc
552.1   Retaining plate
560.1   Pin
811   Motor casing
812.2   Motor casing cover
813   Stator
816   Stator lining
817   Rotor lining
819   Motor shaft

821   Rotor
826   Cable gland
832   Cable inlet
836   Terminal box cover
837.1   Terminal clip
900.2   Stage screw
903.3   Screwed plug
906   Impeller screw
915.1/2   Three square screw
915.4   Three square screw

916.2   Cylindrical screw
920.2   Hexagon nut
930.2/6   Lock washer
930.7   Lock washer
930.10/12   Lock washer
931.1/3   Tabwasher
940.1   Parallel key

3.5.11.3 Vertical side channel pumps with canned motor drive

The space-saving construction which distinguishes side channel pumps with a canned motor drive is further enhanced by their vertical arrangement. In the case of the pump shown in Fig. 3.152, the vertical canned motor is arranged underneath the pump. The pump is designed as an inline pump with an overhung impeller. The rotor is lifted by an appropriate balancing of the pressure difference between the front and rear sides of the rotor. To achieve this, part of the liquid already pumped into the pressure space is passed through the motor-pump shaft and is returned to the inlet point of the product into the impeller, absorbing the greater part of the motor heat losses in the stator-rotor gap as it passes through. An appropriate choice of gap and bore produces an upwards axial thrust which balances the weight of the impeller. The practical effect of the space-saving design of such pumps can be seen in Fig. 3.153. This shows them being used in a tank farm for solvents.

Fig. 3.153 Vertical canned motor side channel pumps as shown in Fig. 3.152, in a tank farm for solvents (Hermetic)

A small pump was developed for laboratory purposes, which could be placed on laboratory tables without the need for securement. This operates on the same principle as that previously described (Fig. 3.154).

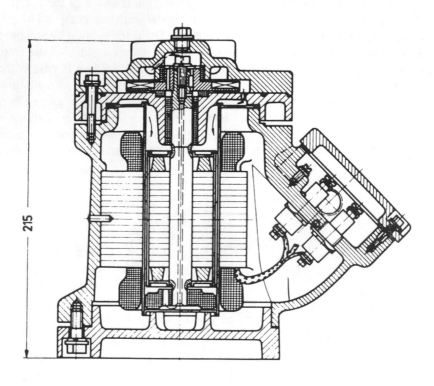

215

Fig. 3.154 Hermetic side channel canned motor pump for laboratory purposes (Hermetic)

3.5.11.4 Self priming canned motor pumps with an impeller cell flushing system

- Single-stage type (Fig. 3.155 and 3.156) -

Side channel pumps have a disadvantage in that they can only be used for fluids which are mechanically pure to turbid. This is due to the relatively narrow clearance between the impellers and the casing wall. Centrifugal pumps with impeller cell flushing can pump both pure and contaminated liquids. Their design is identical to the normal aspirating centrifugal pumps with regard to their hydraulics. The difference is, however, that a guide vane F-S is fitted in the guide volute (Fig. 3.157) which performs the function of jet nozzle similar to a water ejector [3-22].

The rotating impeller generates a flushing flow "S" due to the drag effect of its blade tips. Below the guide surface the flushing flow is drawn into the passing, rotating impeller cells. The impeller cells form a mixture chamber "M" similar to an ejector. The air/gas mixture taken up by the delivery flow "S" mixes with it and is carried to the diffuser. A voluminous separating chamber is located on the pump pressure side, from which the air escapes outwards into the discharge connection whilst the operating fluid falls downwards to again be

264

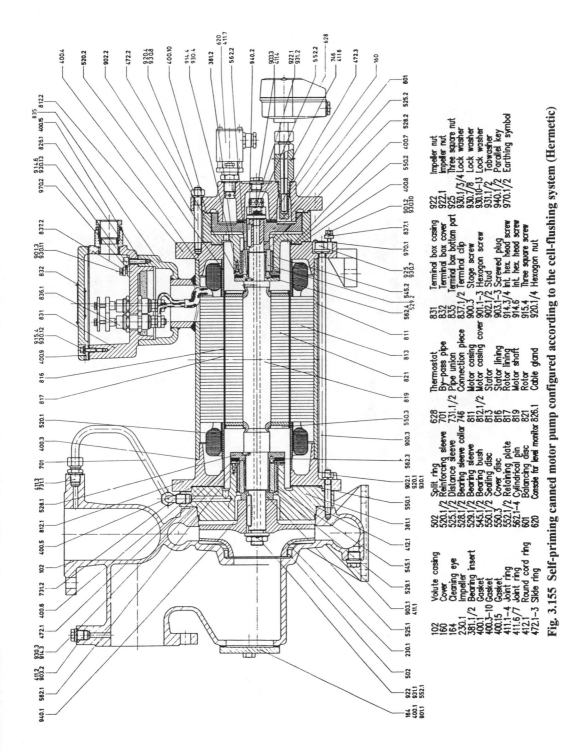

Fig. 3.155 Self-priming canned motor pump configured according to the cell-flushing system (Hermetic)

102	Volute casing	502	Split ring
160	Cover	520.1/2	Reinforcing sleeve
164	Cleaning eye	525.1/2	Distance sleeve
230.1	Impeller	528.1/2	Bearing sleeve collar
381.1/2	Bearing insert	529.1/2	Bearing sleeve
400.1	Gasket	545.1/2	Bearing bush
400.3–10	Gasket	550.1/2	Sealing disc
400.15	Gasket	550.3	Cover disc
411.1–4	Joint ring	552.1/2	Retaining plate
411.6/7	Joint ring	561.1–4	Cylindrical pin
412.1	Round cord ring	601	Balancing disc
472.1–3	Slide ring	620	Console for level monitor

628	Thermostat	831	Termind box casing
701	By-pass pipe	832	Termind box cover
731.1/2	Pipe union	835	Termind box bottom part
746	Connection piece	837.1/2	Termind clip
811	Motor casing	900.3	Stage screw
812.1/2	Motor casing cover	901.1–3	Hexagon screw
813	Stator	902.1/2	Stud
816	Stator lining	903.–3	Screwed plug
817	Rotor lining	914.3/4	Int. hex. head screw
819	Motor shaft	914.6	Int. hex. head screw
821	Rotor	915.4	Three square screw
826.1	Cable gland	920.1/4	Hexagon nut

922	Impeller nut		
922.1	Impeller nut		
925	Three square nut		
930.1/3/4	Lock washer		
930.7/8	Lock washer		
930.10–13	Lock washer		
931.1/2	Tabwasher		
940.1/2	Parallel key		
970.1/2	Earthing symbol		

265

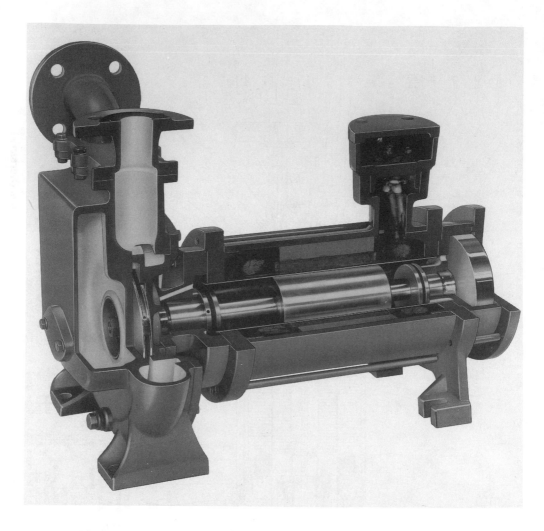

Fig. 3.156 Sectioned self-priming canned-motor pump as shown in Fig. 3.155

come part of the flushing process. This type of pump has the following advantages compared with side channel pumps.

- Improved efficiency when pumping clean liquid, due to a normal radial impeller.
- Lower susceptibility to wear and contamination
- Low sound emission

The pump is particularly suitable for pumping liquids containing gas and solids. The suction capacity is, however, less than that of side channel pumps.

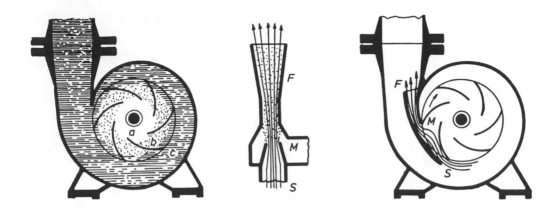

Fig. 3.157 The cell flushing system (Hanibal) [3-22]

- Multistage type using the cell flushing system -

Self priming centrifugal pumps with a cell flushing system are normally used in a single-stage construction for low to medium heads. There are, however, also pumping applications which require a multi-stage design for larger heads and a self priming capacity. This is achieved in the final stage of a multi-stage centrifugal pump (Fig. 3.159) by providing the guide device with a nozzle-type impeller cell flushing (Fig. 3.158), similar to the single-stage type, which flows into a separation chamber.

In this case, the vertical type is particularly advantageous as it can cope with a comparatively small volume of liquid for the priming process. To guarantee that the pump is constantly charged with liquid, the suction line is u-shaped like a siphon (Fig. 3.159). The suction line is connected to the pressure line through a solenoid valve which opens when the pump is shutdown and prevents the suction line being siphoned empty. A solenoid valve is not necessary if the pump is enclosed by a barrel and this has an orifice-type connection to the high part of the suction line (Fig. 3.160).

Fig. 3.158 Cell flushing device of a multi-stage
canned motor centrifugal pump as
shown in Fig. 3.159 (Hermetic)

267

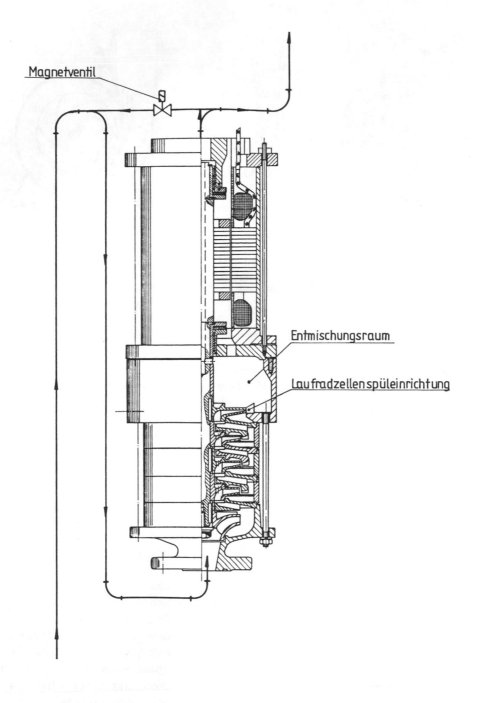

Magnetventil = Solenoid valve;   Entmischungsraum = Separating chamber;
Laufradzellenspüleinrichtung = Impeller cell flushing device

**Fig. 3.159 Multistage canned motor centrifugal pump with a cell flushing system for self
priming**

268

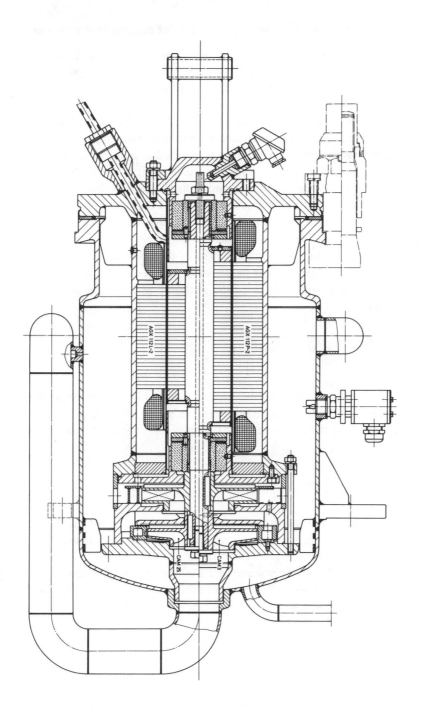

Fig. 3.160 Multistage, vertical, self-priming, canned motor centrifugal pump with a cell flushing system (Hermetic)

3.5.11.5 Vacuum tanks for self-priming operation of canned motor pumps

In those cases where no suitable self-priming pumps are available to draw liquids which are difficult to pump, such as liquids which have severe mechanical contamination or where the suction lines are extremely long, the capability of normal-priming centrifugal pumps can be used to generate suction lift after charging with liquid.

To do this, a vacuum-proof tank, which must be charged with the liquid to be pumped before start-up, is fitted before the suction connection of the pump. As the pump runs up it empties this tank thus generating a vacuum in the tank and the suction pipe connected to it. The liquid is then drawn into this tank via the suction pipe. At the same time, the air present in the suction pipe enters the tank and remains there so that, in contrast to self-priming pumps, the pump does not pump air during the suction process. It is therefore unnecessary to separately evacuate the tank before starting the pump as the centrifugal pump itself generates the vacuum necessary for suction.

The capacity of the vacuum tank is independent of the flow rate of the pump and depends only on the volume of air/gas present in the suction pipe which has to be received by the tank. As we know, the air/gas volume of the suction pipe under atmospheric pressure can be calculated, as follows:

$$V = \frac{D^2 \cdot \pi}{4} \cdot l_s \qquad\qquad (3 - 53)$$

whereby l_s is the developed length of the suction pipe and D is its internal diameter. The air volume calculated in this way expands under the influence of the suction lift or the vacuum which is set up after the tank is evacuated. Conversion is carried out using the Boyer-Mariotte law:

$$p_1 \cdot v_1 = p_2 \cdot v_2 \qquad\qquad (3 - 54)$$

In this case p_1 is the atmospheric pressure which can be used with sufficient accuracy of up to 100 kPa = 1 bar. p_2 is the vacuum pressure which is determined from the geodetic suction lift, the density of the liquid and the friction losses.

The vacuum tank must have this volume and the dimensioning of the tank capacity must allow for leakage and oscillation of the liquid particularly during startup, so that the tank should be adequately dimensioned. For small tanks an oversizing of 3 is adequate and this can be reduced to 2 to 2.5 for large tanks. It follows from this that the suction pipe must enter the tank at as high a level as possible and must not be immersed down to the bottom otherwise the tank could be siphoned empty during shutdown. The pump suction connection is to be connected as low as possible to the tank to ensure that an adequate, effective volume is obtained.

Valves before and after the tank are not necessary and could even be detrimental because of the increased friction losses which they entail.

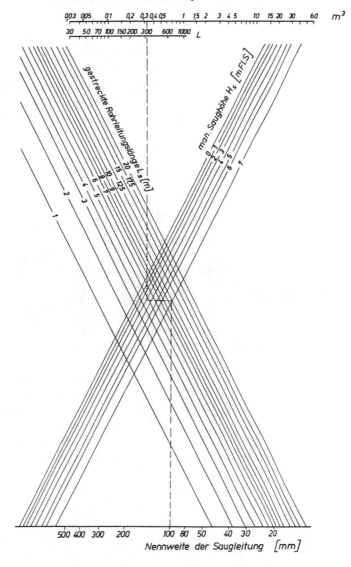

Kesselinhalt der Vakuumanlage = Boiler capacity of vacuum tank
gestreckte Rohrleitungslänge = Developed length of pipeline
man. Saughöhe = Vacuumetric suction lift
Nennweite der Saugleitung = Nominal diameter of suction pipe

Fig. 3.161 Determination of the boiler capacity of a vacuum system

For relatively short suction pipes where the friction losses are negligible, it is sufficient to dimension the vacuum tank in accordance with Fig. 3.161 with whose help the required boiler capacity can be immediately determined using the nominal diameter, developed length of suction pipe and vacuumetric suction lift (note density). In this case it doesn't matter whether the tank is mounted horizontally or vertically to suit local conditions. [3-23]

If the tank is provided with a sight glass, the level in the tank can be observed as the liquid level drops and it can be replenished from the discharge line via an auxiliary pipe. This can, however, only be done when the pump is shut down because when the pump is running only as much liquid can be added as is effectively pumped away on the pressure side. To avoid backflow through the pump, particularly with long discharge lines, a check valve is normally provided between the pump and auxiliary connection (Fig. 3.162).

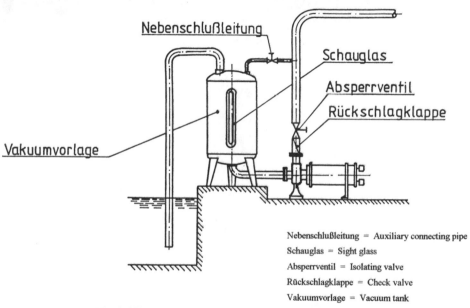

Nebenschlußleitung = Auxiliary connecting pipe
Schauglas = Sight glass
Absperrventil = Isolating valve
Rückschlagklappe = Check valve
Vakuumvorlage = Vacuum tank

Fig. 3.162 Vacuum tank (pump fitted with check valve)

If it is necessary to drain the vacuum tank, thus making repeated refilling necessary, a vessel of the same capacity can be fitted in the pressure line so that with a pump shut down the vacuum tank can be refilled via the auxiliary connecting pipe (Fig. 3.163).

When emptying tank wagons the frequent operation of the charging valve is very tedious. If the check valve is omitted the vacuum above the pump again fills up with the pump shut down so that suction is again possible after a short period. The discharge line must in any case have a drop from the pump or be provided with an automatic vent valve or the pressure vessel should be fitted with automatic venting (Fig. 3.164 and 3.164.1). An arrangement of this kind is also suitable where the pump is automatically switched or operated by remote control.

Canned motor pumps or pumps with permanent magnet couplings which are protected against the maximum permissible rate of flow being exceeded by means of Q_{max} orifices can be started without check valves with the gate valve fully open. Proper operation using vacuum tanks is, however, only possible if air pockets in the suction line between the pump and vacuum tank are avoided. If "sucking over the hill" is unavoidable, the vacuum system must be mounted at the highest point in the suction line.

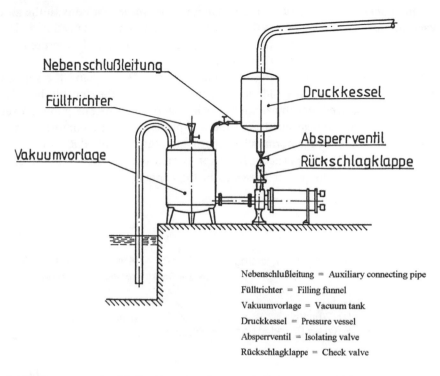

Nebenschlußleitung = Auxiliary connecting pipe

Fülltrichter = Filling funnel

Vakuumvorlage = Vacuum tank

Druckkessel = Pressure vessel

Absperrventil = Isolating valve

Rückschlagklappe = Check valve

Fig. 3.163 Vacuum tank with liquid recharging vessel on the pressure side

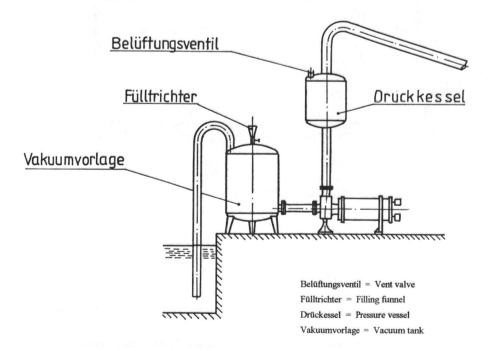

Belüftungsventil = Vent valve

Fülltrichter = Filling funnel

Drückessel = Pressure vessel

Vakuumvorlage = Vacuum tank

Fig. 3.164 Vacuum tank with liquid recharging vessel on the pressure side with a downward discharge line and a vent valve

If air pockets are present in the suction pipe, these vibrate with the air content of the vacuum tank. The pump alters the delivery rate between 0 and maximum delivery in a slow rhythm, whilst the vacuum tank alternately fills and empties. This presents the danger that air will enter the pump which can cause cessation of delivery.

The advantage of vacuum tanks is mainly that on the one hand the suction pipes can be evacuated without further aids and on the other hand they enable heavily contaminated liquids to be sucked, which cannot be dealt with by side channel pumps. Vacuum tanks can be used with advantage even with the self-priming type of pump particularly where the suction pipes exceed the permissible length and thus pose the danger of vaporization of the pumped liquid. This is particularly so for pumps with a canned motor drive or magnetic coupling because the power losses from the canned motor or magnetic coupling lead to additional heating.

Tankwagenentleerug = Tank wagon emptying

Pumpe = Pump

autom. Belüftungsventil = Automatic vent valve

Saug -u. Druckleitung beheizbar = Suction and discharge lines can be heated

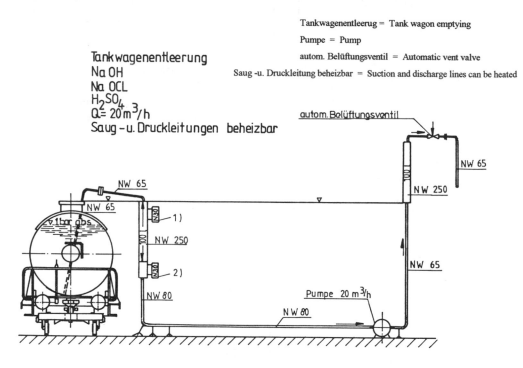

Fig. 3.164.1 **Example of an application for discharging of tank wagons**

For long suction pipes it is not necessary for the vacuum tank to be designed with a safety factor of three and it can in fact be almost completely emptied because it is only necessary to shorten the effective time for which the pump is drawing air. In any case it is then absolutely necessary to recharge the empty vacuum tank for renewed suction.

Fig. 3.165 shows a satisfactory arrangement of the vacuum tank, inlet pipe and connecting pipe [3-22].

This is also particularly suitable for contaminated fluids. The suction vessel (1) in this case is cylindrical and has no internal fittings. The suction pipe (2) enters the top end of the vessel at a tangent to the cylinder wall, the outlet (3) passes to the suction connection of the pump in the same tangential direction. The tangential inlet and outlet has the effect of rotating the liquid in the tank. The particular advantages of design are as follows.

- The kinetic energy (velocity level) is retained.
- Deposits are avoided.
- Fine-bubble air (gas) is included in the flow and carried with it without disturbance. Large bubbles which would cause the flow to break away do not reach the outlet connection and therefore do not enter the pump. Instead they are forced down to the center of the rotation by the weight of the rotating liquid and ascend within this core to re-enter the rotating flow in the upper part of the vessel in the form of fine bubbles. The tank replenishes itself in this way with the pumped liquid after an appropriate running time.
- The pump (4) delivers immediately, suction times are short or zero. The pump only operates on a correspondingly lower output on the first startup, until the vessel is replenished.
- To keep the tank full automatically (including during a long shutdown and for fluids which give off gas) and to recharge automatically after shutdown of the pump, an air exchange valve (5) is fitted between the highest point of the suction pipe and the discharge line. This is fitted with a rubberized valve ball in the lower valve disk and has a cleaning opening. After the pump shutdown the liquid flowing back to the pump forces air or gas through the valve into the discharge line (6).
This is a function of the check valve (7) fitted in the suction pipe. A spindle which passes through to the outside, or a lifter, should be avoided where possible because these could cause ambient air to be entrained.

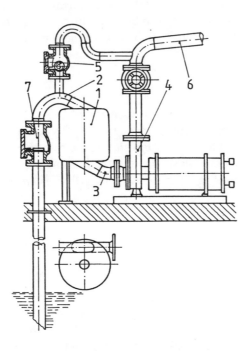

Fig. 3.165 Vacuum system with tangential inlet and outlet flow for the vacuum tank

3.5.11.6 Special measures to protect the canned motor when pumping contaminated liquids using self-priming by means of a vacuum system

Solid particles in the partial flow Q_T through the motor, which could cause damage to bearings or the can, are normally removed by a self-cleaning ring filter. This type of protection is suitable only for occasional impurities in the fluid. Where suspensions are being pumped which have a defined amount of solids a clean liquid is added to the motor which has a separate lubricating-cooling system for this purpose. In addition, it is also necessary to inject a clean liquid into the rotor chamber (refer to Fig. 3.111 Clause 3.5.7 "Pumping suspensions").

3.5.12 Regenerative pumps with a canned motor drive

3.5.12.1 Delivery characteristics

Regenerative pumps have been shown to have relatively good efficiency ($\eta_p = 20$ to 40%) when pumping under extreme delivery pressure/delivery flow conditions in the very low specific speed range ($n_q = 1$ to 10 r.p.m). Although all the characteristics of a flow machine are present on regenerative pumps, their H(Q) behavior is characterized by a steep, straight pattern of the characteristic lines similar to rotating displacement pumps. The hydrodynamic pumping behavior (possibility of throttling) in conjunction with a straight, steep characteristic line pattern makes pumps of this kind particularly suitable for control circuits. These pumps are also used as boiler feed pumps, for high pressure cleaning equipment, atomization, pulverization and extinguishing equipment and also in wide areas of the chemical and pharmaceutical industry. The simple construction and short size give them a good power/weight ratio so that they can be easily fitted to apparatus. For these applications they are not only manufactured from the known metal materials but also from plastic.

3.5.12.2 Constructional description

The regenerative pump is of very simple construction. An impeller with several blades which are always radial and mounted on one or both sides rotates between two plane-parallel housing surfaces. In the area of the blades, the side channel, which is shortened by a short section - the interrupter - extends into the housing around the circumference. The cross-section of the channel is constant with regard to shape and dimensions over the complete extension. The flow of the fluid from the suction connection to the impeller is mainly in the centripetal direction, whereby different connection arrangements are possible (Fig. 3.166).
Several authors [3-24]; [3-25]; [3-26]; [3-27] have dealt with the theory of energy transfer from the impeller to the fluid and also described good methods of mathematically determining the flow processes.
The energy of the fluid, which is imparted to the particular liquid particles by impulse exchange, steadily increases from the inlet into the impeller blades until its exit at the interrupter. The fluid particles entering the blade channel diverts them back into the blade channel. This results in a helical acceleration of the through flow in the circumferential direction to the interrupter (Fig. 3.167). This multiple flow through the impeller gives it an "internal multi-stage" effect.

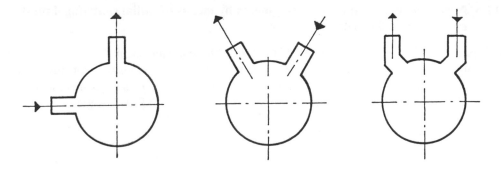

Fig. 3.166 Various suction and pressure connection arrangements for a regenerative turbine impeller

From the type of liquid flow it follows that it is subjected to not inconsiderable shock and friction losses. Furthermore, the unavoidable volumetric losses due to the lateral clearance between the impeller and housing (approximately 0,1 to 0,2 mm) and those which occur at the circumference of the impeller at the interrupter between the fluid inlet and outlet have a decisive influence on the efficiency of these pumps.

Unterbrecher = Interrupter

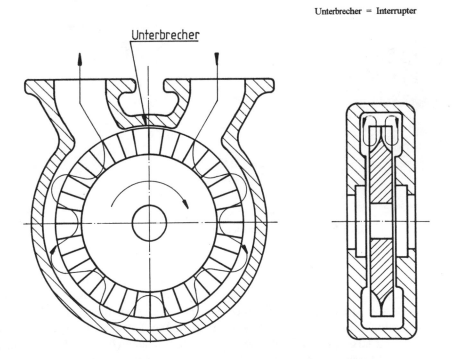

Fig. 3.167 Flow pattern in a regenerative turbine impeller

3.5.12.3 New regenerative pump developments by means of radial centrifugal fluid entry into the impeller

The liquid particles (Fig. 3.166 to 3.167) entering the impeller channel from the suction connection via the impeller outer diameter are subjected to an opposing centrifugal force due to the energy exchange of the impeller. This reduces its energy corresponding to NPSHR thus giving regenerative impeller pumps unsatisfactory NPSHR values.

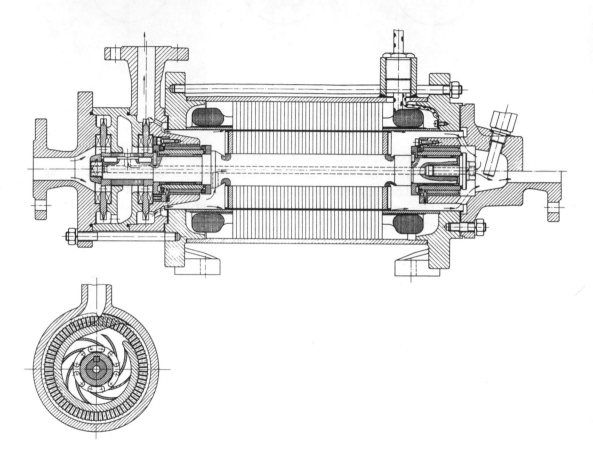

Fig. 3.168 Two-stage canned motor regenerative pump (Lederle)

A design such as shown in Fig. 3.168 is based on a radial impeller with good, normal flow characteristics being positioned before the regenerative impeller (Fig. 3.169) as a booster pump. In this way the comparatively good NPSHR values for this type of impeller can be used and combined at the same time with the advantages which the regenerative impeller has in achieving large pressures at low capacity coefficients (Fig. 3.170). The increase in energy obtained by the front radial impeller also increases the performance of the regenerative impeller pump. Furthermore, this also enables an axial inflow of the liquid with a radial outflow. The radial impellers arranged on both sides of the regenerative impeller to achieve

278

better flow characteristics guide the axial flow entering the pump laterally to the blade inlet of the regenerative impeller through a type of diffuser (Fig. 3.171).

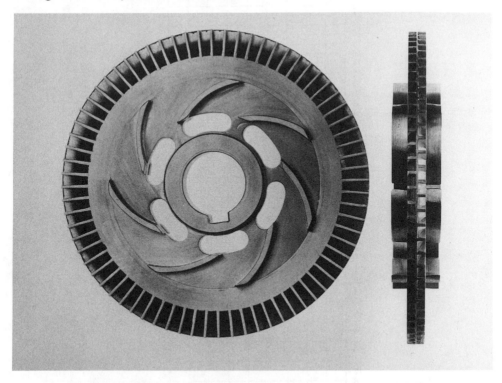

Fig. 3.169 Regenerative impeller with a radial impeller mounted in front (Lederle)

Because of the unequal pressure pattern at the impeller tip, radial transverse forces form which must be taken by the shaft bearing. With the two-stage design the radial force can be largely compensated by offsetting the outlet cross section of the first stage 180° relative to the outlet cross section of the second stage (Fig. 3.168).

The high delivery pressures achievable with regenerative impeller pumps can pose problems with the conventional design using a stuffing box or mechanical seal at the shaft gland. By using a canned motor drive it is possible to build a pump set which meets the requirements of environmental protection, particularly where a harmful substance is being pumped.

The relatively small volume rates of flow also enables an inline design to be achieved (Fig. 3.168 bottom half). In this case the complete flow through the canned motor is split in two. One partial flow passes through the rotor-stator gap whilst the other is passed through a transverse and longitudinal bore in the impeller/rotor shaft to the rotor chamber furthermost from the pump. The two partial flows converge here and flow via bores in the motor rear cover and the plain bearing furthest from the motor to the axially-mounted discharge connection. This arrangement has a further advantage in that the entire flow is used to cool the motor.

279

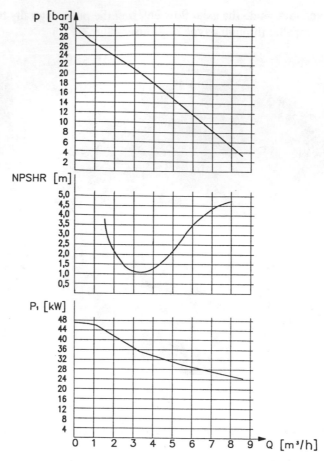

Fig. 3.170 Characteristic curve for a canned motor regenerative impeller pump as shown in Fig. 3.168 (Lederle)

Schnitt = Section

Schnitt A-A

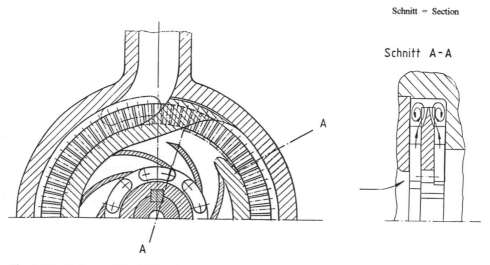

Fig. 3.171 Pattern of liquid flow in the regenerative impeller shown in Fig. 3.168 (Lederle)

3.5.13 Particular characteristics to be noted when operating canned motor centrifugal pumps

3.5.13.1 Characteristic curves which limit the normal working range

It can be seen from Section 3.3 ("Operating range of canned motor centrifugal pumps") that canned motor pumps can only be operated within a defined operating range between the minimum capacity Q_{min} and a maximum capacity Q_{max} if constant, correct operating delivery conditions are to be guaranteed (Fig. 3.48). This applies not only to canned motor pumps or magnetic coupling pumps but also applies in general to all centrifugal pumps with the exception of the limitation for the curve which represents the rising temperature of the partial flow through the rotor. When discussing the various influences which limit the working range the following factors are to be considered:

- The Q_{max}- Q_{min} range is essentially influenced by the NPSHR behavior of the pump. In the overload range and close to the ordinate under partial load delivery there is sometimes a pronounced rise in the NPSHR. If this range is exceeded failure of the pump can result.

- The bending stress on the pump shaft for single-stage radial impeller pumps reaches its maximum value at Q = 0. Together with its rotational forces, this shaft experiences its maximum stress immediately before the bearing on the pump end. Strong radial forces occur not only where there is zero delivery but also in the overload range. The radial pressure which occurs must therefore be allowed for.

 - The increase in load which behaves almost linearly with respect to Q must not lead to overloading of the motor in the overload range. This is for reasons not only of motor life but also because of the associated temperature rise and the fact that this calls into question the explosion protection safety of explosion-proofed motors.

 - For economic reasons therefore, care should be taken to ensure that even if the desired duty point Q_{opt} is not achievable the pump still operates in an acceptable efficiency range.

 - The rise in temperature of the partial flow Q_T when flowing through the rotor-stator gap must not exceed the specified limits, to ensure that on one hand the motor losses occurred are removed without difficulty in the main flow or by the motor cooling liquid where there is external cooling, and on the other hand no vaporization of the partial flow must occur at any point in the rotor chamber or in the pump because otherwise the heat dissipation and the axial and radial thrust balance are endangered and cavitational problems could occur.

The most precise possible compliance with the working range Q_{min}- Q_{max}, established by the manufacturer on the basis of a heat balance calculation and notified to the operator, is a precondition for proper operation of the pump.

3.5.13.2 Measures for compliance with the Q_{max} - Q_{min} limitation

Conditions within the plant are not always such that it can be safely assumed that the flow rates Q_{min} and Q_{max} are not undershot or overshot. For applications of this kind, special precautions can be taken to ensure that the limiting values are met.

- *Compliance with Q_{min}*

To ensure that the minimum rate of flow Q_{min} is maintained, a bypass line to the inlet or suction tank is to be fitted between the pump discharge connection and the first isolating device before any check valve and an orifice which permits minimum rate of flow Q_{min} to pass through is to be fitted in the pipe just before the inlet tank. It is advisable to fit the orifice close to the tank particularly where near-boiling liquids are being pumped, e.g. liquid gases, because the increase in the velocity of the flow through the orifice can cause the liquid to gasify and lead to a two-phase mixture in the bypass pipe, causing a hydraulic cross-section convergence. Fitting a bypass pipe also has a further advantage in that because the system is open to the liquid tank good venting can result. This is particularly important for pumps which handle extremely cold fluids. The pump is "kept cold" by constant revaporization and is therefore always ready for use.

Where a bypass pipe with an integrated Q_{min} is fitted this promotes a steady amount of loss which has a negative effect on the efficiency of the pump set.

This is acceptable where there are small rates of flow but with larger pumps an arrangement using a free-flow check valve is advantageous for reasons of economy (Fig. 3.172) [3-28].

The delivery flow holds the valve disk of the free flow check valve at a specific height. If the flow rate is such that Q_{min} is not undershot the complete rate of flow goes to the consumer (Fig. 3.173, Fig. 1). If the minimum flow rate Q_{min} is undershot, part of the flow goes to the consumer and part is returned to the suction or inlet tank (Fig. 2). If the consumer does not accept any fluid the complete flow Q_{min} is returned to the suction or inlet tank.

This type of liquid flow produces economic operating conditions whilst at the same time protecting the pump if the flow drops below the minimum.

- *Measures to protect against the maximum flow Q_{max} being exceeded*

A Q_{max} orifice fitted immediately on the pump discharge connection (before the junction of the Q_{min} pipe) will ensure that the maximum specified duty point will not be exceeded in any operating condition. Protecting the Q_{min} and Q_{max} values can mean that operating faults such as the opening or closing of isolating valves in the pressure line of the pump do not occur and damage is therefore avoided.

Control of Q_{min} - Q_{max} without losses requires the fitting of a differential pressure valve between the suction and pressure connection of the pump. If the specified Q_{min} - Q_{max} range is exceeded the delivery pressure corresponding to the particular flow switches off the pump. In this case it should be noted that the absence of a Q_{min} bypass pipe can prevent the pump priming, particularly if a check valve is fitted in the pressure line.

The following installation arrangements show how canned motor/magnetic coupling pumps should be correctly installed.

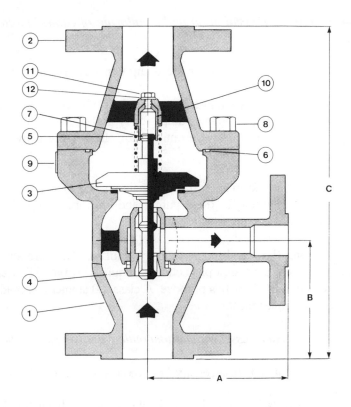

1	Body	5	Spring	9	Identification plate
2	Flange	6	O-ring	10	Guide
3	Check disc	7	Guide ring	11	Bolt
4	Cascade bushing	8	Bolt	12	Lockwasher

Fig. 3.172 Freeflow check valve [3-28]

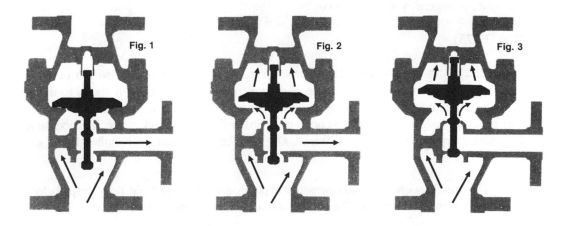

Fig. 3.173 Functional diagram of the freeflow check valve shown in Fig. 3.172

*- Recommendations for installation and monitoring of canned motor/magnetic coupling
centrifugal pumps [3-29]*

The fault-free operation of a hermetic pump depends mainly on the type of installation
and operation.

The following installation arrangements give details for the installation and protection of
canned motor/magnetic coupling pumps for typical applications. Auxiliary pipelines for
draining, flushing or heating were omitted for clarity. Alternative proposals are shown
"dotted".

For all applications the main feature is maintenance of the minimum and permissible
maximum flows required for these pumps and ensuring adequate ventilation or degas-
sing (impairment of bearing function).

Level monitoring (LS-) and temperature monitoring (TS+) are absolutely necessary for
these pumps where explosion protection is required. A type NTS electronic monitoring
device is recommended for this purpose. A class 1,0 ammeter should also be installed for
monitoring the canned motor.

Arrangement 1 : Suction vessel with bottom outlet; positive suction head operation

Minimum flow required. If necessary limit the maximum flow by a suitably dimensioned
Q_{max}.
An additional vent facility must be provided if a check valve is fitted on the outlet side of
the pump.

(Fig. 3.174)

Arrangement 2 : Suction vessel with bottom outlet; operation with minimum flow

Installation in fully automatic systems and when handling liquid gases, (NH_3, Freon,
chlorine, phosgene, vinylchloride etc.).

(Fig. 3.175)

When the pump is shut down the bypass serves as a vent line and when the pump is
running then, in conjunction with a properly dimensioned Q_{min} orifice, it serves to maintain the minimum flow. For higher minimum flow requirements a freeflow checkvalve[1]
can be used instead of the Q_{min} combination of orifice and check valve. If necessary,
limit the maximum flow using a suitable Q_{max} orifice (important for handling liquid gases).

   [1] The freeflow check valve opens the bypass only when the main flow drops below
   the minimum flow rate. The bypass is open when the pump is shut down.

Notes on transferring liquid gases:
- Fully open the shut-off valve in the bypass line (if necessary remove the hand-wheel).
- Install the Q_{min} orifice as close as possible to the suction vessel (avoidance of two- phase flow).
- If vapor pressures are high, provide a gauge to determine the pump differential pressure.
- To avoid cavitation ensure the minimum positive suction head is available.
- Losses in the suction line should be minimized ($V \approx 1$ m/s). Because the NPSHR value of the pump increases with flow it is absolutely necessary to maintain the maximum flow.

Arrangement 3 : Common discharge pipe for several pumps

Parallel operation of several canned motor pumps with a spare pump also installed (standby operation)

(Fig. 3.176)

Reserve pumps must always be kept ready for operation with shut-off valve opened and if possible filled with liquid [2]. To avoid backflow through the reserve pump, a check valve is necessary for each pump. An independent venting device must also be provided for each pump.

   [2] Not recommended when pumping chlorine.

Caution when operating canned motor/magnetic coupling pumps in parallel

Where there are flat H(Q) curves the pump output can be impeded if the pumps have slightly different heads (minimum flow is then not guaranteed). This can be remedied either by fitting orifices [3] or installing separate bypass lines (refer to Arrangement 4 and Section 3.5.13.3).

   [3] This produces a steep curve downstream of the orifices

Arrangement 4 : Common discharge line of several pumps; bypass lines for minimum flow for each pump

Automatic operation of several canned motor pumps. Pumping of liquid gases. Reserve pump also installed (standby operation).

(Fig. 3.177)

For safety reasons each pump should be provided with a bypass line. This is absolutely necessary when pumping liquid gases so that gas which collects in the pump when stationary can be removed. This arrangement also means that the minimum flow of each pump is independent of the differential head of each pump.

Reserve pumps should be kept ready for operation with shut-off valves open and filled with liquid where appropriate. This enables changeover from one pump to another or a second pump to be brought on stream without additional actuation of shut-off valves.

Notes on pumping liquid gases:
- If possible use separate suction lines for each pumps.
- Otherwise the instructions for Arrangement 2 apply
- The limitations in Arrangement 3 regarding parallel operation apply.

Arrangement 5 : Suction vessel with top outlet, priming or siphoning operation with priming tank

The installation requires the following conditions:
Vapor pressure of the liquid less than 0,5 bar.
Tank exposed to atmospheric pressure.
Volume of priming tank two or three times the volume of the siphon pipe. Siphon pipe not higher than 4 m.

(Fig. 3.178)

Before starting the pump the priming tank must be filled either through the filling valve (V_1) or the pressure line. In the latter case the filling valve (V_1) serves as the vent. The pump can only be started when the liquid reaches the higher switching level and it is shut down when it reaches the lower switching level.

If the pump is fitted with a check valve on the outlet side, the priming tank is refilled by a valve V_2, and venting takes place via V_1.

If for technical reasons the volume of the liquid in the pump on the outlet side is not sufficient to fill the priming tank, the use of a re-filling tank with half the volume of the priming tank is recommended. As the pump is shut down the priming tank is replenished by the pump if the shut-off valve V_3 in the pressure line is open.. Shut-off valve V_3 must be closed before starting a new draining cycle. Automatic replenishment of the priming tank can be achieved if valve V_3 is a solenoid valve.

Arrangement 6 : Suction vessel with top inlet, priming or siphoning operation with vacuum controlled system.

This arrangement is recommended for systems with several filling points, whereby all suction and vent pipes are connected through suitable valves to a central vacuum tank. This arrangement also makes automatic priming or siphoning possible.

(Fig. 3.179)

The vacuum pump is switched on or off via two level switches (LS-) and (LS+).
A double-acting check valve on the tank maintains the vacuum when pump 1 is shut down and prevents overflow of the liquid into the vacuum pump if the control system fails.

The lower level switch (LS-) and the vacuum tank should be at least 0,3 m above the highest venting level of the system.

A vent (float valve) fitted in the vent line between the pressure line and vacuum tank prevents liquid from the discharge side of the pump entering the vacuum tank.

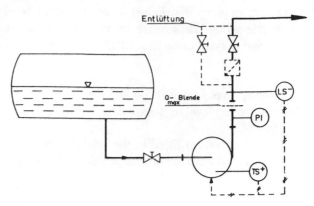

Fig. 3.174

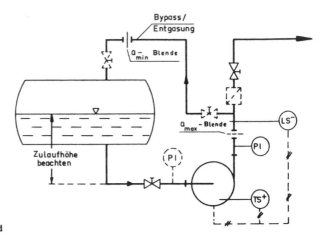

Fig. 3.175

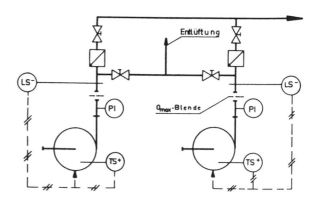

Fig. 3.176

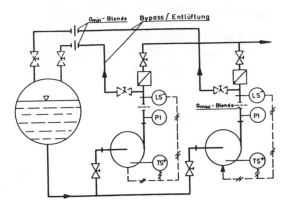

Fig. 3.177

Bypass/Entlüftung = Bypass/venting
Blende = Orifice

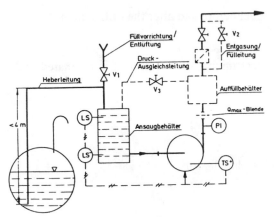

Bild 3.178

Füllvorrichtung/Entlüftung = Filling/venting device
Entgasung/Fülleitung = Degassing/Vent pipe
Druck - Ausgleichsleitung = Pressure balance pipe
Auffüllbhälter = Replenishing tank
Blende = Orifice
Heberleitung = Siphon line
Ansaugbehälter = Priming tank

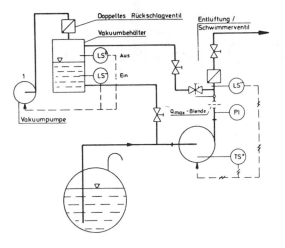

Fig. 3.179

Entlüftung/Schwimmerventil = Vent/Float valve
Doppeltes Rückschlagventil = Double check valve
Vakuumbehälter = Vacuum tank
Vakuumpumpe = Vacuum pump
Blende = Orifice

289

3.5.13.3 Design of orifice plates for setting Q_{min} and Q_{max} [3-8]

The hole diameter d in mm required depends on the desired partial flow Q in m³/h and the pressure difference H in m before and after the orifice plate. It also depends on the ratio of the openings in the orifice and a throttling coefficient f, which are given in the diagram (Fig. 3.180).

According to [3-8] the diameter of a throttling orifice is obtained from:

$$d = f \cdot \sqrt{\frac{Q}{\sqrt{\Delta H}}}$$

(3 - 55)

D	= Diameter of pipe
d	= Diameter of orifice in mm
f	= Throttling coefficient
Q	= Partial flow in m³/h
ΔH	= Difference in pressure level before and after the orifice in m

The opening ratio is initially unknown when determining d. Therefore f is estimated and corrected by repeated calculation.

Example (3-H):

What size must the bore of an orifice plate be for
- D = 100 mm,
- Q = 49 m³/h
- ΔH 16 m ?

f is estimated to be approximately 10, therefore we would get

$$d = 10 \cdot \sqrt{\frac{49}{\sqrt{16}}} = 35 \text{ mm, therefore } \left(\frac{d}{D}\right)^2 = \left(\frac{35}{100}\right)^2 = \underline{0,122} \text{ , in this case } f = 11,6$$

this is then corrected as follows

$$d = 11,6 \cdot \sqrt{\frac{49}{\sqrt{16}}} = 40,6 \text{ mm, therefore } \left(\frac{d}{D}\right)^2 = \left(\frac{40,6}{100}\right)^2 = \underline{0,165} \text{ , in this case } f = 11,55$$

and again corrected as follows

$$d = 11,55 \cdot \sqrt{\frac{49}{\sqrt{16}}} = \underline{40,5 \text{ mm}}$$

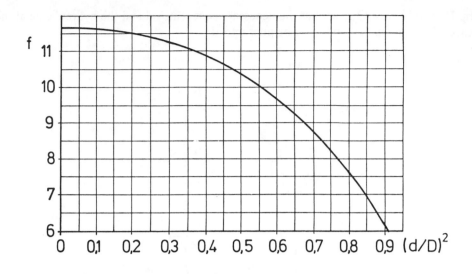

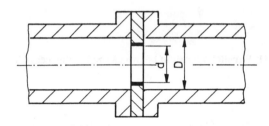

Fig. 3.180 Orifice plate and its throttling coefficients [3-8]

3.5.13.4 Parallel operation of canned motor pumps [3-30]

The curves of two equal pumps which are operated in parallel are seldom identical. They differ, if only slightly, due to manufacturing tolerances.

Even if the curves differ only slightly with regard to the heads, the pump with the lower head can be disadvantaged compared with the other one.

In the case shown in Fig. 3.181, pump II has additional pipe friction losses to overcome compared with pump I.

The two unequal curves are shown schematically in Fig. 3.182 and designated I and II.

It can be seen that during parallel operation pump I can be operated up to point C, without pump II also pumping. From point C onwards both pumps deliver together with the flow rates being added (curves I + II). At the intersection point of the pipe curve with the common pump curve we get duty point B for parallel operation. Pump I contributes the amount Q_I to the total capacity Q_{ges}, pump II in contrast contributes only amount Q_{II}. The situation shown in Fig. 3.182 can now arise where one or even both pumps are working outside the permissible working range Q_{min} and Q_{max}. In the case shown in Fig. 3.182

pump II is still below Q_{min} so that the necessary heat removal and the hydraulic axial thrust balance of this pump can no longer be guaranteed.

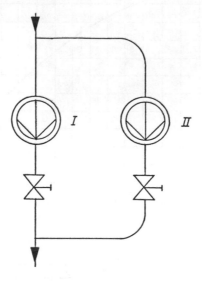

Fig. 3.181 Parallel operation of two canned motor centrifugal pumps

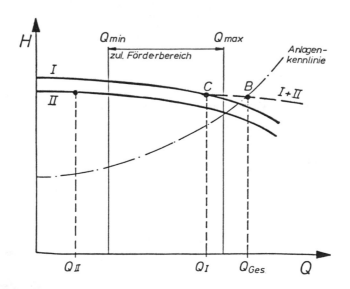

zul. Förderbereich = Permissible working range   Anlagenkennlinie = System curve

Fig. 3.182 H(Q) curves and system curve of two canned motor centrifugal pumps operating in parallel

If at all possible, parallel operation using two canned motor pumps should be avoided, particularly where the curves are relatively flat.

292

3.5.14 Safety and monitoring devices on canned motor pumps [3-31]

The application range of canned motor pumps includes plants in which hazardous, toxic and/or explosive mixtures form or carcinogenic substances have to be moved in pipe systems. Such pumps have to be provided with the best possible safety and monitoring devices. The expenditure required to protect both humans and operating equipment is always justified.

The functioning of canned motor pumps can be monitored by completely electronic means, which contributes substantially to their availability. These include the following

I	Level monitoring
II	Temperature monitoring of winding
III	Excess current release
IV	cos φ - motor load monitor
V	Rotor position indicator
VI	Vibration sensor
VII	Stator pressure monitor transmitter
VIII	Sensitive monitoring of can for tightness

Monitoring measures I and II have already been described in Section 3.2.3 (The use of canned motors in hazardous locations, Explosion protection according to European Standard "EN"). The excess current release III can be regarded as already known.
The monitoring devices named under IV to VIII are further discussed in the following:

IV - cosφ motor load monitor

Monitoring of the inductive load of the canned motor provides a direct source of information on the operating condition of the motor and consequently the operating performance of the pump set. The power consumption of the motor is directly related to the H(Q) behavior of the pump (Fig. 3.183). Canned motor pumps must, however, be operated within a specific pumping range. To the right of pump maximum flow Q_{max} specified by the manufacturer, drastic increases of NPSHR and overloading of the motor may result. It is also possible that the differential in pressure of the partial flow for passing the rotor-stator gap is no longer sufficient. To the left of minimum flow Q_{min} close to the ordinate there is an asymptotic increase of NPSHR, which means that damage to the pump may occur through cavitation. In addition, if the pump flow is less than the minimum Q_{min}, the coolant-lubricant flow is thermally overloaded, which may lead to an excessive increase in motor temperature. The phase angle, which changes with motor loading, can be monitored by a power factor monitor. In this case a red LED is used to indicate that the monitor is ready for operation. The motor load monitor operates on the static current principle; i.e. when the set value (e.g. pumping range between Q_{min} and Q_{max}) is reached, the two output relays for cosφ - min and cosφ - max are energized, and two green LEDs light up. The values for cos φ_{min} and cosφ_{max} which correspond to Q_{min} and Q_{max} can be set on the unit by means of a knob (Fig. 3.184). The set values refer to the rated voltage. If the relevant level drops below or exceeds these values, the corresponding LED lights up. The output relay for cos φ_{min} and cos φ_{max} is de-energized at the same time [3-22].

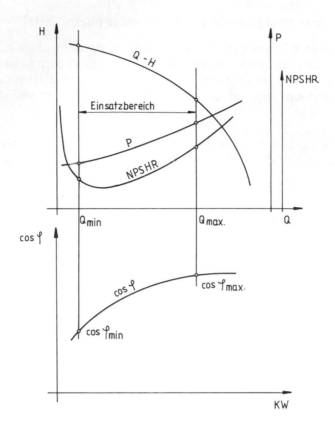

Fig. 3.183 H(Q) diagram with corresponding cosφ curve of canned motor

Fig. 3.184 cosφ motor load monitor in accordance with IEC 255, VDE 0435 (Dold-Söhne)

To prevent the cosφ monitor from reacting on starting the motor, the starting time (which is extremely short) can be set at the unit. The cosφ monitor does not function during this period. A reaction time of 5 seconds is also provided in order to prevent the cosφ monitor from reacting to temporary deviations from the set values. A functional diagram of the cosφ monitor is provided in Fig. 3.185.

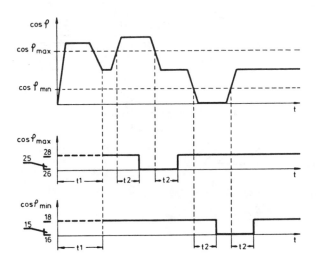

t1 = Start-up bridging time
t2 = Reaction time

Fig. 3.185 Functional diagramm of cos φ - motor load monitor acc. Fig. 3.184

V Rotor position measuring device (RPM) by AEG-Kanis

Hydraulic balancing of axial forces is particularly important in canned motor pumps, as these pumps are not equipped with thrust bearings. The slide rings in the bearings have locating and control functions only; they are not designed to absorb significant axial thrust. If axial thrust balance is disturbed, through cavitation for example, considerable friction damage to the pumps may result. Such damage can be avoided when a rotor position measuring device is mounted on the canned motor pump (Fig. 3.186 and 3.187). The RPM registers changes in the axial position of the pump motor rotor inductively and without contact. When the maximum permissible excursion is exceeded, either an alarm is triggered or the pump is shut down. The device consists of a transducer, with the supply lead and an electronic evaluation system fitted in a pressure housing. A thin shaft is mounted on the end of the rotor. Over the shaft, a support pipe is flanged tightly to the pump casing onto which the transducer is clamped. The support pipe and shaft are made of non-ferromagnetic materials (austenite). The unit is designed for measuring from ± 0,5 mm to ± 1,5 mm, the common range for monitoring of axial clearance.

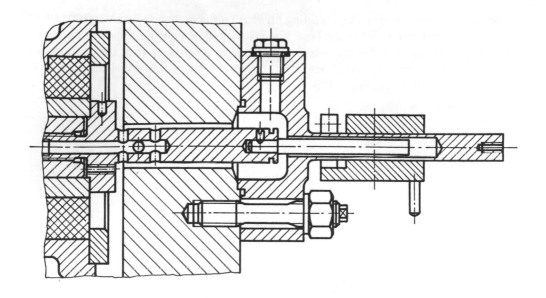

Fig. 3.186 Rotor position measuring device (RPM)

Fig. 3.187 View of an RPM device as shown in Fig. 3.186

296

An alternative design for detecting shaft position is - the VIBROCONTROL 1000 developed by the Schenk Company (Fig. 3.188).

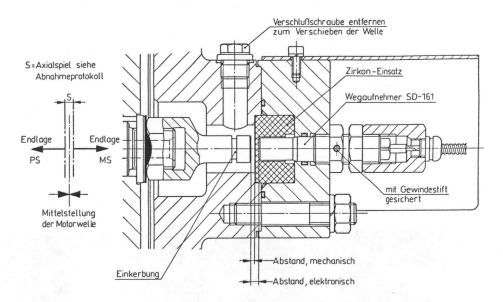

Verschlußschraube entfernen zum Verschieben der Welle = Remove locking screw to adjust shaft;
S = Axialspiel siehe Abnahmeprotokoll = S = Axial clearance, refer to acceptance record; Zirkon - Einsatz = Zircon insert;
Wegaufnehmer = Position pickup; mit Gewindestift gesichert = Locked by grubscrew; Endlage = End position;
Endlage = End position; Mittelstellung der Motorwelle = Mid-position of motor shaft; Einkerbung = Recess;
Abstand, mechanisch = Clearance, mechanical; Abstand, elektronisch = Clearance, electronic;

Fig. 3.188 VIBROCONTROL 1000 rotor position device

This measuring system consists of an electronic system and contactless position pickup. The axial displacement of the rotor about a predetermined zero position is detected by a contactless position pickup and converted into electrical signals. The electronic measuring system forms the excursion s (mm) from these signals, with the correct signs and proportional to the distance.

For visual representation of the measured values;

       - analog indicating instruments and/or
       - dotted-line recorders

can be connected to the standard analog outputs of the electronic measuring system.

VI - Monitoring by measurement of vibration

Canned motor pumps produce very little vibration or noise when working. This is shown by the fact that pump sets of this kind have a sound emission level on average 18 dB lower than the same pumps of conventional design without the drive motor. If during cavitation-free operation the pump exhibits vibration behavior which deviates from the norm this indicates a disturbance. Vibration sensors can therefore be used to obtain a reliable assessment of the condition of bearings and rotating assembly. This can be achieved either by observing the absolute or the relative shaft vibration. The measurement of the relative shaft vibration is more expensive but it provides a better indication than the measurement of the absolute. To do this the relative shaft vibration is measured and monitored as the rotational movement of the shaft in the bearing. To asses the operating condition of the pump it is, however, sufficient in most cases to measure and monitor the absolute bearing vibration using the permanent-dynamic plunger coil measuring method.

Fig. 3.189 High pressure canned motor pump fitted with a vibration sensor (Hermetic)

For this method of measurement, vibration sensors are mounted on the motor housing (Fig. 3.189) and a coil with a flexible diaphragm is arranged in each one in such a way that it is inserted in the air gap of a permanent magnet. If external vibratory movements are now applied to the pickup, the housing of the pickup moves with the part of the pump to be measured and monitored, whilst the plunger coil suspended from the flexible diaphragm remains stationary. This produces a relative movement between the coil and magnetic field thus inducing voltages in the coil which are directly proportional to the velocity of the vibration. The voltage output of the sensor is applied to an electronic circuit for processing, which if a set limiting value is exceeded can trigger an alarm or shut down the pump.

VII - Pressure monitoring in the motor-stator chamber

The terminal box is sealed with cast resin and connected directly to the pressure-proof stator chamber containing a pressure switch (Fig. 3.190). If the motor can is damaged, for example because of damage to the bearings or abrasion from solids in the coolant and lubricant flow, the pumped fluid may escape into the stator chamber. This results in an increase in the internal pressure in the stator chamber. Even relatively low overpressure p_e = 1.0 - 1.4 bar is sufficient to disable the motor by means of the pressure switch. The motor windings are damaged by the pumped fluid in most cases, but serious environmental damage is prevented as the liquid cannot escape into the atmosphere.

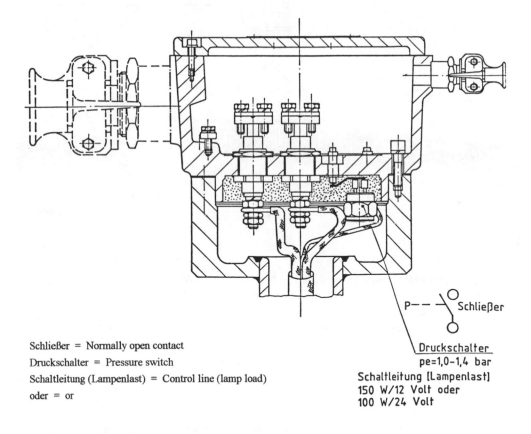

Schließer = Normally open contact
Druckschalter = Pressure switch
Schaltleitung (Lampenlast) = Control line (lamp load)
oder = or

P— —⟋ Schließer

Druckschalter
pe=1,0-1,4 bar
Schaltleitung [Lampenlast]
150 W/12 Volt oder
100 W/24 Volt

Fig. 3.190 Pressure switch integrated into the terminal box

VIII - Sensitive monitoring of the can by means of the "DWS" system [3-34]
(Double-walled Safety) (German patent No. 3639720. "European patent No. 026813; US patent No. 4838763)

The already high safety standard of canned motor pumps with a double sealing casing - canned motor and motor housing - between the product and atmosphere is further enhanced by a second static can thus providing a triple sealing jacket between the fluid and atmosphere.

The double-walled construction of the can enables a sensor to be positioned in the space between both cans which in the event of damage to the inner can in contact with the product will trigger either a visual and/or acoustic signal or shutdown the pump (Fig. 3.191)

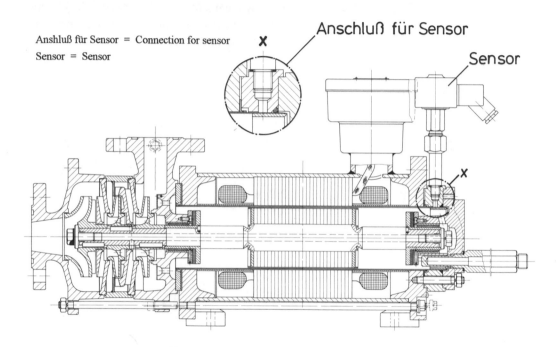

Anshluß für Sensor = Connection for sensor

Sensor = Sensor

Anschluß für Sensor

Sensor

Fig. 3.191 Canned motor with double-walled can (DWS) and signaling sensor

This is of particular interest for highly toxic substances. Apart from this, the DWS system is also an ideal bearing monitoring device which responds in good time in the event of bearing damage. The DWS system is also suitable for reducing repair times and costs of canned motor pumps because the winding is protected from contact with the liquid which previously would normally have led to total failure. Because canned motor pumps are mainly of the explosion-protected type, this substantially increases the ability of the operator to carry out repairs. To further increase the maintenance-friendliness of such pumps, the inner can can be designed in such a way that in the event of the damage it can be removed without having to break welded joints, thus making repair simple for the operator.

3.5.14.1 Specified and recommended monitoring devices in accordance with Section 3.5.14

Re I and II - Level-temperature monitoring
Specified by legislation for all canned motor pumps which work in explosion protection areas, certified by the Eex de II C T1 to T6, Certificate of Conformity. Recommendations by manufacturers to also provide pump sets not subject to explosion protection regulations with the same, to avoid operating errors and consequent damage.
Monitoring of the motor winding temperature by an PTC or NTC thermistor: recommended by manufacturers as the simplest and most effective type of motor monitoring.

Re III - Overcurrent release
Specified by legislation for all canned motors.

Re IV - cosφ - motor load monitor
Recommended by manufacturers if a flowmeter has not been fitted to protect Q_{min} and Q_{max}.

Re V - Rotor position measuring device
Recommendation of the BITC Office (Bureau international technique chlorure) International Association of Chlorine Manufacturers and Processors for all canned motor pumps used to pump chlorine. Recommended by manufacturers for pumps which are expensive and particularly important for the process.

Re VI - Monitoring by measurement and vibration
Recommended by manufacturers for very expensive pumps, particularly in the high pressure area.

Re VII - Pressure switch in the terminal box
Recommended by manufacturers for liquids which are not electrically conductive, e.g. oils, and where they can enter the stator chamber in the event of damage to the can.

Re VIII - The double-walled safety system DWS
Recommended by manufacturers for all canned motor pumps and pumps with a permanent magnet coupling which pump highly-toxic substances.

3.5.15 Simplification of the installation of the power and control connections to the motor

The motor power cable from the motor contacter in the control room terminates in a connection in the terminal box in the motor. The same applies to the control cable for the PTC thermistor in the winding. The cables for the level and temperature monitoring devices are each separately laid from the control room to the particular device. This means that a qualified electrician also has to be present for removal of the pump, in addition to the pump fitter, so that two tradesmen are necessary.

A later development is a system which with the aid of an interlockable four-pole connector which in the CES system corresponds to CEE Publication 17 and to the European Standard EN 50014, provides a movable power cable from a local switch box to the motor. The plug can only be withdrawn after unlocking which cuts off the current. The control cables are provided with separate low-voltage type Exi connectors.

This means that installation and the dismantling can be carried out without a special electrical tradesman, thus saving time and labor. Figures 3.192 and 3.193 show the arrangement of such a connecting system.

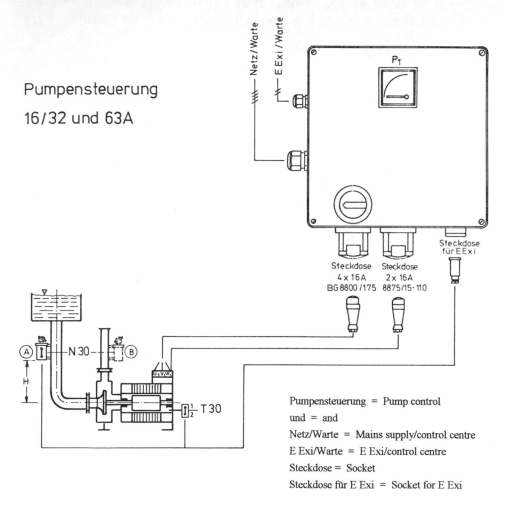

Pumpensteuerung = Pump control
und = and
Netz/Warte = Mains supply/control centre
E Exi/Warte = E Exi/control centre
Steckdose = Socket
Steckdose für E Exi = Socket for E Exi

Fig. 3.192 Cable connection between canned motor and control centre

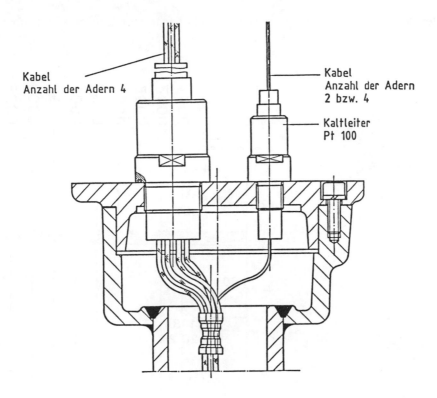

Kabel
Anzahl der Adern 4

Kabel
Anzahl der Adern
2 bzw. 4

Kaltleiter
Pt 100

Kabel Anzahl der Adern 4 = Cable, number of cores 4
bzw. = or
Kaltleiter = PTC thermistor

Fig. 3.193 E Exd cable entry on terminal box

3.5.16 Availability of canned motor pumps

In assessing a pump system for a particular requirement availability also plays a particularly important role in addition to the procurement cost and power input. Shutdown and
repair times cause production failures, and the ensuing repairs affect profitability. Canned
motor pumps have a particularly high availability. Cases are known where pumps have
reached 250 000 running hours in continuous operation without failure. The good performance of canned motor centrifugal pumps with regard to the number of repairs relative to
the total number of pumps used in large chemical plants is shown in a study by Hoechst
AG from 1991 (Fig. 3.194). What is particularly noticeable is the good repair ratio of
7,1% for the canned motor centrifugal pump compared with 20,7% for the conventional
standard chemical pump (refer to columns 11 and 3). Low repair costs save both wage and
equipment costs and brings savings by reducing spares holdings.

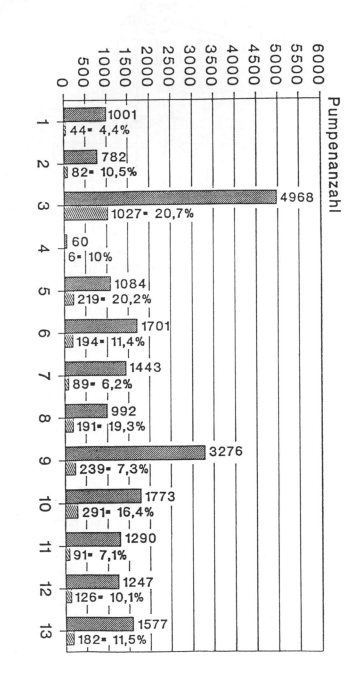

Pumpenanzahl

Pumpenanzahl = Number of pumps

Pumps

Repairs

1001
44 = 4,4%

782
82 = 10,5%

4968
1027 = 20,7%

60
6 = 10%

1084
219 = 20,2%

1701
194 = 11,4%

1443
89 = 6,2%

992
191 = 19,3%

3276
239 = 7,3%

1773
291 = 16,4%

1290
91 = 7,1%

1247
126 = 10,1%

1577
182 = 11,5%

1. Piston metering pump
2. Diaphragm metering pump
3. Standard chemical pump
4. Inline chemical pump
5. Standard chemical pump, magnetic
6. Standard water pump
7. Volute pump

8. Channel impeller pump
9. Side channel pump
10. Liquid ring vacuum pump
11. Canned motor pump
12. Submerged pump
13. Helical rotor pump

Fig. 3.194 Repair statistics for various kinds of pumps in operation in large chemical plants, 1991

3.5.17 Are canned motor pumps economical?

Economic efficiency is an important factor for all working machines used in industry. This must also be the case with canned motor pumps, even though it should not be of first importance because of the environmental-compatibility and operating safety. Particular attention must, however, be paid to economic efficiency with regard to energy and cost saving.

To assess the economic efficiency of a machine, it is necessary for all the relevant cost sources to be taken into account. For canned motor pumps these are as follows:

- Procurement costs
- Availability
- Energy costs
- Service life
- Ease of repair
- Ease of installation
- Maintenance cost
- Space requirement

When assessing the canned motor pump, a conventional type of pump with a double radial shaft seal, coupling, coupling guard, bedplate and explosion-protected pressure-proof encapsulated three-phase motor should be used for comparison in order to begin with a reasonably level basis. The bar charts Fig. 3.195 and 3.196 show the prices of standard chemical pumps and multistage centrifugal pumps with the same hydraulics compared with canned motor pumps, in each case in GGG Spherulitic cast iron and high-grade steel for some sizes. It can be seen from the charts that in most cases the canned motor pumps compare well with regard to price. This is more pronounced with the multistage type than with the standard chemical pump. This assessment, however, considers only the procurement price. Energy consumption is also an important factor and to assess this the efficiency of canned motors must be compared with that of normal three-phase motors. In this case the standard three-phase motor clearly comes out best (refer to the graph in Fig. 3.197 and table 3-IX).

The overall efficiency must be taken into account to assess the power requirement. This includes not only the hydraulic efficiency but also the mechanical efficiency which takes account of the losses due to radial shaft seal and bearing friction. It may also be necessary to allow for the energy expenditure required to pressurize the double mechanical seal. If one compares the overall energy consumption of both pump sets, it can be seen that the canned motor pump is not substantially worse than conventional ones. The canned motor pump has a distinct advantage with regard to availability compared with a pump with mechanical sealing, as can be seen from Section 3.5.16 ("Availability of canned motor pumps") and Fig. 3.194. The repair times and costs are a more important factor in the price performance balance. The ease of installation (no alignment of pumps on foundation) of canned motor pumps must also be allowed for.

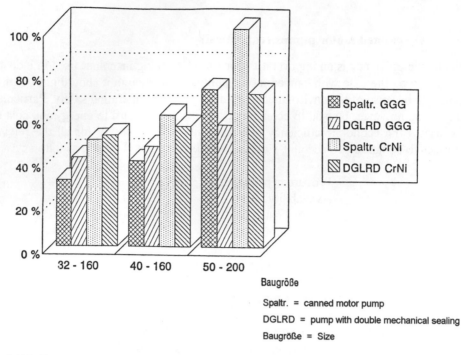

Fig. 3.195 Cost comparison between a canned motor centrifugal pump and a conventional standard chemical pump with a double mechanical seal

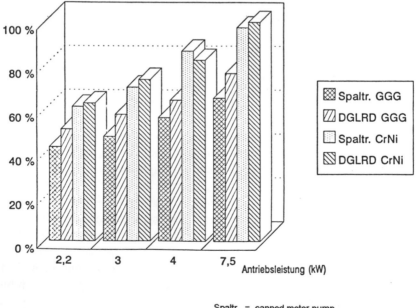

Fig. 3.196 Cost comparison between multistage canned motor centrifugal pumps and conventional centrifugal pumps with double mechanical sealing

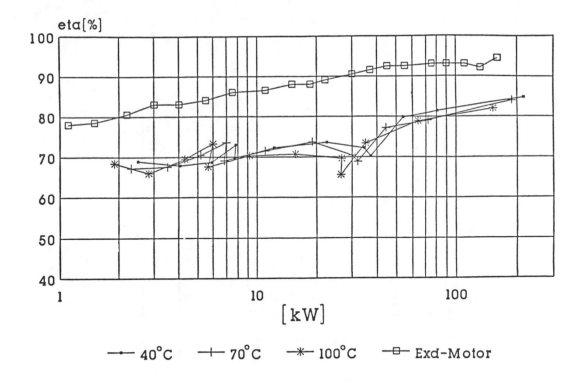

eta[%]

1   10   100

[kW]

—•— 40°C   —+— 70°C   —*— 100°C   —□— Exd-Motor

| Canned motors with different fluid temperatures of product | Normal three-phase motors, IP 55 type of protection "Flameproof enclosure" |

Fig. 3.197 Comparison of the efficiency of canned motors and three-phase motors of protection type IP 55, type of ignition protection "flameproof enclosure"

If all cost factors are summed up it will be seen that the canned motor pump is more cost effective in the final analysis than the radial shaft seal pump and furthermore it does not pollute the environment as it has no leaks which are not only injurious to health but can also incur costs particularly with expensive products.

The decision as to whether standard or canned motor pumps should be used for a specific application can only be made from the point of view of economy after all the aforementioned factors have been examined. In many cases the conclusion will be reached that the possibly higher procurement costs of the canned motor pump are more than made up for by the savings in maintenance and ancillary costs, so that this type of construction is also advantageous from the point of view of economics.

Motoren-Typ	P₁ 40°C [kW]	P₂ 40°C [kW]	Eta 40°C [%]	P₁ 70°C [kW]	P₂ 70°C [kW]	Eta 70°C [%]	P₁ 100°C [kW]	P₂ 100°C [kW]	Eta 100°C [%]
N14L-2	3,74	2,51	67,11	3,44	2,31	67,15	2,78	1,90	68,35
N24L-2	6,00	4,07	67,83	5,20	3,51	67,50	4,30	2,84	66,05
N24N-2	8,60	5,90	68,60	7,40	5,20	70,27	6,20	4,30	69,35
N34L-2	10,70	7,80	72,90	9,50	6,98	73,47	8,20	6,00	73,17
N54E-2	11,10	7,72	69,55	9,90	6,83	68,99	8,40	5,76	67,50
N54P-2	16,90	12,20	72,19	15,40	11,00	71,43	13,10	9,20	70,23
N64r-2	30,60	22,50	73,53	25,80	19,00	73,64	22,10	15,60	70,59
N74n-2	48,00	34,60	72,08	42,80	30,10	70,33	38,40	26,70	69,53
N74rm-2	53,20	37,30	70,11	46,80	32,20	68,80	40,40	26,50	65,59
N80rm-2	67,80	54,00	79,65	57,60	44,40	77,08	48,00	35,20	73,33
N80v-2	99,20	80,60	81,25	91,20	72,00	78,95	80,80	63,70	78,84
CKP85z-2	254,00	215,00	84,65	224,00	188,00	83,93	185,60	152,00	81,90

Table 3-IX Power absorbed (P1) - shaft power (P2) - efficiency (eta) of canned motors for different temperatures

Summary

Centrifugal pumps with integrated canned motors can be used today to solve the most difficult pumping problems in sensitive plants which are environmentally hazardous. Guaranteed zero leakage, greatly reduced noise, low space requirements and, not least, higher availability give these pumps safety and economic efficiency. The supposed contradictions between ecology and economy can be harmonized in this way.

3.6 The permanent-type magnet coupling

An alternative method of hermetically driving centrifugal and rotary displacement pumps is by transferring power from the motor to the pump using a permanent-type magnet coupling. This is basically the same principle as has been shown for the canned motor pumps, but instead of the electromagnet drive (three-phase motor) a coaxial central coupling fitted with permanent magnets is used (Fig. 3.198), which in turn is driven by a normal three-phase motor. Centrifugal pumps with permanent-type magnet couplings have an advantage in that the regulations of the Physical Technical Federal Institute (PTB) for hazardous locations do not apply, although their maximum surface temperature still has to comply with the required temperature class. Permanent-type magnet coupling pumps are not legally required to have temperature level monitoring devices, for safety reasons, however, monitoring of the liquid level on the suction side using a suitable device is recommended to avoid cavitation and possible dry running.

Dauermagnete = Permanent magnets

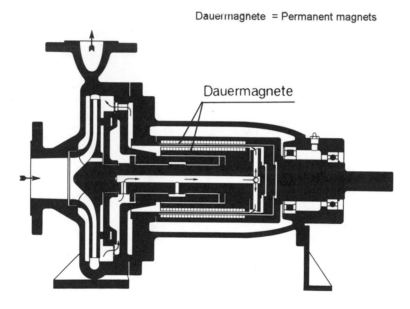

Dauermagnete

Fig. 3.198 Operating principle of a magnetic coupling centrifugal pump

3.6.1 Construction of a permanent magnet coupling

In the case of the magnet coupling, a field of permanent magnets mounted on rotating, outer and inner magnet carriers take the place of the electromagnetic rotating field of the stator (Fig. 3.199). Both are connected to each other through the magnetic field, comparable with a jaw coupling. Whilst with the canned motor the rotor has a certain amount of slip compared with the rotating field, the rotor of a magnetic coupling pump runs in synchrony with the external drive. This synchronism does not produce any additional temperature rise in the inner magnet carrier, in contrast to the canned motor. Fig. 3.199 shown a section view of the magnets. Radially-magnetised permanent magnets with alternating polarity are mounted on the inner and outer magnet carriers. The magnetic flux is deflected by iron cylinders in the manner shown so that the inner and outer magnet carriers are linked together. The fixed can forming the hermetic seal is positioned in between. The rotating outer ring fitted with magnets is shown in Fig. 3.200. Fig. 3.200.1 shows the can which provides the separating wall between the fluid and atmosphere.

Spalttopf = Can
Rotorummantelung = Rotor sheath
getriebener Rotor = Inner torque ring

treibender Rotor = rotating outer magnet ring
Feldlinien = Field lines
Dauermagnete = permanent magnets

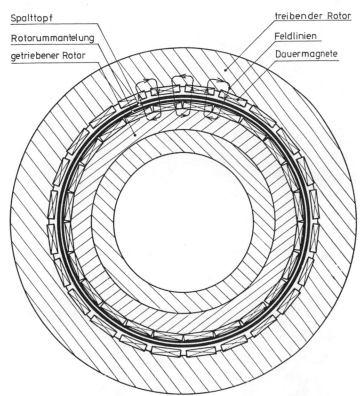

Fig. 3.199 Section of a typical magnetic coupling

Fig. 3.201 will serve to illustrate this process of torque transmission at the coupling. When stationary, the north-south magnetic poles are in opposition and the transmitted torque then = 0. If on startup the outer magnet carrier is rotated by the phase angle φ relative to the inner magnet carrier, a torque with a sinusoidal pattern begins to be established. It reaches its maximum value at the vertex $\frac{\pi}{p}$ (p = number of magnetic poles). Up to this point the rotation of both parts of the coupling is synchronous but when M_{max} is exceeded, poles of the same name increasingly pass over each other and the resulting weakening of the field causes the pump to go 'out of step'. The coupling 'breaks away', i.e. the required torque can no longer be transmitted. Only after the drive is shutdown both halves of the coupling again 'engage' and the pump can be restarted. The magnet material cobalt-samarium (refer to Section 3.6.2 'Magnet material') used almost universally today has the advantage that during this process no demagnetisation of the magnets takes place despite the slipping of the clutch, so that complete torque transmission on re-start is assured. The slipping of the clutch must not, however, last too long because the resulting magnetic

Fig. 3.200 Rotating outer magnet ring of a magnetic coupling

Fig. 3.201 Can of a magnetic coupling

losses with the simultaneous failure of the clutch cooling due to failure of the partial flow leads to overheating of the magnets.

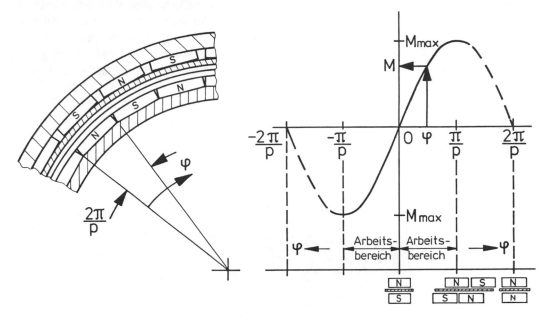

Arbeitsbereich = Working range

Fig. 3.201 Torque characteristic of a magnet coupling [3-33]

The magnetic power transmission using clutches of this kind can also act as an overload protection device. If the pump rotor is blocked the rotating outer ring continues to rotate without overloading the shaft. This is particularly advantageous for rotating displacement pumps.

3.6.2 Magnet materials

As is known, the properties of magnet materials can be illustrated with the aid of a hysteresis loop. It shows the relationship between the outer magnetic field H and the magnetic flux density in the magnet material, the so-called induction B (Fig. 3.202). The magnetisation in the inside of the material first increases with the outer magnetic field until finally with large fields saturation is reached. If the field strength H is again allowed to drop, when H = 0 the so-called remanence point is reached. At this point the magnet is fully magnetised. A field with an opposing direction must now be superimposed in order to bring the magnetisation to 0. The strength of the magnetic field required to do this is known as the coercive field strength. The part of the curve in the second quadrant, the so-called demagnetisation curve is of great practical significance particularly for permanent magnets. Fig. 3.203 shows the demagnetisation curve of normal commercial permanent magnets such as anisotropic ferrite and A1NiCo 500 compared with $SmCo_5$.

The maximum value of the product of B and H is regarded as the quality factor, the socalled energy density. It can be seen that these energy densities are in a ratio of 8 : 1. This means that with $SmCo_5$ the same forces as before are generated with one eighth of the permanent magnet volume, or with an equal magnet volume it can transmit an 8-times larger force. This property makes it possible to fit a powerful magnetic clutch into a smaller space, i.e. where the diameter of the can is small. A second, essential advantage of using cobalt-samarium magnet materials is its particular stability with respect to demagnetising effects.

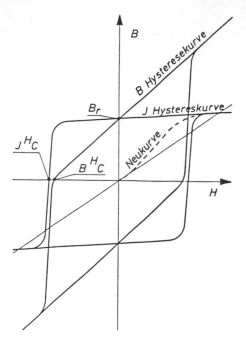

Hysteresekurve = Hysteresis curve

Neukurve = New curve

Fig. 3.202 Hysteresis loop

It could be assumed from the hysteresis curve (Fig. 3.202) that a permanent magnet remains at remanence point B_r after magnetisation, i.e. after the outer magnetic field is switched off. But this applies only to a fully closed magnetic circuit without an air gap. If the air gap is increased from zero, the working point of the magnet moves downwards along the demagnetisation curve. With the magnetic coupling, such shifts in the working point occur when the inner magnet carrier is fitted or removed and also where the outer ring is rotated relative to the inner, perhaps under load but also if the maximum torque is exceeded. If this change, also referred as 'shearing', is reversed i.e. if the air gap is again reduced to zero, the working point does not move back along the demagnetisation curve but instead moves almost in a straight line parallel to the tangent in the remanence point. Where the demagnetisation curve is bent, the magnet is therefore weakened compared with its original state.

Permanent magnets based on SmCo, have flat, almost straight demagnetisation curves with a large coercive field strength where the tangent in the remanence point coincides with the demagnetisation curve itself. After each shearing, or working point shift of equal magnitude, the magnet system returns without weakening to the original condition. Particularly

314

on overload with the associated slippage of the clutch there is no demagnetisation of the material (Fig. 3.203).

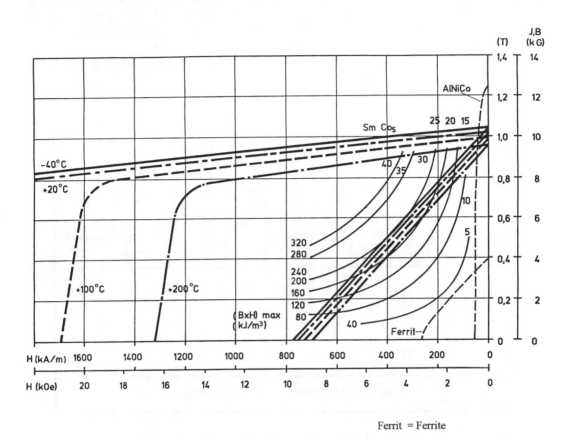

Ferrit = Ferrite

Fig. 3.203 Demagnetisation curves of samarium cobalt magnets compared with Alnico and ferrite magnets (Magnetfabrik Schramberg)

3.6.3 Efficiency of magnetic couplings

Because of the small installation dimensions of new magnets, the eddy current losses, which increase with the square of the can diameter, and the liquid friction losses which increase with $D^4 - D^5$, are strongly reduced. In addition, the slip losses of the rotor are eliminated. This enables efficiencies of $\geq 90\%$ to be obtained from this type of magnetic coupling under full load at 2900 r.p.m. where Hastelloy C is used for the can.

Eddy currents and the resulting losses occur when magnetic field lines are cut by stationary or rotating parts. With the magnetic clutch the field lines are cut by the stationary can. The eddy currents can be minimised if the electrical conductivity is small. This is why the can is made of Hastelloy. With this material the losses remain relatively small and a good efficiency is thus achieved.

The eddy current losses can be completely avoided if a ceramic instead of metal material is used for the can, but this has consequences with regard to strength (tensile stress).

The magnetic coupling pump, however, does not turn out substantially better than a canned motor pump with regard to overall efficiency despite the good efficiency of the coupling. In the following Table 3-X the efficiency for a CAM 3/5 type of canned motor pump with a CKP 74n-2h drive motor is compared with that of a MCAM 3/5 magnetic coupling pump with the same hydraulics, with a size B 4 coupling and a 24 kW Exe drive motor. Although in this case the can is made of 1.4571 instead of the usual high nickel alloy.

Table 3-X shows that the overall efficiency of the magnetic coupling pump is only about 3% better than that of the canned motor pump.

Table 3 - X

Pump (M) CAM 3/5			Drive Canned motor CKP 74n-2h			Drive Three-phase motor B4 / 24 kW Exe-Motor				
Q m^3/h	H m	η_P	η_{mo}	η_{ges}	Loss in produkt kW	η_{mo}	η_K	$\eta_{mo} \cdot \eta_K$	$\eta_{ges} = \eta_{mo} \cdot \eta_K \cdot \eta_P$	Loss in produkt kW
5	165	20	60	12	18,9	76	90	68	14	13,6
10	163	32	62	20	18,7	79	91	72	23	13,2
15	160	41	65	27	18,5	80	92	74	30	12,9
20	155	46	67	31	18,6	82	92	75	34	12,8
25	147	50	69	35	19	83	92	76	38	12,9
Liquid: Vinyl chloride 0 °C, Density = 940 Kg/m^3										

The heat loss given off to the liquid is on the other hand almost 30% less with the magnetic coupling pump than the corresponding heat loss for a canned motor pump. For fluids, which react sensitively to the addition of heat or have a tendency to gasify, the heat balance of the magnetic coupling pump is better.

3.6.4 Starting-torque behaviour of magnetic couplings

The magnetic clutch has substantial disadvantages compared with the canned motor as regards to its starting behaviour. Because of the forces of attraction between the magnets the coupling can only transmit a maximum torque independent of speed. The drive must therefore be dimensioned such that in addition to the rated operating moment the higher acceleration moments occurring when the electric motor has started can also be safely transmitted. Otherwise the coupling breaks down and is no longer in step. This can be compared with towing a car. The towrope can only transmit a maximum force, in the same way as a magnetic coupling. If the front vehicle accelerates too fast or the rear vehicle is too heavy the rope breaks when pulling away. After the size of the magnetic coupling and drive motor has been decided, it is necessary to establish whether on the basis of the given mass moments of inertia and on the starting and stalling torques of the electric motor the

3.6.4.2 Starting conditions

After the size of the coupling and drive motor have been decided, a check should be made to ensure that the condition of non-separation of the magnetic clutch upon direct switch-on of the motor is fulfilled. For this purpose it is necessary to form the moment of inertia on the drive and take power off.

$I_1 =$ Moment of inertia, outer magnet carrier
    + Moment of inertia of coupling
    + Moment of inertia of motor

$I_2 =$ Moment of inertia, inner magnet carrier
    + Moment of inertia of the impeller
    + Shaft

and thus the design/safety factors:

$$S_{erf} = K \frac{I_2}{I_1 + I_2} \qquad (3 - 58)$$

$$K = 4,3 \text{ (from tests)}$$

$$S_{vorh} = \frac{P_{MM}}{P_{mo}} \qquad (3 - 59)$$

The condition on non-separation of the magnetic drive on direct switch-on of the three-phase motor (Fig. 3.204) is then:

$$S_{vorh.} \geq S_{erf.} \qquad (3 - 60)$$

getrieben = driven; treibend = driving; Laufräder = impellers; Magnetkupplung = Magnetic coupling
Kupplung = Coupling; Motor = Motor

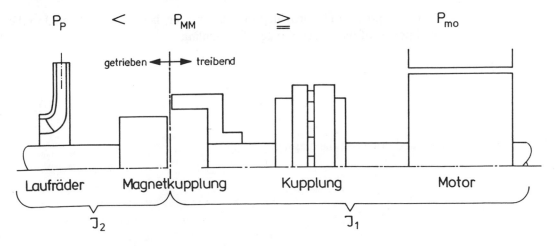

Fig. 3.204 Moments of inertia of driving and driven rotor

conditions for non-separation of the magnetic clutch are met when the electric mo
directly switched on.

Important: The drive motor of a magnetic pump must not be too large. For this reas
technical specification must also include the maximum drive power of the electric mo

3.6.4.1 Selection criteria for magnetic couplings

To ensure problem-free transfer of power from the drive motor to the pump the d
and driven torques must be harmonised.
This means the following.

During operation the following conditions must be fulfilled:

- The sum of the pump power requirements and the power loss of the magnetic
  pling must be smaller than the rated motor power.

Therefore:

$$P_p + P_{vk} \leq P_{mo}$$

where

P_p = Power requirements of the pump relative to Q_{max} at a corresponding de
ty and viscosity of the fluid

P_{vk} = Power loss of the magnetic coupling

P_{mo} = Rated motor power

- The motor power to be provided must be smaller than or at most equal to
  rated power of this selected magnetic coupling.

therefore we get:

$$P_{mo} \leq P_{MM}$$

where P_{MM} is the rated power of the magnetic coupling.

3.6.5 Influence of temperature on the magnets

A rise in temperature of a magnet is combined with a change in its magnetic properties and at higher temperatures leads to noticeable losses in magnetisation. The distinction is made between reversible losses i.e. those which can be reversed on cooling to the initial temperature and irreversible losses which are those which lead to a permanent weakening of the magnets even after cooling to the initial temperature. The reversible losses includes those which cause a change in the magnetic structure due to higher temperature; these are only repairable by re-magnetisation of the material. The irreversible losses occur due to metallurgical and chemical changes in the material. These are not repairable even by remagnetisation

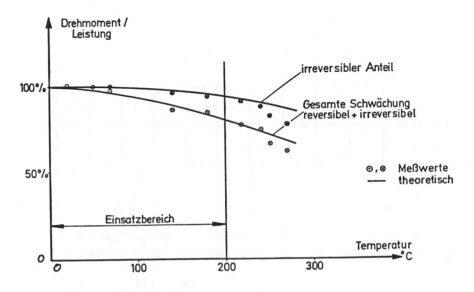

Drehmoment/Leistung = Torque/power; irreversibler Anteil = irreversible portion;
Gesamte Schwächung reversibel + irreversibel = total weakening, reversible + irreversible
Meßwerte = measured value; theoretisch = theoretical; Einsatzbereich = working range
Temperatur = Temperature

Fig. 3.205 Reduction in performance of magnetic coupling with increasing temperature

Fig. 3.205 shows the torque losses of a magnetic clutch relative to temperature for $SmCo_5$ magnets, with curves representing the irreversible and reversible losses to be expected. The points are measured values obtained from a series of tests. It can be seen from this illustration that the weakening of the maximum torque at 250°C is already 30% and at 200°C it is still 18%, whereas at 100°C it is only 8% of the value at room temperature. This means that the economic working range of permanent magnets is from -100 °C to + 200°C, without having to make special allowance for the reduction in performance of the magnets. Today there are already magnet materials with a permissible temperature range up to 350°C. The reduction in performance relative to temperature of a series of magnets is shown in a Fig. 3.206, whilst Fig. 3.207 shows the temperature rise of a magnetic clutch where the drive has separated.

319

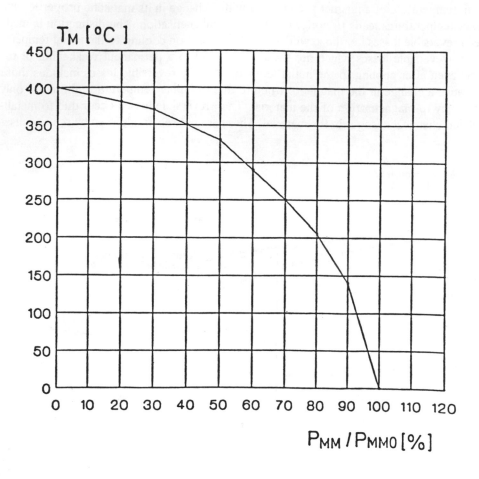

Fig. 3.206 Magnetic transmission power relative to P_{MM} = f (Tm) °C

P_{MM} = Magnet power
P_{MM0} = Magnet power at 0°C
T_M = Temperature of magnets

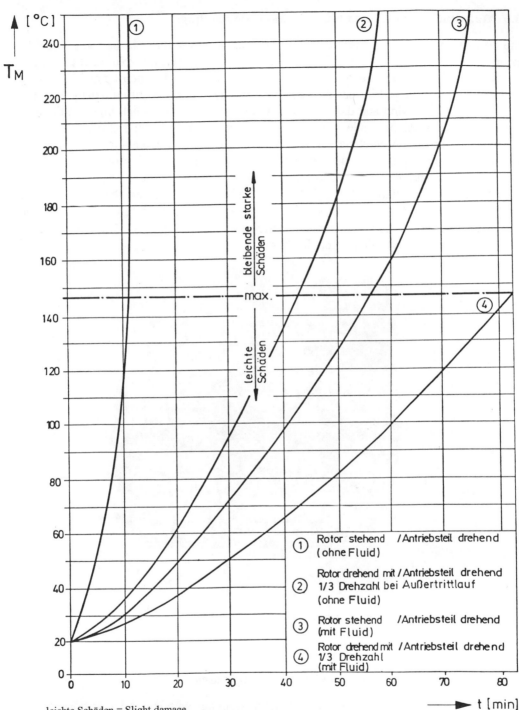

Fig. 3.207 Temperature rise of a magnetic clutch in the separated condition °C = f(t min)

leichte Schäden = Slight damage
bleibende starke Schäden = Permanent severe damage
1 = Rotor stationary/driving part rotating (without fluid)
2 = Rotor rotating with drive part rotating 1/3 speed in out of step condition (without fluid)
3 = Rotor stationary/driving part rotating (with fluid)
4 = Rotor rotating with drive part rotating 1/3 speed (with fluid)

3.6.6 Magnitude of power and losses of magnetic couplings

The coupling volume is a measure for the amount of power that can be transmitted. To arrive at the smallest possible coupling size for a series of pumps the volume of the magnet can be varied in both the axial and radial direction on the inner torque ring and outer magnet ring. This is achieved in that the required magnetic plates are fitted axially in one or more rows in the driving and driven part of the coupling. In the radial direction the magnets can, for example, be single or double on the inner torque ring and also single or double on the outer ring. A combination of double magnets on the outer ring and single magnets on the inner ring is also possible. Fig. 3.208 shows an inner ring with four different arrangements of magnets in the axial direction. Use of these different possible combinations enables a differentiation in the power harmonisation on one and the same coupling. Fig. 3.209, 3.209.1, 3.210, 3.210.1 show the output and losses of different couplings sizes with a variety of magnet arrangements relative to speed, whilst Fig. 3.211 and 3.211.1 show the stalling powers of different clutches.

Fig. 3.208 Torque ring with examples of magnet arrangements

3.6.7 Power losses in magnetic couplings with high viscosity fluids

The increasing power requirements of magnetic couplings with increasing viscosity shows up directly as a power loss in the power balance and has to be carefully considered.
The increased power requirement relative to viscosity and speed for two series of couplings can be determined using the diagrams in Fig. 3.212 and 3.213.

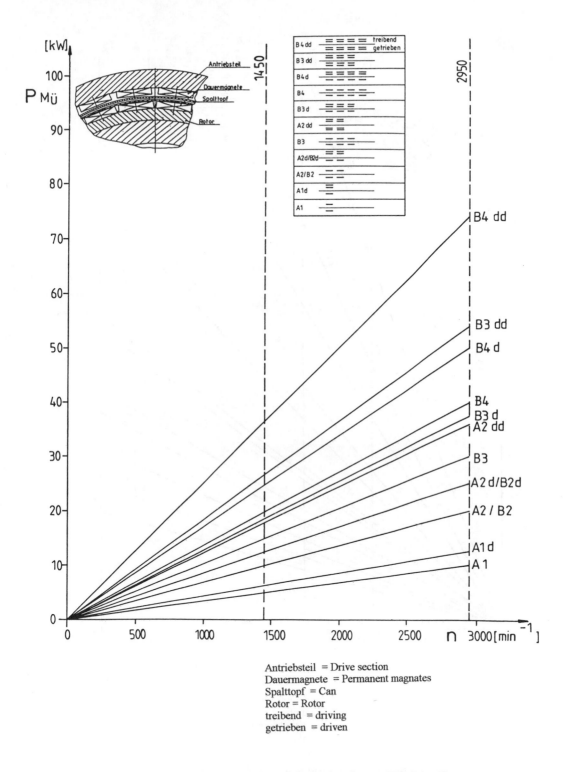

Fig. 3.209 Transmission power P_{M0} of size A/B magnetic couplings

Antriebsteil = Drive section
Dauermagnete = Permanent magnates
Spalttopf = Can
Rotor = Rotor
treibend = driving
getrieben = driven

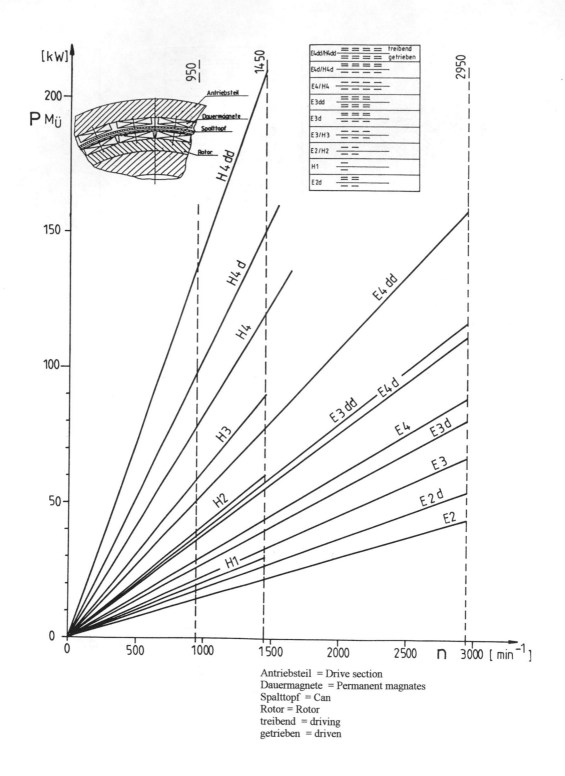

Antriebsteil = Drive section
Dauermagnete = Permanent magnates
Spalttopf = Can
Rotor = Rotor
treibend = driving
getrieben = driven

Fig. 3.210 Transmission power $P_{M\ddot{U}}$ of size E/H magnetic couplings

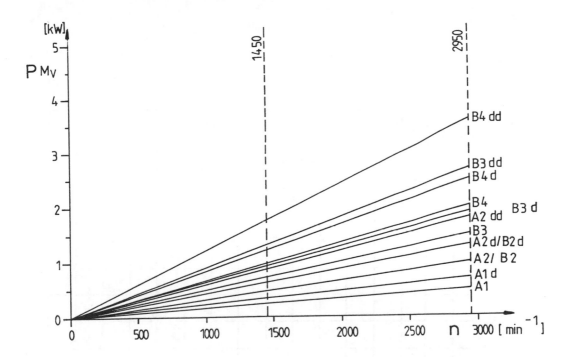

Fig. 3.209.1 Power losses P$_{MV}$ for size A/B magnetic coupling in accordance with Fig. 3.209

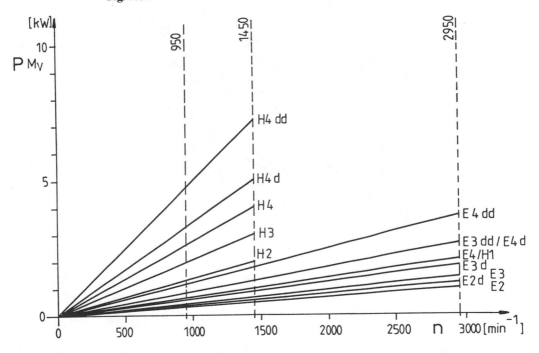

Fig. 3.210.1 Power losses P$_{MV}$ for size E/H magnetic coupling in accordance with Fig. 3.210

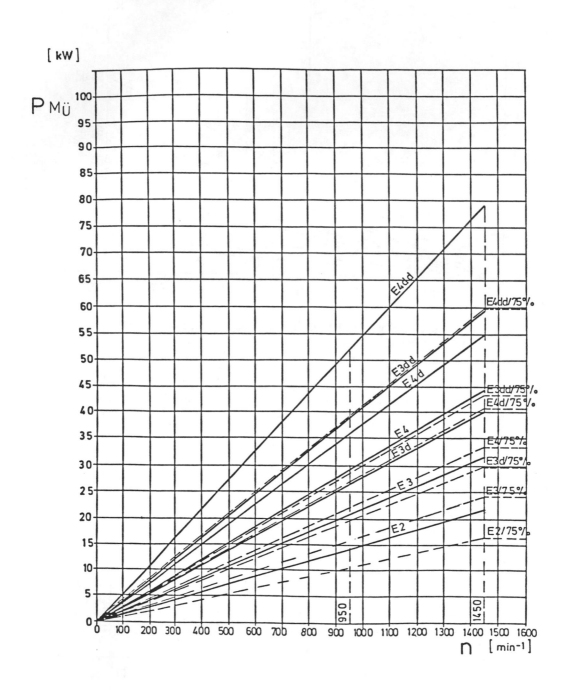

Fig. 3.211 Determination of stalling power [kW] = 100% of a type E coupling relative to speed n = f (KP$_M$) at 20 °C

326

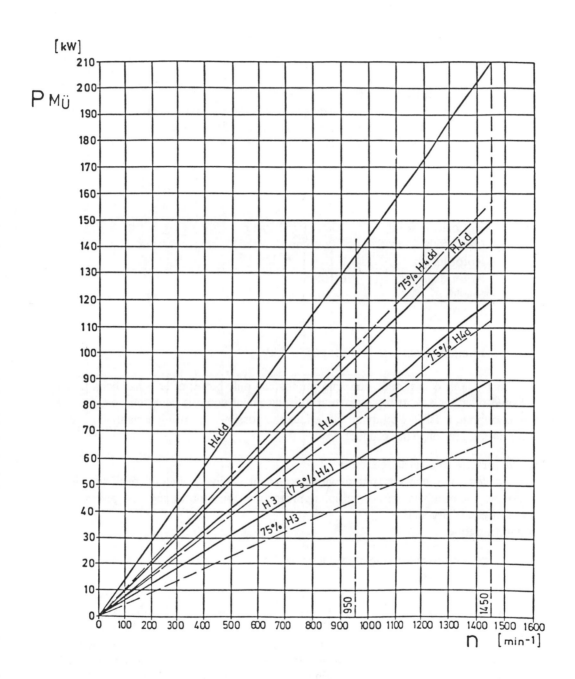

Fig. 3.211.1 Determination of stalling power [kW] = 100% transmission power = 75% of a type H magnetic coupling relative to speed n = f (KP$_M$) at 20 °C

327

At higher viscosities, such as occur in the area of rotary displacement pumps, the friction losses of the magnet coupling can be approximately determined using the equation (3-61)

$$P_V = \frac{\eta \cdot A \cdot U^2}{S}$$

(3 - 61)

where:

η = Viscosity in Pa $\cdot$ s
A = Developed (unrolled) rotor jacket area m^2
U = Circumferential velocity of the rotor m/s
S = Gap width at the rotor in m

example [3 - I]

Given:

Viscosity of fluid:		1000 cP = 1 Pa $\cdot$ s
Gap width:	S	0,0006 m
Rotor length:		0,1 m
Rotor diameter:		0,108 m
Speed:		1450 r.p.m.

Results:

Friction power loss in rotor gap 3,8 kW

The viscosity limit of magnetic coupling pumps can only be set after it has been properly established whether a centrifugal or rotary displacement pump is being chosen. Only then can an economic feasibility study be made into the possibility of using a magnetic coupling with direct integration in the product flow.

Three factors are to be included in this:
- Prime cost of pump set (pump with permanent magnet coupling, resilient coupling between the pump and motor, coupling protection, drive motor and base plate),
- cost of electricity and
- operating time

It is therefore not possible to clearly set a general maximum viscosity limit and this must be determined case by case.

If the anticipated friction losses are unacceptable, the fluid contains solids, fibres or gas or has a tendency to solidify, products can still be pumped without difficulty using the same design measures as for the canned motor, i.e. by injecting a sealing fluid into the rotor chamber. A precondition for this is a throttling gap or a shaft seal to form the separation from the working space of the pump and the magnetic clutch. The sealing fluid is best injected using a piston diaphragm metering pump, to guarantee on one hand that the complete unit is hermetically sealed and on the other hand to be able to determine and regulate the quantity of liquid injected. In addition to the sealing function, the injected fluid also carries the heat loss to the delivery flow of the pump.

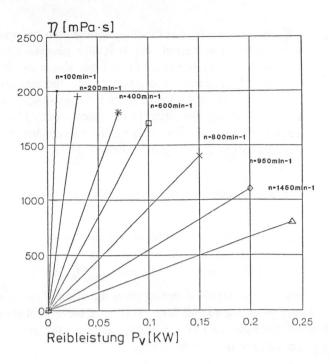

Fig. 3.212 Friction power loss P_V in the rotor/stator gap as a function of speed n and viscosity η at 0,6mm gap distance (type A_{00})

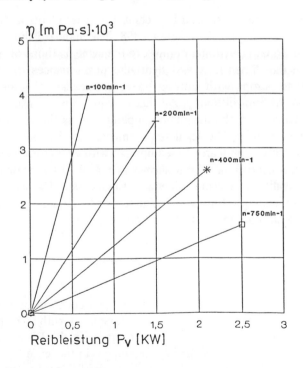

Fig. 3.213 Friction power loss P_V in the rotor/stator gap as a function of speed n and viscosity η at 1 mm gap distance (type E_2)

The pump working space and the rotor space of the magnetic clutch can also be satisfactorily separated using a single-acting mechanical seal. It is then necessary for the coupling rotor space to be provided with a pressure system (1 to 2 bar above the delivery pressure of the pump). The coupling heat loss can be removed either by a thermosiphon effect using a pressure tank provided with a cooling device or a cooler consisting of a nest of boiler tubes with the aid of a centrifugal impeller fitted in the rotor space. The sealing-cooling system to be chosen depends on the operating parameters, the condition of the fluid, its temperature and the compatibility of the sealing fluid with the product.

3.7 Operating principle and examples of construction of permanent magnet coupling pumps

The operation of hermetic centrifugal pumps provided with a permanent magnet coupling is the same as equivalent canned motor pumps. All the observations made for these pumps with regard to axial thrust, radial pressure, NPSHR and Q_{min}/Q_{max} limitations also apply.

The power loss of the permanent magnet coupling takes the place of the motor power loss with its effect on the heating of the partial flow. This is, however, substantially less than that of the canned motor because the electrical losses at this point are omitted and only appear again in the power balance of the overall unit.

3.7.1 Standard chemical pumps to DIN 24256 / ISO 2858

The magnetic coupling pump most used in technical process plants is the standard single-stage centrifugal pump to DIN 24256 / ISO 2858. The standard dimensions and performance grid of conventional centrifugal pumps are the same as those of the magnetic coupling type. Because the dimensions and hydraulic performances of magnetic coupling pumps are identical to pumps with stuffing boxes or mechanical seals, replacement can usually be carried out without difficulty. All that is required is to ensure that the power of the drive motor is matched to the magnetic coupling, and as this is so in most cases, replacement is relatively simple. The legal requirements for environmental protection can also often be fulfilled more simply after re-equipment with a magnetic coupling type.
The standard chemical centrifugal pump shown in Fig. 3.214 represents a development initiated by the Association of German Chemical Engineers (VCI) which took account of all these relevant requirements. Fig. 3.215 is a sectioned model and Fig. 3.216 shows a hermetic standard chemical pump in a technical process plant.

3.7.2 Single-stage magnetic coupling centrifugal pumps of monobloc construction

Magnetic coupling pumps are generally driven by type B3 three-phase motors. In addition to the magnetic coupling this requires an additional jaw coupling between the pump and motor and also a base plate. Most of these can be done away with if the pump is driven via a motor pedestal connected directly to the drive motor in the B3/B5 arrangement (Fig. 3.217). The shorter construction is a further advantage. In this case the hydraulics are the same as for DIN 24256 / ISO 2858. Of late, a variation of this type, the inline model (Fig. 3.218), is receiving increasing attention because of its ease of installation.

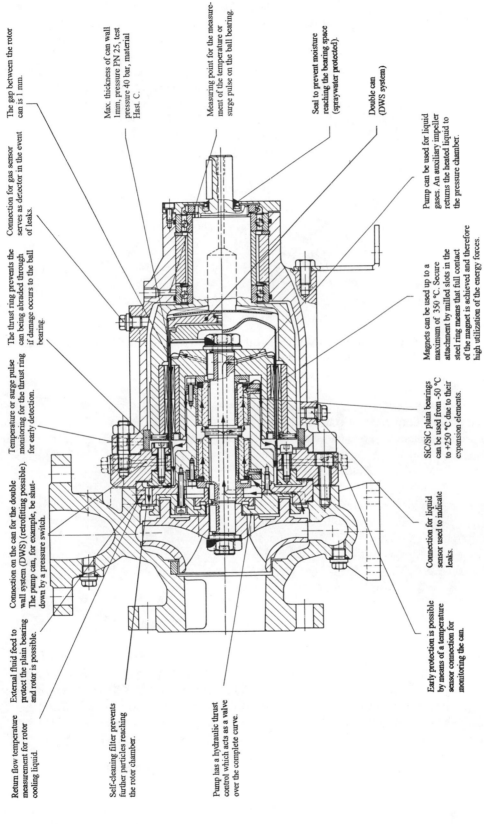

The gap between the rotor can is 1 mm.

Max. thickness of can wall 1mm, pressure PN 25, test pressure 40 bar, material Hast. C.

Measuring point for the measurement of the temperature or surge pulse on the ball bearing.

Seal to prevent moisture reaching the bearing space (spraywater protected).

Double can (DWS system)

Connection for gas sensor serves as detector in the event of leaks.

Pump can be used for liquid gases. An auxiliary impeller returns the heated liquid to the pressure chamber.

The thrust ring prevents the can being abraded through if damage occurs to the ball bearing.

Temperature or surge pulse monitoring for the thrust ring for early detection.

Magnets can be used up to a maximum of 350 °C. Secure attachment by milled slots in the steel ring means that full contact of the magnet is achieved and therefore high utilization of the energy forces.

Connection on the can for the double wall system (DWS) (retrofitting possible). The pump can, for example, be shut-down by a pressure switch.

External fluid feed to protect the plain bearing and rotor is possible.

SiC/SiC plain bearings can be used from -50 °C to +250 °C due to their expansion elements.

Connection for liquid sensor used to indicate leaks.

Return flow temperature measurement for rotor cooling liquid.

Self-cleaning filter prevents further particles reaching the rotor chamber.

Pump has a hydraulic thrust control which acts as a valve over the complete curve.

Early protection is possible by means of a temperature sensor connection for monitoring the can.

Fig. 3.214 Standard chemical centrifugal pump to DIN 24 256/ISO 2858 with a magnetic coupling to VDI Recommendation (Lederle)

Fig. 3.215 Cut model of a Standard chemical magnetic coupling centrifugal pump in accordance with Fig. 3.214 (Lederle)

Fig. 3.216 Standard chemical magnetic coupling centrifugal pump in a process plant (Lederle)

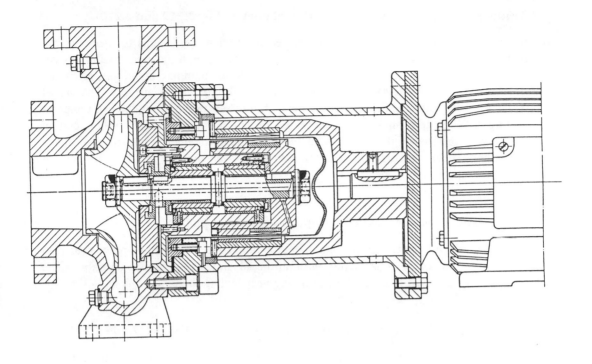

Fig. 3.217 Single-stage magnetic coupling centrifugal pump of monobloc construction
(Lederle)

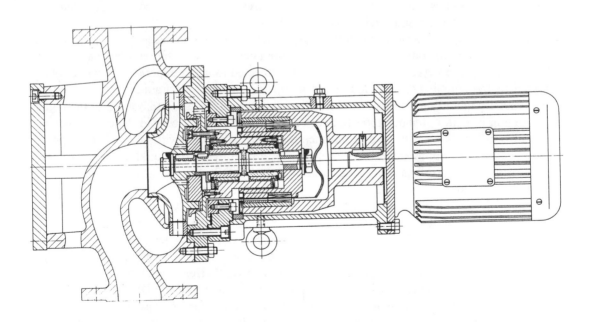

Fig. 3.218 Magnetic coupling centrifugal pump of inline construction (Lederle)

3.7.3 Multi-stage magnetic coupling centrifugal pumps (Fig. 3.219 and 3.220)

Magnetic coupling pumps are also manufactured in multi-stage designs to meet the hydraulic requirement and to obtain satisfactory specific speeds with the associated improvement in efficiency. These pumps are particularly suitable for pumping liquid gas, as can be seen from Fig. 3.219. A suction impeller ensures that NPSHR values are low. The partial flow Q_r is returned after the first impeller. Temperature sensors monitor the can temperature and the DWS system, shown here, increases the safety of the pump set. Where certain tasks require an external feed, this can be achieved through the inlet bore provided for the purpose. Shock pulse sensors and temperature control sensors are fitted in the bearing housing to monitor the roller bearings.

3.7.4 Multi-stage monobloc magnetic coupling centrifugal pumps

As already shown with the standard chemical pump, multi-stage centrifugal pumps can also be designed in a monobloc construction with all the advantages previously described. The pump shown in Fig. 3.221 is suitable for pumping near-boiling fluids.

3.7.5 Magnetic coupling centrifugal pumps for heat transfer (Fig. 3.222 and 3.223)

Pumping heat transfer oil requires particular attention to be paid to the tightness of the pump set. Magnetic coupling pumps are also used in addition to canned motor pumps for this purpose.

Although present-day magnets can be used up to temperatures of 350 °C, it is advisable to keep the temperatures in the magnetic coupling to a minimum because a rise in temperature in the coupling can lead to a reversible, or irreversible, power reduction as stated in Section 3.6.5 ('Temperature influences on the magnet').

A spatial separation between the pump and magnetic coupling is necessary to prevent the high operating temperatures of the circulating fluid coming in contact with the magnets.

This means that no creep heat from the pump subjected to high temperatures can be transferred to the coupling. The coupling and pump are, however, hydraulically connected by an annular gap and this means that where there are temperatures up to 350 °C in the liquid in the pump section, temperatures between 160 °C and 200 °C occur in the can. This type can also be provided with an external tube bundle or air cooler with an auxiliary impeller installed in the coupling can be used to circulate the rotor chamber liquid through the rotor and cooler, as described for canned motor pumps in Section 3.5.5.1 ('Canned motor pumps with externally-cooled motors') Fig. 3.83. The design of this type also makes it suitable for pumping suspensions because the secondary circulation can be either via a self-priming cleaning filter or by means of an auxiliary impeller with integrated external infeed (Fig. 3.224, 3.225 and 3.226). All the monitoring devices such as temperature control on the can, shock pulse and bearing temperature sensing devices at the roller bearings and the introduction of a second can are of course possible (refer to Section 3.7.3 'Multi-stage magnetic coupling centrifugal pumps'). The installation of the DWS system is particularly recommended for pumping heat transfer oil. If the liquid escapes to the atmosphere from a single-wall type coupling, fires, with all their consequences, can occur due to self-ignition of the heat transfer fluid.

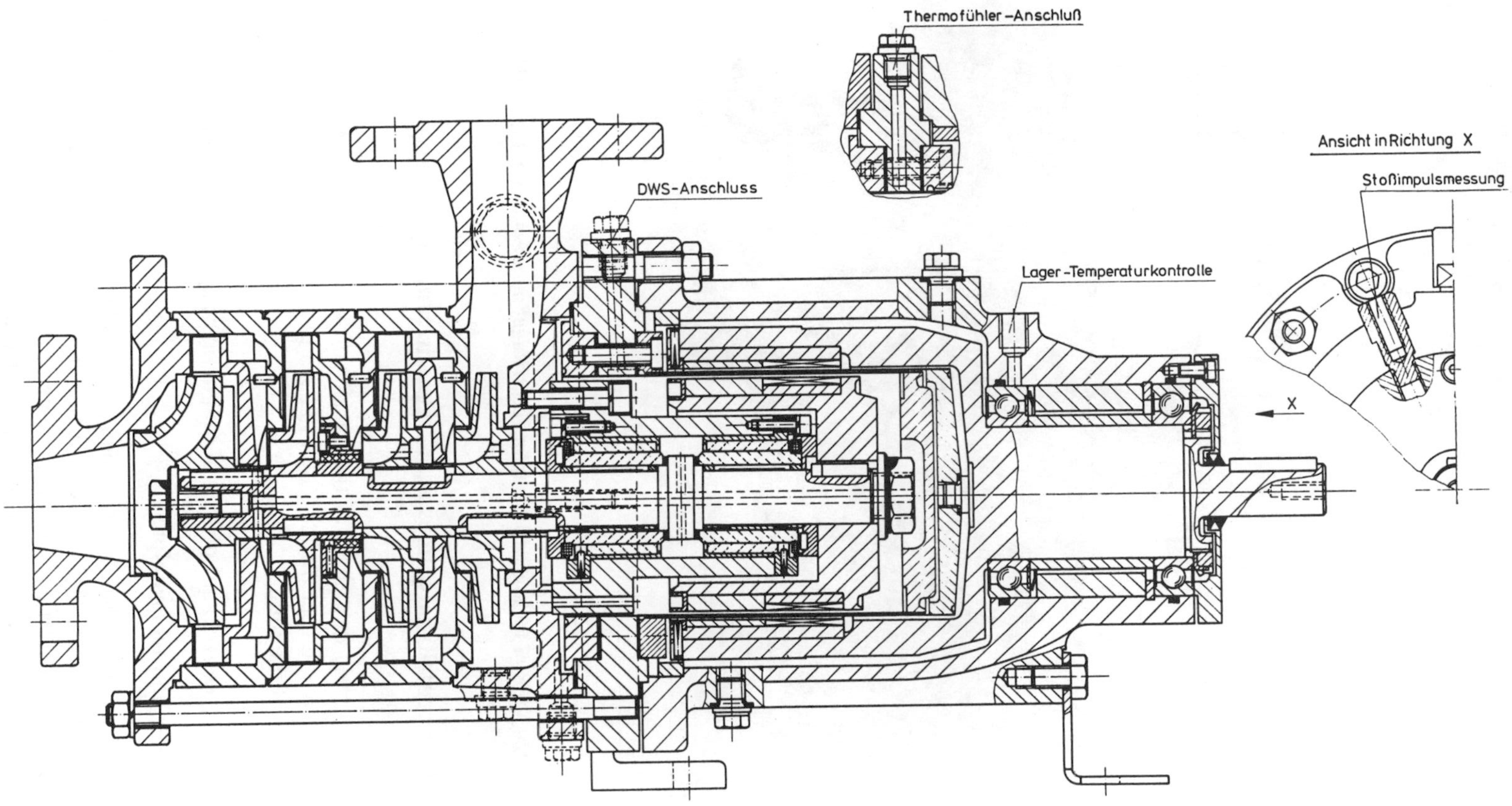

Ansicht in Richtung X

Stoßimpulsmessung

X

Thermofühler –Anschluß

Lager –Temperaturkontrolle

DWS –Anschluss

Fig. 3.219 Multi stage magnet coupling centrifugal pump with double-wall can (Lederle)

Fig. 3.220 View of the magnetic coupling centrifugal pump shown in Fig. 3.219 (Lederle)

3.7.6 Monobloc centrifugal pumps with magnetic coupling, for liquids at high temperature

The pumps previously described for the transfer of heat can also be designed in monobloc construction. To avoid heat transfer from the can to the motor bearing an additional fan is fitted on the hub of the rotor.

3.7.7 Self-priming centrifugal pumps of magnetic coupling construction

3.7.7.1 The side channel principle

Large numbers of known side channel pumps are installed in chemical plants for pumping liquid gas and hydrocarbons. Driving by means of a magnetic clutch is a particularly good ecological solution, because the large numbers of these pumps of conventional construction represent a not inconsiderable environmental hazard.

336

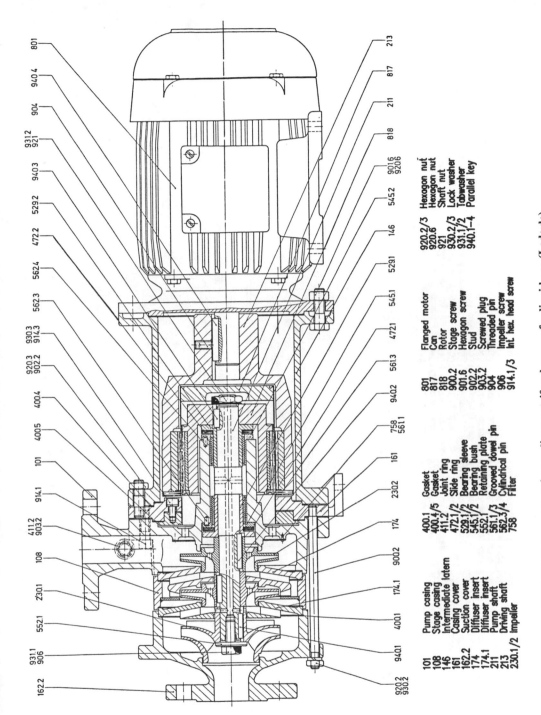

801		920.2/3	Hexagon nut	
817		920.6	Hexagon nut	
818		921	Shaft nut	
900.2	Flanged motor	930.2/3	Lock washer	
901.6	Can	931.1/2	Tabwasher	
902.2	Rotor	940.1–4	Parallel key	
903.2	Stage screw			
904	Hexagon screw			
906	Stud			
914.1/3	Screwed plug			
	Threaded pin			
	Impeller screw			
	Int. hex. head screw			

101	Pump casing	400.1	Gasket
108	Stage casing	400.4/5	Gasket
146	Intermediate latern	411.2	Joint ring
161	Casing cover	472.1/2	Side ring
162.2	Suction cover	529.1/2	Bearing sleeve
174	Diffuser insert	545.1	Bearing bush
174.1	Diffuser insert	552.1	Retaining plate
211	Pump shaft	561.1/3	Grooved dowel pin
213	Driving shaft	562.3/4	Cylindrical pin
230.1/2	Impeller	758	Filter

Fig. 3.221 Multistage magnetic coupling centrifugal pumps for liquid gas (Lederle)

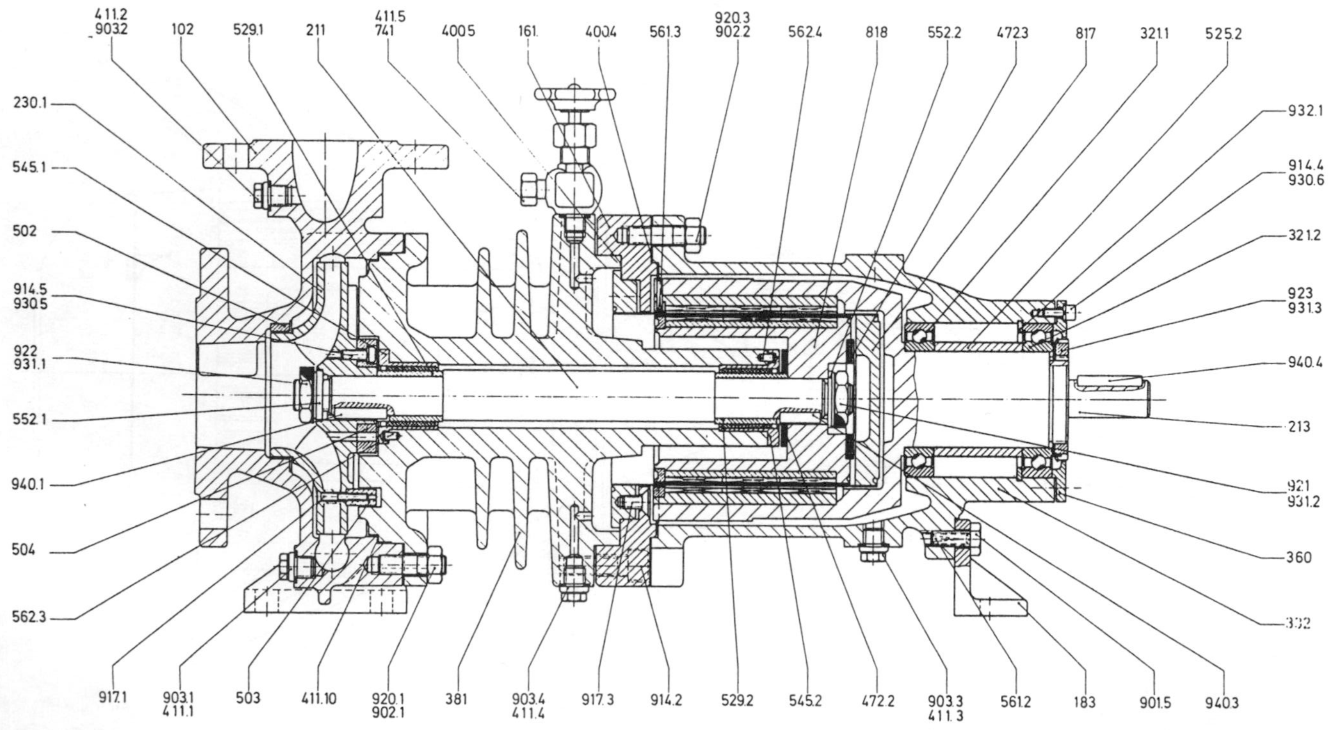

Fig. 3.222 Magnetic coupling centrifugal pump for heat transport (Lederle)

102 Volute casing	400.4-6 Gasket	543 Distance bush	818 Rotor	921 Shaft nut
161 Casing casing	410 Profil gasket	545.1/2 Bearing bush	900 Ring screw	922 Impeller nut
183 Support foot	411.3/4 Joint ring	545.4 Safety bearing bush	901.5 Hexagon screw	923 Bearing nut
211 Pump shaft	411.10 Joint ring	550.1 Sealing plate	902.1/2 Stud	930.2/3 Lock washer
213 Driving shaft	412.3/4 Round cord ring	552.1/2 Retaining plate	903.4 Screwed plug	930.6/7 Lock washer
230.1 Impeller	473.1/2 Slide ring	553.1/3 Equalizing disk	904 Threaded pin	930.9/10 Lock washer
321.1/2 Radial ball bearing	513 Slide ring collar	561.3 Grooved dowel pin	914.2/4 Int. hex. head screw	931.1-3 Tabwasher
332 Bearing bracket	525.2 Distance sleeve	673 Balance screw	914.6 Int. hex. head screw	932.1 Circlip
360 Bearing cover	526.1/2 Equalizing sleeve	741 Valve	914.8/9 Int. hex. head screw	940.1/3 Parallel key
381 Bearing insert	529.1/2 Bearing sleeve	817 Cam	920.1/2 Hexagon nut	940.4 Parallel key

Fig. 3.223 Section of a model of a magnetic coupling centrifugal pump for heat transport (Lederle)

Naturally, all the aforementioned safety devices can be used on the pumps shown in Fig. 3.227 and 3.228. The functional principle of these pumps is described in Section 3.5.11.1 ('Side channel pumps').

3.7.7.2 The impeller cell flushing principle (Fig. 3.229)

Whilst the side channel type can be used only for mechanically pure liquids or those which are only lightly contaminated, fluids containing solids can be pumped with the impeller cell flushing type. Section 3.5.11.4 ('Self-priming canned motor pumps with impeller cell flushing') gives details of the operating principle of such pumps.

3.7.8 Regenerative pumps with a magnetic coupling (Fig. 3.230 and 3.231)

Regenerative pumps designed for high pressure with low delivery coefficients place particular demands on the shaft seal due to their high achievable heads. The magnetic coupling provides a simple solution to this problem. This also makes these pumps suitable for both liquids and gases.

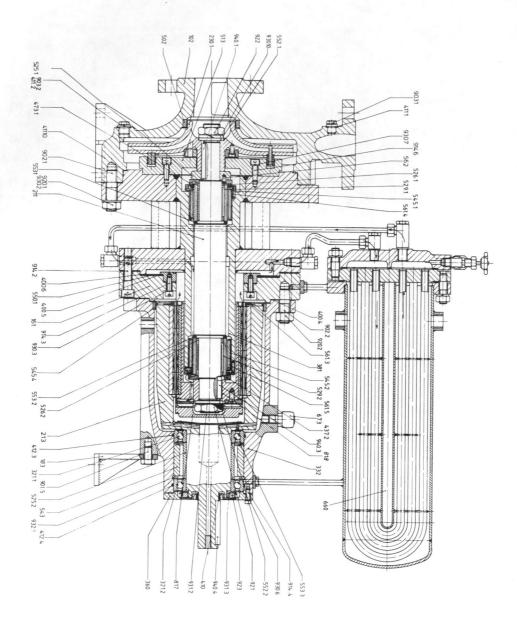

Fig. 3.224 Magnetic coupling centrifugal with external cooling (Lederle)

102	Volute casing
161	Casing cover
183	Support foot
211	Pump shaft
213	Driving shaft
230.1	Impeller
321.1/2	Radial ball bearing
332	Bearing bracket
360	Bearing cover
381	Bearing insert
400.4/6	Gasket
410	Profil gasket
411.1/2	Joint ring
411.1b	Joint ring
412.3/4	Round cord ring
473.1/2	Side ring collar
502	Wear ring
513	Wear ring
525.1/2	Distance sleeve
528.1/2	Equalizing sleeve
529.1/2	Bearing sleeve
543	Distance bush
545.1/2	Bearing bush
545.4	Safety bearing bush
550.1	Sealing plate
552.1/2	Retaining plate
553.1/3	Equalizing disk
561.3/5	Grooved dowel pin
562	Cylindrical pin
660	Cooler
673	Balance screw
817	Can
818	Rotor
901.5	Hexagon screw
902.1/2	Screwed plug
903.1/2	Stud
914.2/4	Int. hex. head screw
914.6	Int. hex. head screw
920.1/2	Hexagon nut
921	Shaft nut
922	Impeller nut
923	Bearing nut
930.2/3	Lock washer
930.6/7	Lock washer
930.10	Lock washer
931.1/3	Tabwasher
932.1	Circlip
940.1/3	Parallel key
940.4	Parallel key

340

Fig. 3.225 A view of a magnetic coupling centrifugal with external cooling as shown in Fig. 3.224 (Lederle)

341

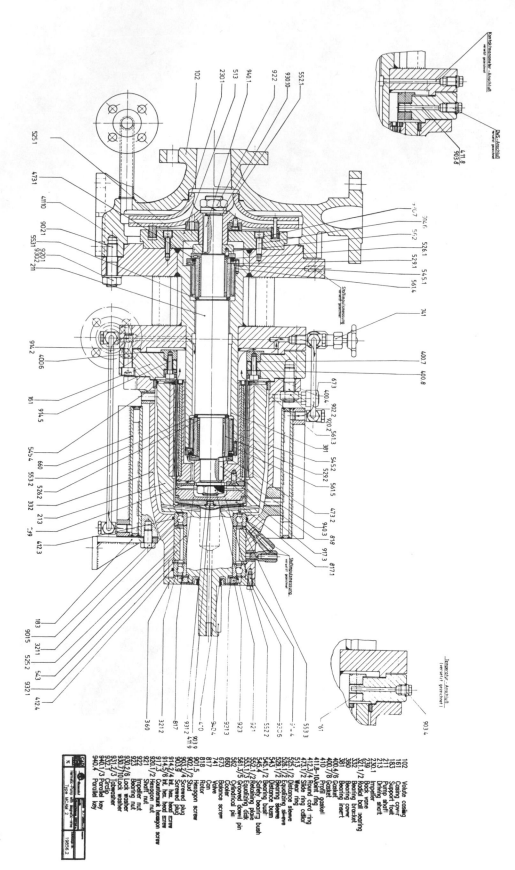

Fig. 3.226 Magnetic coupling centrifugal pump with self-cooling (convection cooler) (Lederle)

342

102	Volute casing
161	Casing cover
183	Support foot
211	Pump shaft
213	Driving shaft
230.1	Impeller
239	Back vane
321.1/2	Radial ball bearing
332	Bearing bracket
360	Bearing cover
381	Bearing insert
400.1/6	Gasket
400.7/8	Gasket
410	Profil gasket
411.8	Joint ring
412.3/4	Round cord ring
473.1/2	Slide ring collar
513	Wear ring
525.1/2	Distance sleeve
528.1/2	Equalizing sleeve
528.1/2	Bearing sleeve
545.1/2	Bearing bush
545.4	Safety bearing bush
552.1/3	Distance bush
553.1/3	Grooved disk
561.3/3	Cylindrical pin
562	Grooved dowel pin
660	Cooler
673	Balance screw
741	Valve
818	Rotor
902.1/5	Stud
903.2/4	Screwed plug
914.2	Hexagon socket screw
914.3/6 Int. hex. head screw
917.3 Countersunk hexagon screw
920.1/2	Hexagon nut
921	Hexagon nut
923	Bearing nut
930.2/6	Lock washer
930.7/10 Lock washer
931.1/3	Lock washer
932.1/3	Circlip
940.1/3	Toewasher
940.4	Parallel key
	Parallel key

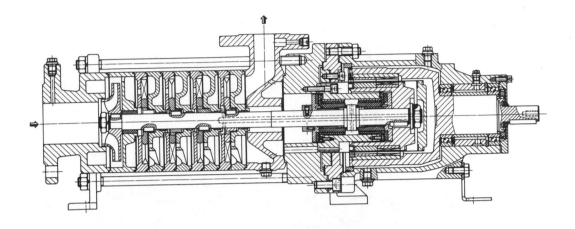

Fig. 3.227 Self-priming side channel pump with magnetic coupling (Lederle)

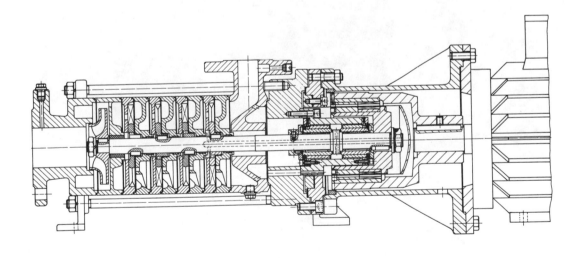

Fig. 3.228 Self-priming side channel pump with magnetic coupling in monobloc construction
(Lederle)

343

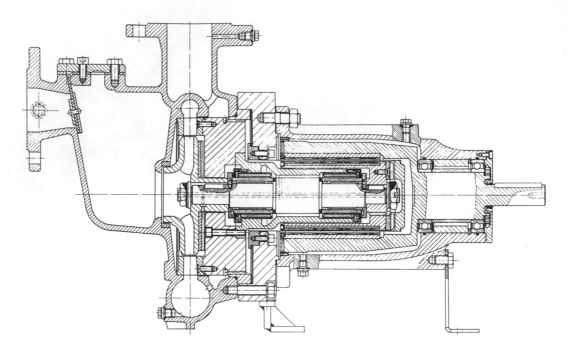

Fig. 3.229 Self-priming centrifugal pump using impeller cell flushing and a magnetic coupling (Lederle)

Fig. 3.231 View of the regenerative pump as shown in Fig. 3.230

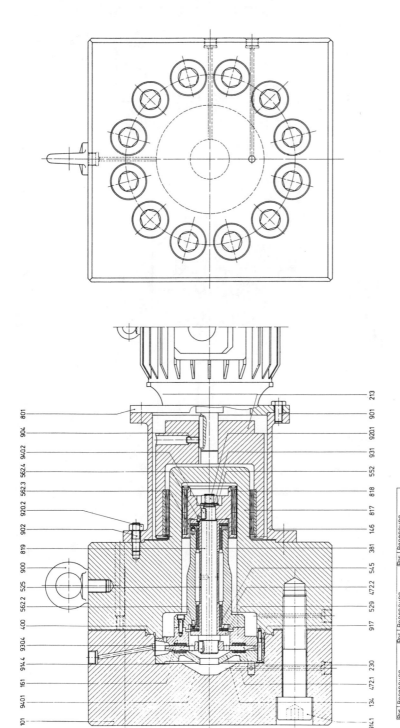

HERMETIC-Pumpen
G.m.b.H
Hermetic-Pumpe mit Magnet -
Antrieb Typ MCHB 6
1:1   221311

Pos	Benennung	Pos	Benennung	Pos	Benennung
101	Pumpengehäuse	562.2	Zylinderstift	9002	Sechskantmutter
134	Profilwand	562.3	Zylinderstift	9004	Federring
146	Motorlaterne	562.4	Zylinderstift	931	Sicherungsblech
161	Gehäusedeckel	801	E-Motor	9201	Paßfeder
213	Antriebsteil	817	Spalttopf	9202	Paßfeder
230	Laufrad	818	Rotor	917	Senkschraube
381	Lagerensatz	819	Welle		
400	Flachdichtung	900	Traghaken		
4721	Anlaufscheibe	902	Sechskantschraube		
4722	Gleitring	904	Stiftschraube		
525	Distanzhülse	9041	Gewindestift		
529	Lagerhülse	914.1	Innensechskantschr		
545	Lagerbuchse	914.4	Innensechskantschr		
552	Spannscheibe	9201	Sechskantmutter		

Fig. 3.230 Regenerative pump in monobloc construction with magnetic coupling, PN 700 (Lederle)

345

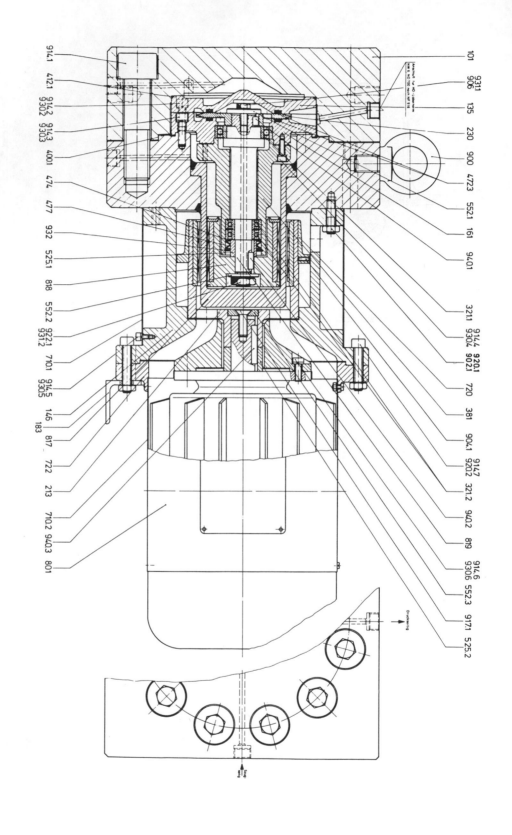

Fig. 3.232 High pressure regenerative pump with magnetic coupling (gas circulator) (Lederle)

346

Fig. 3.233 Hermetic gas circulation in a technical process plant using a regenerative pump as shown in Fig. 3.232 (Lederle)

The possible application of a pump of this kind can be demonstrated using an example taken from practice. The magnetic coupling pump shown in Fig. 3.232 is used to circulate molecular hydrogen under high pressure. The rotor runs on roller bearings permanently packed with grease.

The following parameters are specified as the operating conditions:

Operating temperature   20 °C
Product   Molecular hydrogen
Density at operating
  temperature   27 kg/m^3
Rated pressure   320 bar
Design pressure   320 bar
Test pressure   500 bar
Motor output   1,1 kW
Speed   2900 min^{-1}

Fig. 3.233 shows this pump installed in the plant.

The removal of eddy current losses in the relatively thick-walled can is achieved by the airflow of the drive motor, which passes through special feed channels between the can and outer magnet carrier. If such machines are used to pump liquids they are then fitted with plain bearings instead of roller bearings. Regenerative magnetic coupling pumps are also made of plastic, as shown in Fig. 3.234.

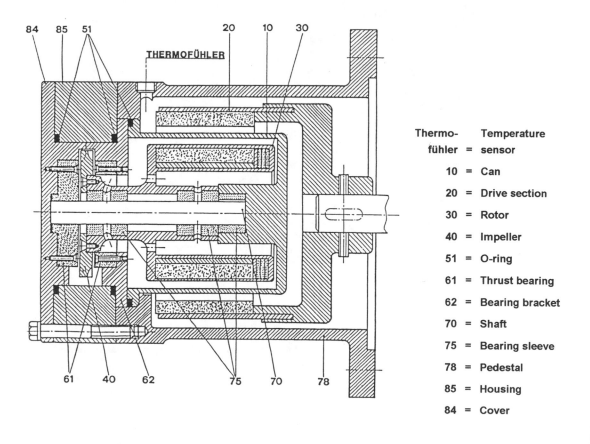

Thermo-fühler	=	Temperature sensor
10	=	Can
20	=	Drive section
30	=	Rotor
40	=	Impeller
51	=	O-ring
61	=	Thrust bearing
62	=	Bearing bracket
70	=	Shaft
75	=	Bearing sleeve
78	=	Pedestal
85	=	Housing
84	=	Cover

Fig. 3.234 Regenerative pump made of plastic, with magnetic coupling (schematic)

3.7.9 Magnetic coupling centrifugal pumps made of plastic (Fig. 3.235 and 3.236)

A series of standard chemical pumps to DIN 24256/2858 was developed to handle particularly aggressive fluids. The hydraulics are similar in design to that of stainless steel pumps, so that as well as good efficiencies equal NPSHR values are also achieved. The casing is of thick plastic protected by an outer steel jacket. The following materials are used depending on the aggressiveness of the product:

- The can is made of Hastelloy C lined with 4 mm thick diffusion-proof and universally corrosion-resistant PTFE.

- The shaft and pump bearings are made of SiC.

- The casing, impellers and the jackets of the magnet rotors are made in either PP or PVDF as required.

The family of characteristics shown in Fig. 3.237 clearly shows the large application range of plastic pumps. The temperature limits for the product are from -20 to + 100 °C.

Schmier- und Kühlflüssigkeitsführung	= Lubricating and cooling flow
Kühlstrom	= Cooling flow
Hauptstrom	= Main flow

Schmier- und Kühlflüssigkeitsführung

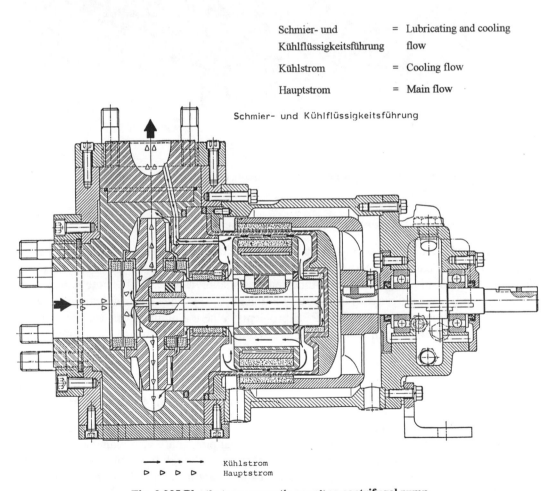

→ → → Kühlstrom
▷ ▷ ▷ ▷ Hauptstrom

Fig. 3.235 Plastic-type magnetic coupling centrifugal pump

Fig. 3.236 View of the plastic-type magnetic coupling centrifugal pump shown in Fig. 3.235

3.7.10 Vertical magnetic coupling centrifugal pumps

Vertical magnetic coupling pumps are particularly suitable for draining ground-level tanks or vessels which cannot have a ground drain for safety reasons. To do this the pumps are fitted in the tank in a so-called wet (immersed) installation (Fig. 3.238). These pumps can be either single or multi-stage (Fig. 3.239, 3.240 and 3.241). They are mainly used in a dry type of installation where there are NPSH problems. Where a pump is vertically immersed in a shaft an additional suction lift can be created by a direct feed line to the suction connection (Fig. 3.242).

3.7.11 Hyphenate magnetic coupling centrifugal pumps (Fig. 3.243 and 3.244)

In the same way as the high pressure canned motor pumps (Section 3.5.9 'Canned motor pumps in high pressure systems'), hermetic centrifugal pumps are also manufactured with permanent-magnet couplings for high pressure systems, although the application limits for these are substantially narrower than those for canned motor pumps. This applies particularly to the pressure, temperature and performance parameters.

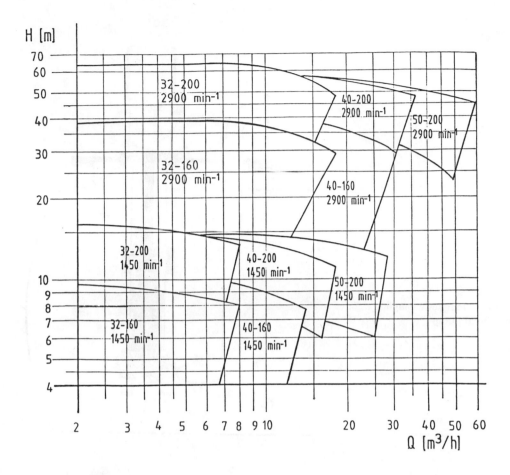

H [m]

32-200
2900 min⁻¹

40-200
2900 min⁻¹

50-200
2900 min⁻¹

32-160
2900 min⁻¹

40-160
2900 min⁻¹

32-200
1450 min⁻¹

40-200
1450 min⁻¹

50-200
1450 min⁻¹

32-160
1450 min⁻¹

40-160
1450 min⁻¹

Q [m³/h]

min⁻¹ = r.p.m.

Fig. 3.237 Characteristic curves for magnetic coupling centrifugal pump as shown in Fig. 3.235 and 3.236 (Lederle)

The restricted use results from the substantial increase in the thickness of the can wall and the resulting higher eddy current losses. An optimum has to be found between the rotor diameter and the thickness of the can wall which still results in a corresponding torque. The pump shown in Fig. 3.243 is designed for a system pressure of 620 bar.

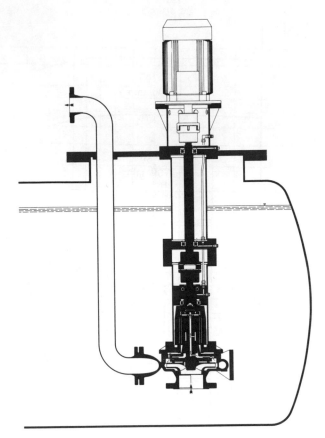

Fig. 3.238
Immersed magnetic coupling
centrifugal pump

Fig. 3.239
Vertical, single-stage magnetic centrifugal
pump (view) in accordance with (Fig. 3.238)
(Lederle)

352

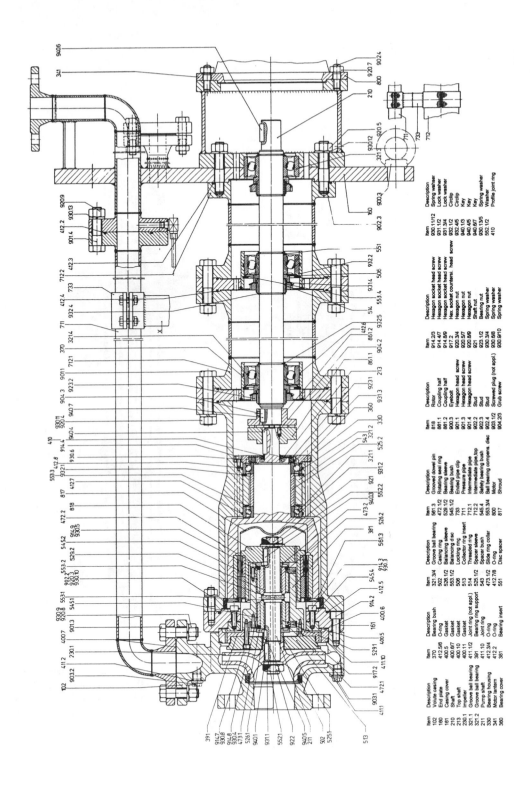

Fig. 3.240 Vertical, single-stage magnetic centrifugal pump (Lederle)

Item	Description
102	Volute casing
160	End plate
161	Casing cover
210	Shaft
213	Top shaft
230.1	Impeller
321.1	Groove ball bearing
321.2	Groove ball bearing
211	Pump shaft
330	Bearing housing
341	Motor lantern
360	Bearing cover

Item	Description
370	Bearing bush
412.5/6	O-ring
400.5	Gasket
400.6/7	Gasket
400.10	Gasket
400.11	Gasket
411.1/2	Joint ring (not appl)
391	Bearing ring support
411.10	Joint ring
412.3/4	O-ring
412.2	O-ring
381	Bearing insert

Item	Description
321.3/4	Groove ball bearing
502	Casing ring
528.1/2	Balancing sleeve
553.1/2	Gasket
506	Balancing disc
513	Collector ring
514	Locking ring
525.1/2	Spacer sleeve
543	Spacer bush
473.1/2	Slide ring collar
412.7/8	O-ring
551	Disc spacer

Item	Description
561.3	Grooved dowel pin
472.1/2	Rotating seal ring
528.1/2	Balancing sleeve
545.1/2	Bearing bush
733	Ended pipe clip
711	Threaded ring
712.1	Collector ring insert
712.2	Intermediate pipe
545.4	Intermediate pipe top
553.3/4	Safety bearing bush
800	Ball bearing compens. disc
817	Shroud

Item	Description
818	Rotor
861.1	Coupling half
861.2	Coupling half
900.3	Eyebolt
901.1	Hexagon head screw
901.3	Hexagon head screw
901.4	Hexagon head screw
921	Shaft nut
902.2	Stud
902.3	Stud
903.1/2	Screwed plug (not appl)
904.2/3	Grub screw

Item	Description
914.2/3	Hexagon socket head screw
914.47	Hexagon socket head screw
914.69	Hexagon socket head screw
917.2	Hex. socket countersi. head screw
920.34	Hexagon nut
920.57	Hexagon nut
920.89	Hexagon nut
921	Shaft nut
923.1/2	Bearing nut
930.34	Stud
930.68	Stud
930.910	Spring washer

Item	Description
930.11/12	Spring washer
931.1/2	Lock washer
931.34	Lock washer
932.12	Circlip
932.45	Circlip
940.13	Key
940.45	Key
940.67	Key
930.135	Spring washer
552.1/2	Spring washer
410	Profile joint ring

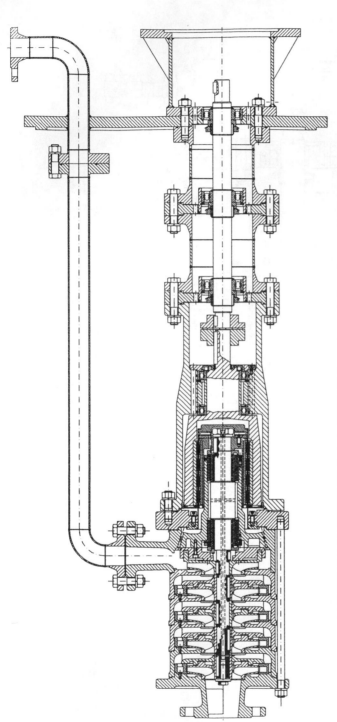

**Fig. 3.241 Immersible multi-stage magnetic
coupling centrifugal pump
(Lederle)**

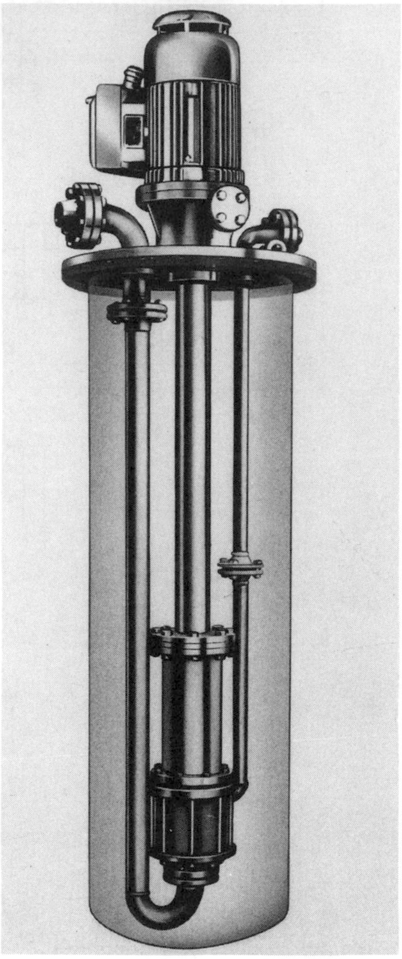

**Fig. 3.242 Multi-stage magnetic
coupling centrifugal
pump for dry installa-
tion (Lederle)**

354

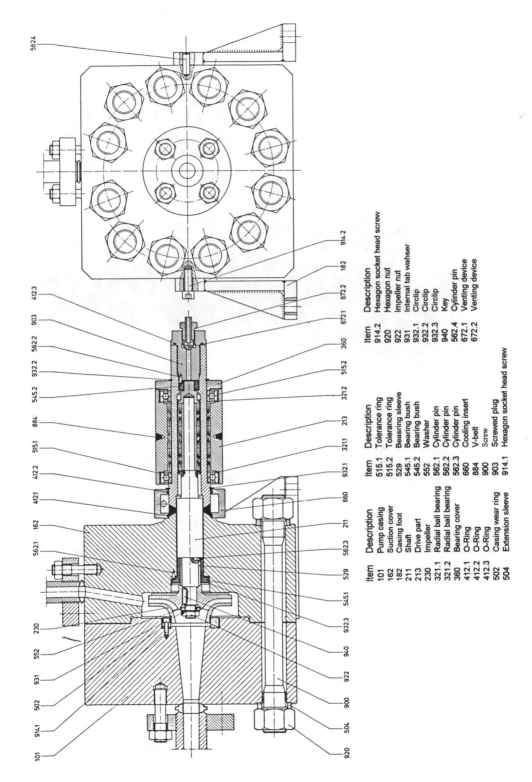

Item	Description	Item	Description	Item	Description
101	Pump casing	515.1	Tolerance ring	914.2	Hexagon socket head screw
162	Suction cover	515.2	Tolerance ring	920	Hexagon nut
182	Casing foot	529	Beasring sleeve	922	Impeller nut
211	Shaft	545.1	Bearing bush	931	Internal tab wahser
213	Drive part	545.2	Bearing bush	932.1	Circlip
230	Impeller	552	Washer	932.2	Circlip
321.1	Radial ball bearing	562.1	Cylinder pin	932.3	Circlip
321.2	Radial ball bearing	562.2	Cylinder pin	940	Key
360	Bearing cover	562.3	Cylinder pin	562.4	Cylinder pin
412.1	O-Ring	660	Cooling insert	672.1	Venting device
412.2	O-Ring	884	V-belt	672.2	Venting device
412.3	O-Ring	900	Screw		
502	Casing wear ring	903	Screwed plug		
504	Extension sleeve	914.1	Hexagon socket head screw		

Fig. 3.243 Multi-stage magnetic coupling centrifugal pump in high pressure specification for 620 bar system pressure (Lederle)

355

Fig. 3.244 View of a high pressure magnetic coupling centrifugal pump as shown in Fig. 3.243

3.7.12 Shaftless magnetic coupling centrifugal pumps [3-14]

The assumption that the rotor carrying forces (refer to 3.2.10.1 'The hydraulic carrying force F_L') should be a multiple of the weight of the impeller was the reason for the development of a series of shaftless pumps. The first test models produced convincing evidence of the correctness of this hypothesis. Fig. 3.245 shows the construction of such a pump. It is different from traditional constructions in that it is noticeably simple and easy to manufacture. The bedding of the inner magnet carrier is almost exclusively hydrodynamic and therefore the carrying capacity is increased by the omission of a shaft and normal shaft bearings.

The casing wear ring on the discharge side of the impeller as well as the axial hydraulic balancing device in the can are used as guide or guard bearings during starting and stopping the pump. Casing wear ring and sliding parts of the balancing device are made of highly abrasion-resistant silicium carbide and ensure that the magnet carrier does not contact the inside of the can. The bending moment due to transverse forces (radial force) is substantially reduced with this design, as the effective bending arm is greatly shortened.

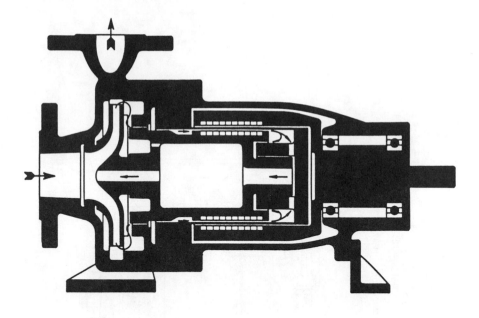

Fig. 3.245 Functional diagram of a shaftless magnetic coupling centrifugal pump (Lederle)

3.7.12.1 Venting

The construction of hermetic pumps means that complete venting is an absolute requirement during starting and running. This is relatively easily achieved during shutdown by
means of suitable bores, but other criteria have to be taken into account when the pump is
running. Gases which escape from the fluid or those which do not fully escape from the
pressure or vent line during shutdown venting collect as a concentric ring of gas in and
around the shaft areas due to the rotational movement of the impeller.
This can lead to damage to bearings. The shaftless pump on the other hand avoids this
disadvantage because the gases are drawn off through the hollow bore of the impeller hub
into the suction part of the impeller.

3.7.12.2 Axial hydraulic balance and balancing flow Q_T

A radial hydrodynamic bearing can only be effective if hydraulic balance is also provided in
the axial direction. To safely avoid thrust damage, the design of a balancing device must
assume that the forces are adequately balanced over the complete H (Q) curve. The pumps
shown in Fig. 3.246 and 3.247 have a control mechanism of this type. To eliminate the
thrust forces acting on the impeller and at the same time remove the heat losses arising in
the magnetic coupling, a partial flow Q_T is taken from the pressure chamber formed by the
rear side of the impeller and the flange of the can. This partial flow passes through the
rearward gap seal of the impeller and the rotor gap (Fig. 3.245).

357

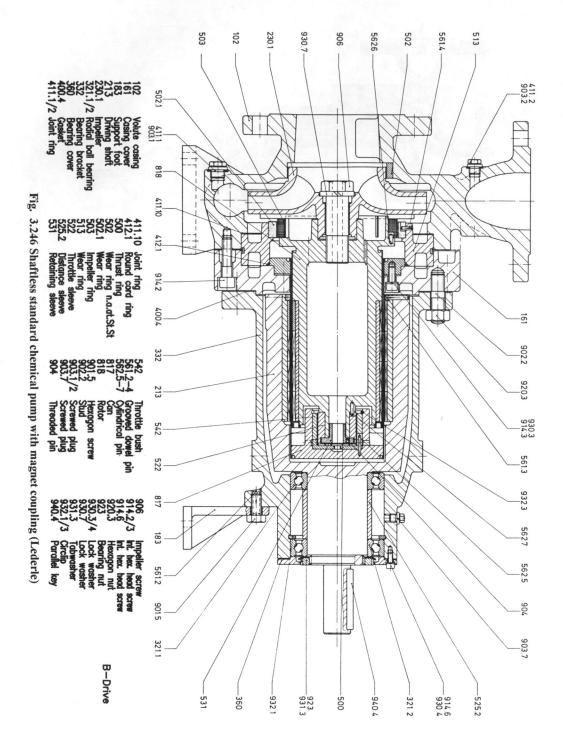

102	Volute casing	411.10	Joint ring	542	Throttle bush	906	Impeller screw
161	Casing cover	412.1	Round cord ring	561.2-4	Grooved dowel pin	914.2/3	Int. hex. screw
183	Support foot	500	Thrust ring	562.5-7	Cylindrical pin	914.6	Int. hex. head screw
213	Driving shaft	502	Wear ring	817	Can	920.3	Int. hex. head screw
230.1	Impeller	502.1	Wear ring n.a.ct.St.St	818	Rotor	923	Hexagon nut
321.1/2	Radial ball bearing	503	Wear ring	901.5	Hexagon screw	930.3/4	Bearing nut
332	Bearing bracket	513	Impeller ring	902.2	Stud	930.7	Lock washer
360	Bearing cover	522	Throttle sleeve	903.1/2	Screwed plug	931.3	Lock washer
400.4	Gasket	525.2	Distance sleeve	903.7/2	Screwed plug	932.1/3	Circlip
411.1/2	Joint ring	531	Retaining sleeve	904	Threaded pin	940.4	Parallel key

Fig. 3.246 Shaftless standard chemical pump with magnet coupling (Lederle)

B—Drive

358

Fig. 2.347 Section view of a model of a standard chemical centrifugal pump with a magnetic clutch as shown in Fig. 3.246 (Lederle)

The partial flow Q_T passes through bores in a guide bush in the bottom of the can and then through the vent gap S_h into the rotor space, from where it then passes through the hub of the impeller to the suction side. It should also be possible initially to quantify the anticipated balancing flow Q_T.

3.7.12.3 Initial calculation for thrust forces and balancing flow (4-K)

Designations
Refer to Fig. 3.248 and 3.249 and also to 'symbols'
Circular areas at D_2, D_s etc are designated A_2, A_s etc.

Pressure drops through the impeller pressure side and can
The pressure drop takes place in several stages.
Between D_2 and D_r the liquid rotates at half the angular velocity of that of the impeller.
The following generally applies to any angular velocity.

$$\Delta H = \frac{\omega^2}{2 \cdot g \cdot \pi} \cdot (A_{externd} - A_{internal}) \tag{3 - 62}$$

in particular the following applies between D_2 and D_r where $\omega = \frac{1}{2} \cdot \Omega$ and $\alpha = \frac{\Omega^2}{2 \cdot g \cdot \pi}$

$$H_0 - H_1 = \frac{1}{4} \cdot \alpha(A_2 - A_r) \tag{3 - 63}$$

Between the back shroud blade $\varnothing \, D_r$ and casing wear ring $\varnothing \, D_3$ the vortex on the impeller shroud side rotates at almost the same angular velocity as the impeller. Therefore we get:

$$H_1 - H_2 = \alpha(A_r - A_3) \tag{3 - 64}$$

The pressure drop in the casing wear ring and rotor gap can be calculated according to Yamada as follows:

$$\Delta H = \left(\lambda \cdot \frac{L}{2 \cdot s} + 1,5\right) \cdot \frac{\overline{V}_x^2}{2 \cdot g} = E \cdot Q_T^2 \tag{3 - 65}$$

where $Re_{su} > 10^4$ the following applies (refer to Section 3.2.8 and Fig. 3.28)

$$\lambda = 0,26 \cdot Re^{-0,24} \cdot \left[1 + \left(\frac{7}{16}\right)^2 \cdot \left(\frac{U}{\overline{V}_x}\right)^2\right]^{0,38} \tag{3 - 66}$$

where $\quad Re = \frac{2 \cdot s \cdot \overline{V}_x}{\nu}, \; Re_{su} = \frac{2 \cdot s \cdot V}{\nu}$

and $\quad E = \dfrac{\left(\lambda \frac{L}{2 \cdot s} + 1,5\right)}{(2 \cdot g \cdot D^2 \cdot \pi^2 \cdot S^2)} \tag{3 - 67}$

In this case $\overline{V}_x$ is the mean axial velocity of the gap flow and U is the circumferential velocity of the impeller and rotating wear ring.

It will be noted that λ over Re depends on Q_T!

The drop in head between the inlet wear ring and the end of the impeller is therefore:

$$H_2 - H_3 = [E_1 + E_2] \cdot Q_T^2 \qquad (3 - 68)$$

where E_1 represents the casing wear ring dimensions and E_2 the rotor dimensions.

The liquid ring on the rear side of the rotor again rotates at $\omega = \frac{\Omega}{2}$ and we therefore get the following:

$$H_3 - H_4 = \frac{1}{4} \cdot \alpha [A_{Rot} + A_{ax}] \qquad (3 - 69)$$

The following formula applies to the axial vent gap at the end of the rotor.

$$Q_T = \mu \cdot A_{Sp} \cdot \sqrt{2g \cdot \Delta H_{ax}} \qquad (3 - 70)$$

with the gap area $A_{sp} = \pi \cdot D_{ax} \cdot S_h$ depending on the gap width S_h and therefore the impeller position.

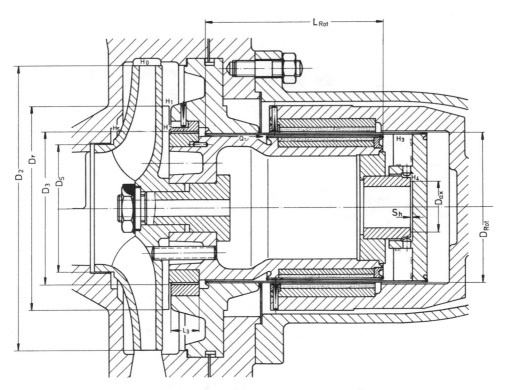

Fig. 3.248 Impeller set with axial thrust device

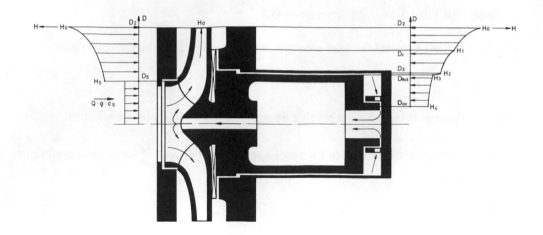

Fig. 3.249 Axial thrust forces on impeller

If release takes place towards ΔH_{ax} we get:

$$\Delta H_{ax} = \frac{E_3}{S_h^2} \cdot Q_T^2 \qquad (3 - 71)$$

where

$$E_3 = \frac{1}{2g \cdot \mu^2 \cdot \pi^2 \cdot D_{ax}^2} \qquad (3 - 72)$$

- Calculation of balancing flow Q_T

The total of the pressure drops from (3-63; 3-64; 3-68; 3-69; 3-71;) gives the total head H_0 of the pump.

$$H_0 = \frac{1}{4} \cdot \alpha[A_2 - A_r] + \alpha[A_r - A_3] + \left[E_1 + E_2 + \frac{E_3}{S_h^2}\right] \cdot Q_T^2 + \frac{1}{4} \cdot \alpha[A_{rot} - A_{ax}] \qquad (3 - 73)$$

The balancing amount Q_T for the given gap S_h can be iteratively determined from the equation (E_1 and E_2 also depends on Q_T).

- Thrust forces in the direction of the suction end

The forces on the surface $(A_2 - A_r)$, $(A_r - A_3)$ and $(A_3 - A_{ax})$ are obtained from the corresponding pressure paraboloids (refer to Fig. 2.249) as follows:

$$\overleftarrow{K} = \rho \cdot g \cdot \left[\frac{H_0 + H_1}{2} \cdot [A_2 - A_r] + \frac{H_1 + H_2}{2} \cdot [A_r - A_3] + \frac{H_3 + H_4}{2} \cdot [A_{Rot} - A_{ax}]\right] \qquad (3 - 74)$$

362

This made use of the fact that the volume of a paraboloid ($\int pdF$) is equal to the volume of a cylinder of the same external diameter and with half the height.

Because a calculation can be made to the given S_h and Q_T from (3-73), the heights H_1 - H_4 can be calculated using the formulae (3-63; 3-64; 3-68; 3-69) and therefore also K using formula $\overleftarrow{K}$ (3-74).

- Forces in the direction of the drive end

From the pressure paraboloid on the impeller shroud we therefore get the following where $\omega = \dfrac{\Omega}{2}$.

$$H_5 = H_0 - \tfrac{1}{4} \cdot \alpha \cdot [A_2 - A_s] \qquad\qquad (3\text{ - }75)$$

The deflection of the flow Q from the axial to the radial direction produces an impulse force $\rho \cdot Q \cdot V_s$, with V_s representing the inlet velocity on the suction side.

The total axial force in the direction of the drive end is therefore:

$$\overrightarrow{A} = \rho \cdot g \cdot \frac{H_0 - H_5}{2} \cdot [A_2 - A_s] + \rho \cdot Q \cdot v_s \qquad (3\text{ - }76)$$

Example of calculation (3-K)

Given:

Pump

Impeller diameter	$D_2 = 169$ mm
Back shroud blade diameter	$D_r = 120$ mm
Casing wear ring diameter suction	$D_s = 70$ mm
Nominal diameter of suction connection	$D = 50$ mm

Drive

Casing wear ring diameter	$D_3 = 112$ mm
Gap clearance	$S_3 = 0,15$ mm
Gap length	$L_3 = 18$ mm
Rotor diameter	$D_{Rot} = 106,2$ mm
Rotor gap	$S_{Rot} = 0,5$ mm
Rotor gap length	$L_{Rot} = 85$ mm
Average throttling diameter	$D_{ax} = 38$ mm

Calculated:

These problems were solved using formula (3-74) for the thrust forces in the direction of the suction end and formula (3-76) for the direction of the drive end, and plotted in the form of a graph (Fig. 3.250) for the duty points at zero delivery $Q = 0$ m³/h at head $H_0 = 40$ m and flow $Q = 16$ m³/h at head $H = 33$ m.

The formula (3-73) give the values for the balancing (loss) flow $Q_T = f(S_h)$. The relationship

$K = f(S_h)$ indicates the stability of the balancing device. Equilibrium of the forces and therefore a zero force state exists when:

$$\overrightarrow{G} = \overleftarrow{K} \qquad\qquad (3 - 77)$$

In this example:

Equilibrium at zero delivery $Q = 0$ m³/h		G	= 6790 N
		K	= 6790 N
where		S_h	= 0,32 mm
with		Q_T	= 0,93 m³/h
Equilibrium at flow	$Q = 16$ m³/h	G	= 5590 N
		K	= 5590 N
at gap width		S_h	= 0,26 mm
with gap flow		Q_T	= 0,68 m³/h

Its easy to see that the equilibrium situation is stable because the force $\overleftarrow{K}$ in the direction of the suction end reduces when deflected in the direction of the suction end (increase in S_h) and vice versa it increases when deflected in the direction of the drive end.

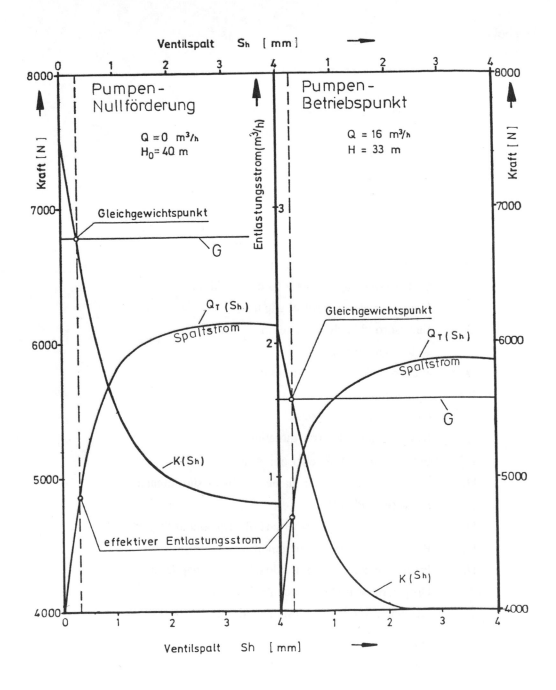

Ventilspalt	=	Valve gap	Kraft	=	Force
Pumpen-Nullförderung	=	Pump zero delivery	Spaltstrom	=	Gap flow
Pumpen-Betriebspunkt	=	Pump duty point	effektiver Entlastungsstrom	=	Effective balancing flow
Gleichgewichtspunkt	=	Equilibrium point	Ventilspalt	=	Valve gap

Fig. 3.250 Hydraulic axial forces and partial flows for specimen calculation (3-K)

Symbols

A_2	Circular area D_2 m^2
A_3	Circular area of D_3 m^2
A_{ax}	Circular area of D_{ax} m^2
A_r	Circular area of D_r m^2
A_{Rot}	Circular area of D_{Rot} m^2
A_S	Circular area of D_S m^2
A_{SP}	Annular ring area of gap m^2
α	Flow coefficient
D	Nominal width of suction nozzle m
D_2	Impeller external diameter m
D_3	Casing wear ring diameter (pressure side) m
D_{ax}	Mean diameter of control gap m
D_r	Back shroud blade diameter m
D_{Rot}	Rotor diameter m
D_s	Wear ring diameter suction side m
E	Proportionality factor
g	Gravity acceleration m/s^2
$\vec{G}$	Resulting force in drive direction N
H_0	Pressure head at impeller outlet m
H_1	Pressure head on outer diameter of back shroud blade m
H_2	Pressure head at D_3 m
H_3	Pressure head on pressure side of rotor end at D_{Rot} m
H_4	Pressure head on control gap m
H_5	Pressure head on suction side of impeller gap D_S m
ΔH_{ax}	Drop in pressure head in control gap m
$\overleftarrow{K}$	resulting force in direction of suction end N
L	Gap length m
L_3	Gap length (sealing gap - rear side of impeller) m
L_{Rot}	Rotor gap length m
λ	Flow coefficient
μ	Friction coefficient
ν	Kinematic viscosity m^2/s
ω	Angular velocity l/s
Q	Flow rate m^3/s

Q_T	Balancing (partial) flow m³/s
Re	Reynolds number in axial direction
Re_{su}	Reynolds number in circumferential direction
ρ	Density of fluid kg/m³
S	Gap width m
S_3	Gap clearance (one-sided casing wear ring - pressure side) m
S_h	Control gap width m
U	Circumferential velocity m/s
$\bar{V}$	Flow velocity in gap m/s
V_s	Flow velocity at impeller inlet m/s

3.7.12.4 Multi-stage design (Fig. 3.251)

There are multistages designs similar to the single-stage types. In this case the coupling itself is shaftless and the pump is provided with a shaft corresponding to the number of stages. Because this is connected directly to the rotor of the pump and provided with a hollow bore, the multi-stage pump vents in the same way as the single-stage pump. It should also be noted that the shaftless design of this coupling is also successfully used on hydrostatic, rotating displacement pumps.

3.7.13 Safety and monitoring systems on magnetic coupling pumps

In principle the safety aspects for magnetic coupling pumps are the same as those as for canned motor pumps with the exception of the regulations for explosion protection.

3.7.13.1 Dry-run protection

The bearing in the working chamber must be protected from running dry and therefore require level monitoring on the inlet side. This can be achieved by float switches, capacitive or optoelectronic sensors depending on the density, conductivity or internal pressure.

3.7.13.2 Partial flow temperature monitoring (Fig. 3.214)

The requirement for nonvaporisation of the partial flow liquid, particularly for fluids close to boiling point, e.g. liquid gases, can be met using a monitoring thermostat (PT 100) fitted in a bore provided in the casing for that purpose.

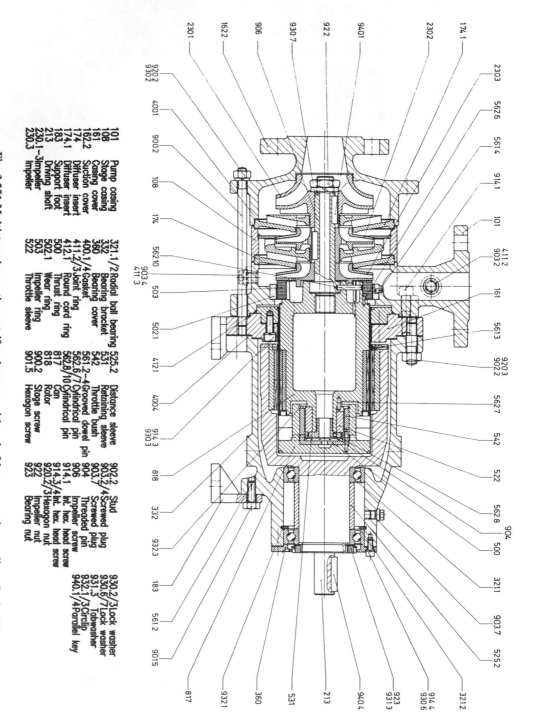

101	Pump casing	321.1/2	Radial ball bearing	525.2	Distance sleeve	902.2	Stud	930.2/3	Lock washer
108	Stage casing	332	Bearing bracket	531	Retaining sleeve	903.2/4	Screwed plug	930.6/7	Lock washer
161	Casing cover	360	Bearing cover	542	Throttle bush	903.7	Screwed plug	931.3	Tabwasher
162.2	Suction cover	400.1/4	Gasket	561.2	Grooved dowel pin	904	Threaded pin	932.1/3	Circlip
174	Diffuser insert	411.2/3	Joint ring	562.6/7	Cylindrical pin	906	Impeller screw	940.1/4	Parallel key
174.1	Diffuser insert	412.1	Round cord ring	562.8/10	Cylindrical pin	914.1	Int. hex. head screw		
183	Support foot	500	Thrust ring	817	Can	914.3/4	Int. hex. head screw		
213	Driving shaft	502.1	Wear ring	818	Rotor	920.2/3	Hexagon nut		
230.1–3	Impeller	503	Impeller ring	900.2	Stage screw	922	Impeller nut		
230.3	Impeller	522	Throttle sleeve	901.5	Hexagon screw	923	Bearing nut		

Fig. 3.251 Multistage hermetic centrifugal pumps with a shaftless magnetic coupling (Lederle)

368

3.7.13.3 Monitoring the can temperature (Fig. 3.214)

This can be achieved by fitting a temperature sensor (PT 100) directly on the can, so that it will also provide an indication of the temperature of the partial flow. If the pump runs dry, the temperature of the can rises rapidly and the can temperature sensor can therefore be used as a warning device.

3.7.13.4 Roller bearing monitoring (Fig. 3.214)

Damage to the external bearing system can be detected in good time and further damage to the pump prevented by temperature and impact pulse measurement at the ball bearings of the bearing bracket.

3.7.13.5 Double wall security system (DWS) [3.34]

A disadvantage of magnetic coupling pumps is that the wall separating the product from the atmosphere, the can, has to have a thin wall (approximately 0,5 to 1 mm) in order to minimise eddy current losses. This means that if damage to the can occurs due to corrosion or abrasion the fluid escapes to atmosphere. This can lead to problematic malfunctions with dangerous substances. Abrasion damage can occur, for example, if the ceramic bearing materials, mainly used today, break away due to impact or temperature shock and are then caused to rotate and penetrate the can wall (Fig. 3.252). Because of their higher density these parts rotate under the effect of centrifugal force Z against the inner wall of the can. If they are greater in size than S (gap width) they cannot pass through the annular gap between the rotor and stator, but are instead forced by the gap flow in the axial direction with force L_K against the outer circumference of the rotor. Force R then results from Z and L_K and after a relatively short period this leads to the can being abraded through.
Ferrite particles entrained in the product are attracted by the sychronous rotating magnets and carried with them, which can lead to damage to the sealing shroud. A further danger is if the pump unintentionally runs dry, which would damage the plain bearing and cause the inner rotor to be in contact with the can. Impact damage is then unavoidable. The double wall safety system (DWS) is designed to protect against such dangers. The can which forms the dividing wall between the fluid and atmosphere is encased by a second one with a small clearance between them (Fig. 3.253). If the inner can is damaged, the second outer can performs the sealing function and prevents leakage of the product. For static reasons both cans have the same wall thickness. The outer can may be easily retrofitted to pumps which have only one can because the bore dimension of the magnetic clutch is designed such that this can be achieved at any time. This is of particular interest for standard chemical pumps to DIN 24256. The space between both cans can easily be connected to a monitoring system and a wide range of measuring and control techniques such as switching, signaling and monitoring facilities are available. These could be, for example, pressure switches, humidity sensors or level sensors or sensors which react to conductivity, or a switching system based on analytical measurement. For example, in the case of the optical fibre level sensor, an optical fibre conductor is immersed in the operating liquid, which in the event of a leak enters the space between the inner and outer can, and uses the difference in the refractory index between liquid and gas (Fig. 3.254). The particular advantage of these sensors is their

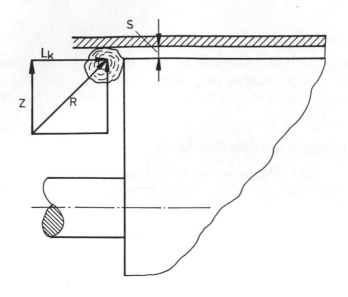

Fig. 3.252 Damage to can due to solids in the fluid

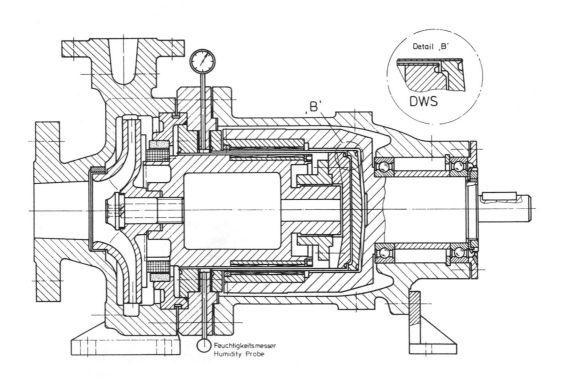

Feuchtigkeitsmesser
Humidity Probe

Detail ‚B'

DWS

Fig. 3.253 Magnetic coupling centrifugal pump with the DWS system

small dimensions which are available in thread sizes from M5 upwards. They are also resistant to high pressures, have shorter response times, can be used in hazardous areas and, if made of quartz glass, are resistant to aggressive fluids.

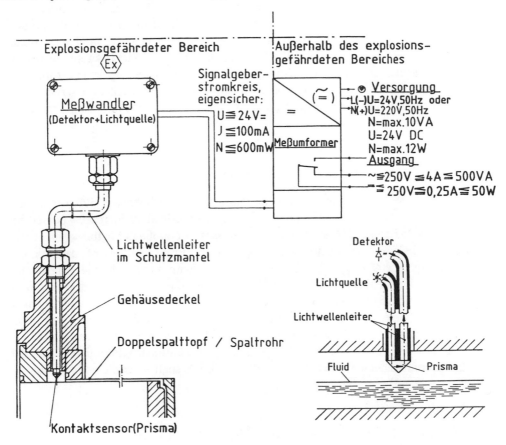

Explosionsgefahrdeter Bereich	=	Hazardous area
Außerhalb des explosionsgefährdeten	=	Outside the hazardous area
Versorgung	=	Supply
oder	=	or
Ausgang	=	Output
Signalgeberstromkereis, eigensicher	=	Intrinsically safe sensing element circuit
Meßwandler (Dedektor + Lichtquelle)	=	Instrument transformer (detector and light source)
Meßumformer	=	Transducer
Lichtwellenleiter	=	Optical fibre conductor in protective sheath
Gehausedeckel	=	Housing cover
Doppelspalttopf /Spaltrohr	=	Double can enclosure/can
Detektor	=	Detector
Lichtquelle	=	Light source
Lichtwellenlleiter	=	Optical fibre conductor
Fluid	=	Fluid
Kontaktsensor (Prisma)	=	Contact sensor (prism)

Fig. 3.254 Electronic monitoring of double can

The sensor receives light from a specially developed LED (light source) through a flexible optical fibre conductor (diameters from 0,5 to 1 mm). If fluid comes into contact with the sensor the light is decoupled in the liquid so that the incoming light intensity at the detector is far lower compared with when in contact with gas or air. An electronic amplification circuit is placed after the detector to provide a suitable alarm or to switch off the drive motor. Because they take up little space these sensors are particularly suitable for double can monitoring. These sensors enable visual and acoustic signals to be triggered which shut down pumps and also activate the closure of valves in associated pipes. Interruptions in operation can be avoided if a standby pump is automatically brought into service in the event of such a fault and the damaged pump is shut down. The monitoring and control facilities can be expanded as required.

3.7.13.6 Additional losses due to the DWS system

The addition of a second sealing shroud (double can) causes further eddy current losses which reduce the efficiency of the coupling and therefore lead to increased power consumption. Fig. 3.255 and 3.256 show the power losses of couplings for the series of magnetic couplings discussed in Section 3.6.6 ("Power and loss quantities of magnetic couplings" Fig. 3.209, 3.209.1, 3.210, 3.210.1) when fitted with a metal double can. It can be seen that there is an additional reduction of 4 to 5 percentage points in the efficiency of the coupling. The designer or operator must consider whether the additional safety, considering the material being handled, afforded by the double wall is either necessary or appropriate. If the answer is yes then safety must take precedence over increased power consumption. It is in any case also possible to avoid the additional eddy current losses by making the second outer can of a non-metallic material, such as carbon fibre reinforced plastic.

3.7.13.7 Transport of molten materials using the double wall system

In addition to the facility for monitoring malfunctions, the double wall can also enable intensive heating of the magnetic clutch. Whereas previously the use of hermetic pumps with permanent magnet couplings for substances which have to be heated either before pumping or during pumping was unsatisfactory, the DWS system today offers a particularly good solution to this problem (Fig. 3.257). A double wall can and can base acting as a heating jacket mean that the can no longer acts as a thermal insulator against heating of the inner bearings. It is now an active element of a heater which is in direct thermal contact with the bearing area. Adequate heating of the bearing and the liquid surrounding the rotor is therefore guaranteed. This means that full hydrodynamic lubrication can be established on startup and the starting torque held within normal limits.

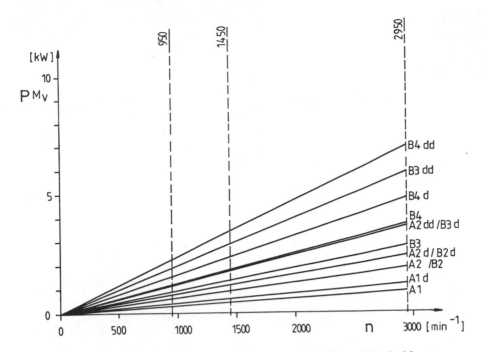

Fig. 3.255 Power loss P_{MV} of a size A/B magnetic coupling with a double can

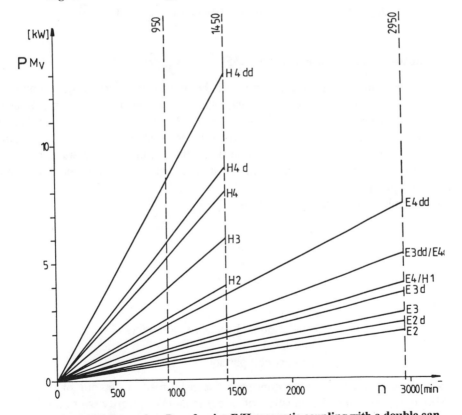

Fig. 3.256 Power loss P_{MV} of a size E/H magnetic coupling with a double can

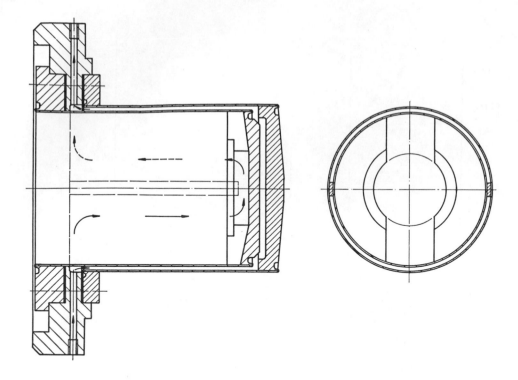

Fig. 3.257 Double can using the DWS system for fluid heating

3.8 Canned motor or magnetic coupling [3.35]

Now that both systems of leak-free pumping of the liquids using centrifugal pumps have been presented, this naturally raises the question for the engineer engaged in planning or operating as to which is the most suitable, safest and also the most economical for his task. This question cannot be answered clearly in favour of one system or the other. It requires a thorough analysis of the task, operating conditions, the required safety standard and the financial situation. The following considerations should help to decide on the most suitable type of hermetic pump.

To do this, the following initial considerations:

>   - safety
>   - explosion protection
>   - maintainability
>   - temperature of the fluid
>   - pressure range
>   - efficiency
>   - installation and space requirements
>   - noise level
>   - and cost comparison

have to be balanced to decide what concessions can be made in favour of one or other system. Above all, however, the question of safety must take priority over all other considerations.

3.8.1 Safety

The primary difference between canned motor and permanent magnet coupling systems is that the canned motor is equipped with an additional safety shell (Fig. 3.52) which is sealed to the atmosphere. The terminal boxes and terminal connections are also gas and liquid tight and are designed for the nominal design pressure of the pump. Should damage occur to the motor can resulting from damage to the bearings or corrosion, dangerous substances cannot escape to the atmosphere. In contrast such damage in the magnetic coupling system constitutes a considerable safety hazard.

This risk can also be minimised to a great extent in magnetic coupling pumps by using a double wall drive "can" (Fig. 3.253). However, this is possible only in conjunction with continuous monitoring of the gap between the cans.

Canned motor pumps are therefore recommended when the pumped fluid has a high hazard potential. Either the lower toxic limit value of the fluid or a classification number (hazard rating diamond according to Hommel "Manual of hazardous substances") can be used to assess this risk.

Thus, canned motor pumps should be used under all circumstances for highly dangerous liquids (health hazard degree 4), and not magnetic coupling pumps. Such hazardous liquids include hydrocyanic acid, phosgene, hydrogen fluoride, monochloracetic acid, acrylonitrile etc.

With limitations, this also applies to liquids of health hazards degree 3 which are still classified as extremely dangerous. Examples of such liquids are chlorine, ammonia, chlorosulfonic acid, diamide, amines, perchloric acid etc.

Both types of hermetic drive systems can be used equally well for fluids with lower hazard classifications.

The canned motor type also proves superior to the magnet drive type when highly combustible liquids (fire hazard degree 4 according to Hommel) are to be pumped. This applies, for example, to compressed gases such as ethylene, ethane, propylene, vinyl chloride etc.

Also included in this category are thermal oils, for example diphyl, which have a relatively high vapour pressure (6.6 bar abs at 360 °C) and a flash point of 115 °C. If the can of a magnetic pump is damaged, allowing hot diphyl to escape, a fire hazard always exists. Such accidents can be reliably prevented only with the use of a canned motor pump.

3.8.2 Explosion protection

An argument commonly used in favour of magnetic coupling machines is that such pumps need not comply with explosion protection conditions and therefore do not require a permit from the PTB (Federal Institute for Physics and Technology). In addition, monitoring equipment as prescribed for canned motor pumps are not required.

Such arguments must, however, be viewed with caution. Magnetic coupling pumps may also act as an ignition source for external explosive atmospheres through hot surfaces or mechanically generated sparks. Sparks may be caused by dry running, overheating of bearings, or by grazing the outer magnet carrier on the coupling casing.

The canned motor pump, designed as an integral unit consisting of pump and electricmotor, offers a substantially higher degree of safety regarding explosion protection. The liquid level and temperature monitoring devices required for this pump type prevent dry operation and overtemperatures in both the motor and the pump. To attain an equal degree of safety in a magnetic pump, the pump set must also be monitored for dry operation and overtemperatures.

3.8.3 Maintainability

Magnetic coupling pumps clearly offer advantages over canned motor pumps in this respect, as only mechanical and no electric parts must be repaired in case of breakdown.

In a canned motor pump, damage to the motor can usually result in damage to the windings, which must then be either replaced or re-wound. Many users with large chemical or petrochemical plants often carry out their own stator repairs without problems. The equipment required for winding motors or welding motor cans (TIG welding) is normally available.

In contrast, certain reservations must be made regarding the repair of magnetic coupling pumps. If the pump is operated when dry, eddy current losses may result in a rapid increase in temperature which in turn leads to overheating of the permanent magnet. If this is not recognised in time, partial or complete demagnetisation of the permanent magnets may result. In this case repair is no longer possible. The inner, and at worst the outer, magnet carrier must be completely replaced.

3.8.4 Temperature of the fluid

Standard magnetic coupling pumps can be operated for liquid temperatures up to 200 °C without an additional cooling system. When a spacer is installed between pump and drive the operating range may be extended to up to 400 °C (without a cooling system) (Fig. 3.222).

The operating temperature range for canned motor pumps is primarily determined by the insulation class of the motor. In pumps with a specified insulation class, a functional relationship exists between the maximum permissible temperature of the liquid and the output of the motor. For example, the permissible load in a canned motor is 8 kW at a liquid temperature of 40 °C, 7 kW at 70 °C and only 5.8 kW at 100 °C. For temperatures above 100°C, canned motors with external cooling systems or motors of insulation class C must be used.

No temperature restrictions other than those pertaining to the material are stipulated for canned motors with external cooling systems (Fig. 3.83). The temperature in the motor cooling circuit remains almost constant regardless of the operating temperature on the pump side. For this reason, canned motor pumps with external cooling systems are also suitable where extreme fluctuations in temperature may occur. One particular example of this is combined heating-cooling circuits.

The use of magnetic coupling pumps at low temperatures (below -10 °C) is restricted, as ice may form on the outside of the coupling can. As a result the outer magnet carrier may lock, especially when starting the pump. Magnetic coupling pumps are therefore not commonly used for refrigeration liquids. Canned motor pumps are preferred for these applications.

3.8.5 High pressure applications

Magnetic coupling pumps have two physical air gaps, one on the inside and one on the outside of the coupling can. In addition, these pumps are not equipped with the reinforcement rings normally used for canned motor pumps. For this reason, and due to the thin walls of the can, the use of magnetic coupling pumps in high pressure systems is restricted. Magnetic coupling pumps designed for higher system pressures are available but require extremely thick cans. As the air gap dimensions and wall thickness increase the efficiency of the pump set decreases rapidly, especially when combined with higher drive powers and speeds.

This problem can be solved easily and extremely efficiently with the canned motor pump (Fig. 3.121). The motor can is supported from the outside of the stator and beyond the stator by the reinforcement rings. This permits a thin walled design of the motor can, which results in minimum eddy current losses. Canned motor pumps for system pressures of up to 1200 bar have been constructed with up to 80% efficiency.

3.8.6 Efficiency

The magnetic coupling pump generally achieves 5 - 10% higher efficiency than the corresponding canned motor pump, resulting in lower temperature increases in the partial flows. This can be of particular advantage when liquids with a tendency to polymerize are pumped.

Despite higher efficiency, the magnetic coupling pump (including drive motor) does not rate better in overall efficiency than the canned motor pump and therefore offers no power saving advantages.

3.8.7 Starting behaviour

Magnetic couplings have serious drawbacks in starting behaviour in comparison to the canned motor.

The magnetic coupling can transfer only a maximum torque. If this maximum torque is exceeded, for example when the motor is started directly, the magnetic coupling disengages and cannot fall back into step. Magnetic couplings must be carefully designed and provided with the required torque reserves in order to withstand the torque peaks attained during starting and normal operation. In many cases the maximum torque can be increased only by enlarging the balance weight at the driving end or by installing smooth start devices.

3.8.8 Installation and space requirements

Because the pump and motor form an integral unit, canned motor pumps generally require neither baseplates nor special foundations. The units are not at all sensitive to distortions caused by the piping.

This is of special importance if heat transfer oils are to be pumped at high temperatures.

In magnetic coupling pumps, torsion and thermal expansion of the pipework may lead to misalignment of the mechanical coupling and in turn result in damage to the roller bearings. Because the gap between the outer magnet carrier and the coupling can is only 0,5 - 1 mm, damage to the can from the outside by the magnet carrier cannot be excluded. Under such circumstances the pumped liquid would escape through the ball bearings. This constitutes a safety hazard which is not to be underestimated if explosive, toxic liquids are pumped.

Canned motor pumps may easily be installed without foundations, for example using Avenarious screws (Fig. 3.56). In some cases this may mean substantial savings in installation costs in comparison with the costs required for magnetic drive pumps.

Canned motor pumps are characterised by extreme compactness and consequentially low space requirements. This allows the pump to be installed in otherwise inaccessible places, for example in ships or in nuclear power stations.

3.8.9 Noise level

Overall noise emission of magnetic coupling pumps is substantially higher than that of canned motor pumps. This is above all due to the electromotor (fan), the roller bearings (normally four roller bearings) and the coupling.

For example, the sound pressure level of an 11 kW standard magnetic coupling pump, size 50-200 was measured at 70 dB(A), and the sound pressure level of a canned motor pump with the same output and size at 53 dB(A).

Thus one magnetic coupling pump has the same sound pressure level as 50 canned motor pumps.

Regarding noise emission the canned motor pump is considerably more environmentally friendly while fulfilling the most stringent regulations stipulated by the authorities.

3.8.10 Cost comparison

Canned motor pumps designed for operating temperatures of up to approx. 100 °C hold a price advantage regarding initial costs if the fact is taken into consideration that a magnetic coupling pump requires a base plate, a coupling and a drive motor in order to operate.

For canned motor pump installations in hazardous locations, additional costs must be included for monitoring equipment.

More and more operators require temperature and dry-run monitoring systems in magnetic coupling pumps for increased pump reliability and availability. At the same time, however, initial costs also increase.

If a double wall magnetic coupling system is required for reasons of safety, the space between the two cans must be equipped with a continuous leakage monitoring system, also a factor in the overall acquisition costs.

3.8.11 Conclusion

Fig. 3.258 illustrates the scope of application of the two pump systems. The diagram has three axis representing the parameters: temperature, pressure and hazard potential. The inner tetrahedron represents the operating range of pumps with packing; the next larger tetrahedron represents the range of application for pumps with mechanical seals. The second largest tetrahedron represents the operating range of magnetic coupling pumps while the outer tetrahedron, representing high hazard potential, high temperatures, high pressure, or a combination of these, outlines the typical range in which canned motor pumps are used.

The decision in favour of either a magnetic coupling or canned motor pump should be based on the requirement profile. Both systems offer advantages depending on application, pressure rating, temperature, hazard degree of the pumped fluid, explosion protection and safety requirements.
Due to the more stringent health and safety regulations, it may be assumed that mechanical seals are being replaced by hermetic systems at an increasing rate; whereby

   - magnetic drive systems offer an alternative to single mechanical seals and

   - canned motor drive systems offer an alternative to double mechanical seals.

Both hermetic systems can be used to solve problem pumping applications with few and simple parts. Also both provide leakage-free, tight systems even under extreme temperatures, thus fulfilling all conditions required for environmental protection. In addition, if installed and operated properly, these pumps can render technical processes maintenance-free and safe, and also contribute to increased availability and reduced maintenance costs.

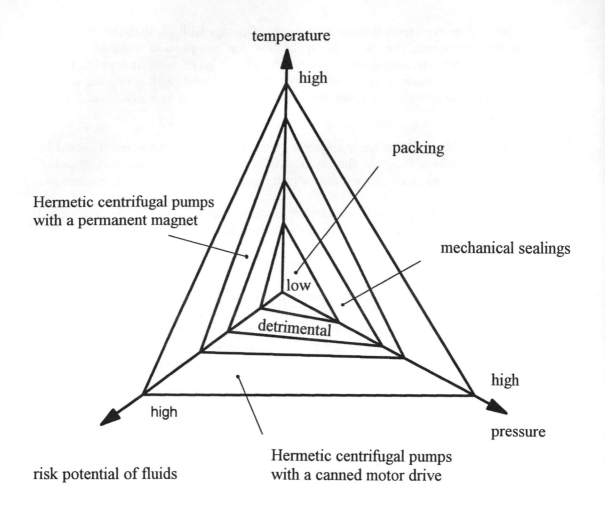

temperature

high

packing

Hermetic centrifugal pumps
with a permanent magnet

mechanical sealings

low

detrimental

high

high

risk potential of fluids

pressure

Hermetic centrifugal pumps
with a canned motor drive

Fig. 3.258 Operating parameters of pumps

3.9 Hermetic stirrers and isolating devices

Environmentally-hazardous fluids which are being pumped in circuits always cause problems if moving shafts or spindles pass out of the system. Where centrifugal or rotary displacement pumps are used, a canned motor or magnet coupling can provide absolute technical tightness. However, it serves little purpose to provide a hermetic seal for pumps, which clearly represent the most frequent causes of leaks, if the stirrer in the system and the isolating valves are sealed by stuffing boxes, radial shaft seal or mechnical sealing of some other kind which represent an "open" connection between the product and atmosphere. To completely seal the system, stirrers are provided both with canned motors (Fig. 3.259) and also magnetic couplings (Fig. 3.260, 3.26l, 3.262). Shut-off valves can also be rendered leak-free with the aid of magnetic couplings (Fig. 3.263).

With the extended applications of canned motors or magnetic couplings technical process plants can therefore be fully hermetically sealed.

Fig. 3.259 Hermetic stirrer with a canned motor drive in a technical process plant

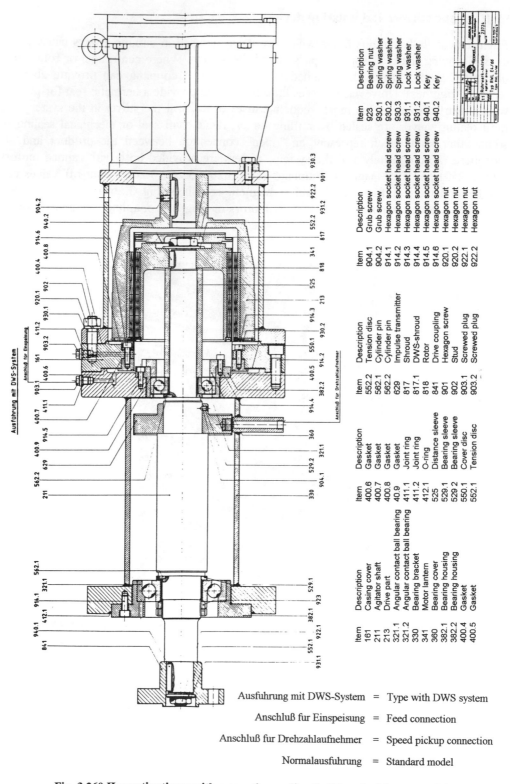

Item	Description
161	Casing cover
211	Agitator shaft
213	Drive part
321.1	Angular contact ball bearing
321.2	Angular contact ball bearing
330	Bearing bracket
341	Motor lantern
360	Bearing cover
382.1	Bearing housing
382.2	Bearing housing
400.4	Gasket
400.5	Gasket

Item	Description
400.6	Gasket
400.7	Gasket
400.8	Gasket
40.9	Gasket
411.1	Joint ring
411.2	Joint ring
412.1	O-ring
525	Distance sleeve
529.1	Bearing sleeve
529 2	Bearing sleeve
550.1	Cover disc
552.1	Tension disc

Item	Description
552.2	Tension disc
562.1	Cylinder pin
562.2	Cylinder pin
629	Impulse transmitter
817	Shroud
817.1	DWS-shroud
818	Rotor
841	Drive coupling
901	Hexagon screw
902	Stud
903.1	Screwed plug
903.2	Screwed plug

Item	Description
904.1	Grub screw
904.2	Grub screw
914.1	Hexagon socket head screw
914.2	Hexagon socket head screw
914.3	Hexagon socket head screw
914.4	Hexagon socket head screw
914.5	Hexagon socket head screw
914.6	Hexagon socket head screw
920.1	Hexagon nut
920.2	Hexagon nut
922.1	Hexagon nut
922.2	Hexagon nut

Item	Description
923	Bearing nut
930.1	Spring washer
930.2	Spring washer
930.3	Spring washer
931.1	Lock washer
931.2	Lock washer
940.1	Key
940.2	Key

Ausfuhrung mit DWS-System = Type with DWS system
Anschluß fur Einspeisung = Feed connection
Anschluß fur Drehzahlaufnehmer = Speed pickup connection
Normalausfuhrung = Standard model

Fig. 3.260 Hermetic stirrer with magnetic coupling (ball bearing) for cover drive

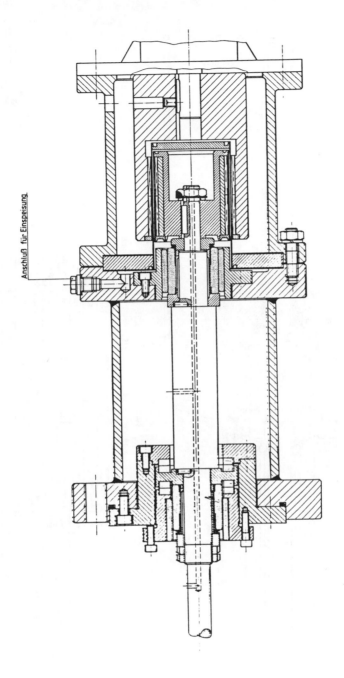

Anschluß fur Einspeisung   = Feed connection

Fig. 3.261 Hermetic stirrer with magnetic coupling (plain bearing)

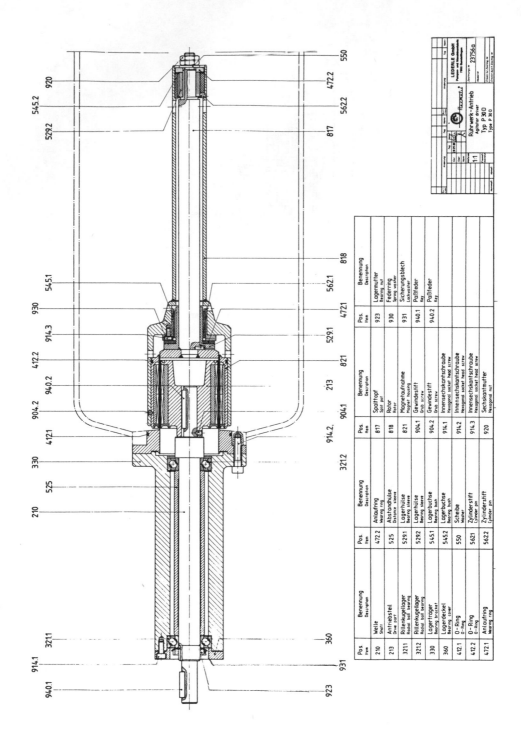

Pos. Item	Benennung Description	Pos. Item	Benennung Description	Pos. Item	Benennung Description	Pos. Item	Benennung Description
210	Welle Shaft	472.2	Anlaufring Wearing ring	817	Spalttopf Split pot	923	Lagermutter Bearing nut
213	Antriebsteil Drive part	525	Abstandhülse Distance sleeve	818	Rotor Rotor	930	Federring Spring washer
3211	Rillenkugellager Radial ball bearing	529.1	Lagerhülse Bearing sleeve	821	Magnetaufnahme Magnet housing	931	Sicherungsblech Lockwasher
3212	Rillenkugellager Radial ball bearing	529.2	Lagerhülse Bearing sleeve	904.1	Gewindestift Grub screw	940.1	Paßfeder key
330	Lagerträger Bearing bracket	545.1	Lagerbuchse Bearing bush	904.2	Gewindestift Grub screw	940.2	Paßfeder key
360	Lagerdeckel Bearing cover	545.2	Lagerbuchse Bearing bush	914.1	Innensechskantschraube Hexagonal socket head screw		
412.1	O-Ring O-Ring	550	Scheibe Washer	914.2	Innensechskantschraube Hexagonal socket head screw		
412.2	O-Ring O-Ring	5621	Zylinderstift Cylinder pin	914.3	Innensechskantschraube Hexagonal socket head screw		
472.1	Anlaufring Wearing ring	5622	Zylinderstift Cylinder pin	920	Sechskantmutter Hexagonal nut		

Fig. 3.262 Hermetic stirrer with magnet coupling for bottom drive

384

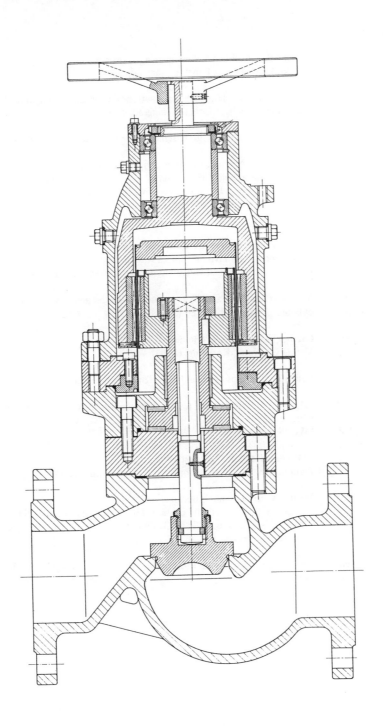

Fig. 3.263 Hermetic shut-off valve with magnet coupling

References for Chapter 3

[3-1] Neumaier, R.:   Handbuch neuzeitlicher Pumpenanlagen (Manual of modern pump systems)
                              Verlag Alfred Schütz Lahr/Schwarzwald 1966

[3-2] Tscherenousav,   Germetischeskije chimiko-technologitscheskije maschiny i aparaty Verlag
       N.P.;   Maschinostrojenije, Moskau-Leningrad, 1965
       Kutin, V.F.:
[3-3] AEG   Allgemeine Elektrizitätsgesellschaft Werk Esslingen, Company News 1966

[3-4] Schilling, R.;   Strömung und Verluste in drei wichtigen Elementen radialer Kreiselpumpen -
       Siegele, H.;   eine Literaturübersicht (Flow and losses in three important elements of radial
       Stoffel, B.:   centrifugal pumps, literature overview) Report of the Institute of Science of Flow
                              and Flow Machines, University (Technical University) of Karlsruhe 16th July
                              1974

[3-5] STAMPA, B.:   Experimentelle Untersuchungen an axial durchströmten Ringspalten
                              (Experiments on radial clearances with axial flow)
                              Dissertation, Technical University Braunschweig (1971)

[3-6] KAYE, J. and   Modes of adiabatic and diabatic fluid flow in an annulus with an inner rotation
       E.C. ELGAR: cylinder. Trans. ASME, 80 (1958), 753.

[3-7] Schlichting, H.:   Grenzschicht- Theorie, (Boundary layer theory)
                              Verlag G. Braun, Karlsruhe (1964)

[3-8] KSB   Centrifugal Pumps Lexicon Klein-Schanzlin and Becker AG, Frankenthal 1980

[3-9] Allaire, E. Paul;   "Design, construction and test of magnetic bearings in an industrial canned motor
       Imlach, J.;   pump
       McDonald, J.P.:   "World Pump, September 1989"

[3-10]   VDI Guideline 2204, August 1968
                              Plain bearing calculations for hydrodynamic plain bearings subject to stationery
                              load
                              Beuth - Vertrieb GmbH, Berlin and Cologne
[3-11] Taylor G.I.:   Stability of a viscous liquid contained between two rotating cylinders Phil. Trans.
                              Series A., vol. 223, p. 289 (1923) Fluid Friction between rotating cylinders,
                              Proc.roy.Soc.A, vol. 157, p. 546 (1936)

[3-12] Constantinescu   On turbulent Lubrication Proc. Instn. Mech. Engrs. vol. 173, Nr.38, p. 881
       V.N.:   (1959)

[3-13] Freese H.D.:   Querkräfte in axialdurchströmten Drosselspalten (Transverse forces in throttling
                              clearances with axial flow)
                              Pump Conference, Karlsruhe '78 Sect. K 6 (1978)

[3-14]   Krämer, R.;   Kreiselpumpen und rotierende Verdrägerpumpen hermeticscher Bauart (Hermetic
         Neumaier, R.:   centrifugal pumps and rotating displacement pumps) Company Literature.
                             Lederle-Hermetic GmbH 1986

[3-15   Lehmann, W.;   Untersuchung der Geräuschemission von Kreiselpumpen in Abhängigkeit der
         Mehlhorn, P.:   konstruktiven Ausführungen und Betriebsbedingungen (Investigation of noise
                             emission of centrifugal pumps relative to design and operating conditions).
                             Pump Conference, Karlsruhe '78, Sect. K 9 Noise

[3-16]   Gordzielik, H.:   Versuch zur Vorherbestimmung der Schalleistungspegel von einstufigen
                             Kreiselpumpen und -aggregaten (Test to predetermine the sound power level of
                             single-stage centrifugal pumps and sets) Pumpen und Verdichten (Pumps and
                             Compressors), Information I (1975)

[3-17]   Krämer, R.;   Entwicklung und Bau einer Hermetic-Umlaufpumpe für hohen Systemdruck
         Neumaier, R.;   (Development and construction of a hermetic recirculation pump for high system
         Schließer, K.:   pressures) Verfahrenstechnik (Process Technology) 6 (1973)

[3-18]   Neumaier, R.:   Hermetische Kreiselpumpen in Hochdruckausführung (High-pressure hermetic
                             centrifugal pumps) Pump Conference, Karlsruhe '73, Sect. K 7 Centrifugal
                             Pumps (1973)

[3-19]   Dittmer, H.;   Weiterentwicklung und Betriebserfahrungen mit einer Spaltrohrmotorpumpe
         Rejman, A.;   unter Druckwasserreaktorbedingungen (Further development and operating
         Reinecker, H.:   experience with a canned motor pump under pressurized water reactor
                             conditions))Pump Conference, Karlsruhe '78, Sect. A Canned Motor Pumps.

[3-20]   Pihowicz, W.:   Entwurf, Konstruktion und Sicherheit von Kernenergieschiffsantrieben (Design,
                             construction and safety of nuclear energy propulsion systems for ships)
                             Papers by the Association of Friends and Supporters of The GKSS
                             Forschungszentrums Geesthacht e.V. (GKSS Research Centre Geesthacht), 1989

[3-21]   Neumaier, R.:   Seitenkanal-Strömungsmaschinen (kollektiv) Leckfreie Seitenkanalpumpen mit
                             Spaltrohrmotor-Antrieb (Side channel flow machines (collective), leak-free side
                             channel pumps with canned motor drive)
                             Verlag und Bildarchiv W.H. Faragallah Sulzbach / Ts 1992

[3-22]   Hanibal GmbH, Düsseldorf, Works Literature

[3-23]   Hilsebein, G.:   Vacuumvorlagen für nicht selbstansaugende Kreiselpumpen (Vacuum
                             specifications for non self-priming centrifugal pumps)
                             Chem. Technik (Chemical Technology), year 17, vol. 4, April 1965

[3-24] Grabow, G.:   Seitenkanal-Strömungsmaschinen (Kollektiv) Übersicht über die Theorien zur Klärung der Strömung von Seitenkanalmaschinen (Side channel flow machines (collective). Overview of theories for explanation of flow in side channel machines)
Verlag und Bildarchiv W.H. Faragallah Sulzbach / Ts 1992

[3-25] Surek, D.:   Kenngrößen von Peripheralradpumpen (Characteristic variables of regenerative pumps) Maschinenbautechnik (Mechanical Engineering Technology) Berlin 39 (1990) 5

[3-26] Pfleiderer, C.:   Die Kreiselpumpen für Flüssigkeiten und Gase (Centrifugal pumps for liquids and gases)
Springer Verlag Berlin/Göttingen/Heidelberg 1955

[3-27] Faragallah, W.H.: Seitenkanal-Strömungsmaschinen (Kollektiv) Peripheralradpumpen (Side channel flow machines (collective). Regenerative pumps)
Verlag und Bildarchiv W.H. Faragallah Sulzbach/Ts 1992

[3-28]   Keystone-Yarway-Produkte Mönchengladbach Low pressure automatic recirculation valve, Company Literature
[3-29])   Hermetic-Pumpen GmbH, Gundelfingen (Works Literature

[3-30])   Hermetic-Pumpen GmbH, Gundelfingen (Works Literature

[3-31] Neumaier, R.:   Spaltrohrmotorpumpen - ein wesentlicher Beitrag zur leckfreien Förderung (Canned motor pumps, a substantial contribution to leak-free delivery) (aus leckfreien Pumpen and Verdichter (leak-free pumps and compressors)
Publisher G. Vetter, Vulkan- Verlag Essen 1992)

[3-32]   E. Dold + Söhne KG; Furtwangen, Works Literature

[3-33] Lehmann, W.:   Leckfreie Pumpen und Verdichter (Kollektiv). Eigenschaften und Auslegungskriterien von Magnetantrieben (leak-free pumps and compressors (collective). Properties and design criteria of magnet drives)
Publisher G. Vetter, Vulkan- Verlag Essen 1992

[3-34] Krämer, R.;   Die doppelwandige Sicherheit (Double-wall safety)
        Neumaier, R.:   Chemische Industrie (Chemical Industry), vol. 3, March 1987

[3-35] Krämer, R.:   Gegenüberstellung Spaltrohrmotor-Magnetantrieb Leckfreie Pumpen und Verdichter (Comparison of canned motor pumps and magnet drive leak-free pumps and compressors)
Publisher G. Vetter, Vulkan- Verlag Essen 1992

Part III

Rotary displacement pumps

Basic Principles

4. Rotary displacement pumps

4.1 Introduction

The third part of this book which covers leak-free rotary displacement pumps, first deals with conventional pumps of this type. The author considers this appropriate because relatively little information is given in technical literature on this subject, particularly on rotary piston and helical vane pumps. This provides a better basis for understanding the other information.

The high level of importance which is nowadays attached to the energy consumption and environmental aspects of the operation of industrial plant and machinery is on a par with considerations of functionability. The selection of fluid machinery which satisfies these criteria thus represents a particular challenge to the planning engineer. As long as the fluids in question are watery fluids of low viscosity, centrifugal pumps of various designs are used accordingly, unless the specific speed falls below $n_q = 10$ r.p.m. and the nature of the delivery product does not permit centrifuging or any degree of speed fluctuation due to its admixtures/inclusions or shearing sensitivity.
As viscosity increases, the centrifugal pump's power demand also shaply Increases, such that (from approximately 300 mPa·s upwards), the pumping duty can be handled better and more cost-effectively by hydrostatic (positive displacement) pumps. For this purpose, we may resort to two different types of pumping systems: oscillating pumps and rotary displacement pumps. Oscillating pumps are mainly used for solving metering and pumping problems in the high-pressure range, whilst rotary pumps are employed for transport and circulation duties.

4.2 Rotary piston pumps [4.1]

In the class of rotary positive-displacement pumps, the rotary piston pump is a type which nowadays covers a wide spectrum as regards both the delivery product and potential delivery parameters. Nowadays these pumps are mainly used in the chemical, petrochemical, cosmetic, foodstuffs, paper processing and bitumen processing industries. Their ability to continuously deliver single and multi-phase mixtures with and without solid ingredients or gas inclusions as well as fibrous admixtures in high concentration opens up a wide area of application for pumps of this type. Within this "rotary piston pumps" family, moreover, there is a seemingly infinitely wide variety of displacer shapes in terms of geometry and kinematics. Again and again, inventors have been inspired to create new types of displacer. This is an indication that this type of pump has been "thought-out" to a degree uncommon among operational machines. In his description of the "Classification of the Rotary Piston Machine", Wankel lists no fewer than 332 possible designs for parallel-axis rotary piston machines with working chamber walls made of rigid materials.

It is understandable that not all of these designs have been able to make their mark - neither would this be reasonable in pure economical terms. The following will therefore only deal with displacer systems which have proved themselves to be expedient and economical as regards their use and their manufacture. First, however, we shall take a short look at the history of the development of the rotary piston pump.

4.2.1 History

Our knowledge of the progress of technical development up to the beginning of the 15th century is fairly vague. It is only the invention of printing that gave us access to reliable documents reporting the invention of machines and appliances and this enables us to trace the history of the rotary piston pump up to the present. The Italian engineer and mathematician Agostino Ramelli, who published the first drawings of these machines in his book "Del arteficiose macchine" in 1588 in Geneva which he dedicated to his supreme commander Henry III of France, is generally regarded as the inventor of the rotary piston machine. In his book he describes his rotary piston water pumps (Fig. 4.1).

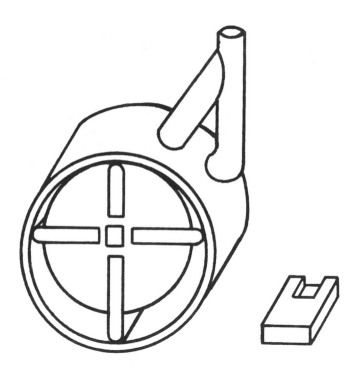

Fig. 4.1 Rotary piston water pump designed by Agostino Ramelli in 1588

It was, however, to be quite a while before the theory was put in practice, for the necessary technical production conditions as regards manufacturing accuracy, sealing and pipeline problems were not present at the time. Around 1600, the astronomer Johannes Kepler (Fig. 4.2) invented the gear pump. In 1602 he commissioned an Augsburg fountain builder to manufacture a pump model, but this was not satisfactory (Fig. 4.3).
A further model made by the Prague watchmaker Jost Buergi in 1604 is described as an "amusing little fountain" which produces a water jet "of quite some height". Kepler was

392

not however able to find a commercial use for his pump. On the other hand, literature on the subject names the German engineer Pappenheim as the inventor of the gear pump; in 1636 he created a fully functional pump which is shown in Fig. 4.4 .

The rotary piston pump as we know it today, however, was not invented until 1867, by Behrens (Fig. 4.5), when industrialisation made it possible to achieved the necessary production tolerances. During the period of rapid industrial expansion in Germany at the beginning of the 1870s, several inventors took an interest in this pump type. They produced numerous usable models, such as

Fig. 4.2 Johannes Kepler (1571-1630)

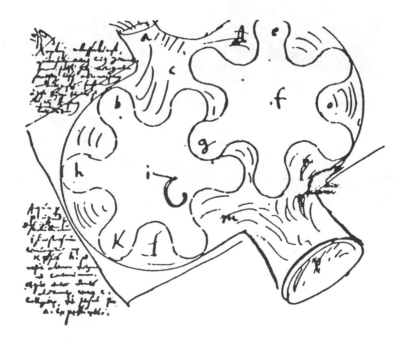

Fig. 4.3 Gear pump designed by Johannes Kepler

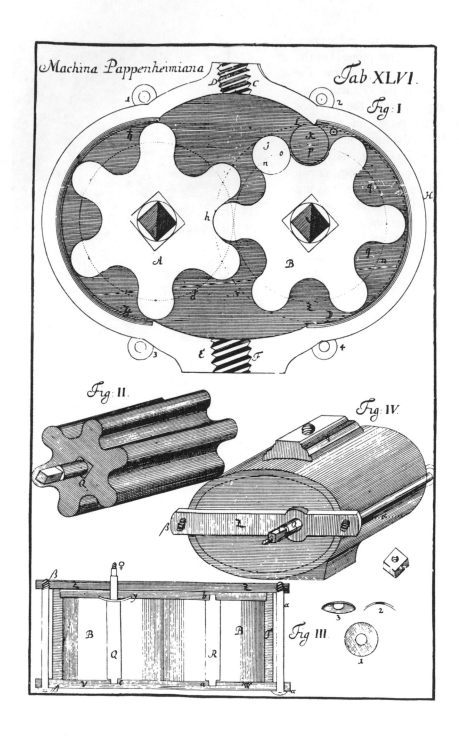

Fig. 4.4 Gear pump designed by Pappenheim 1636

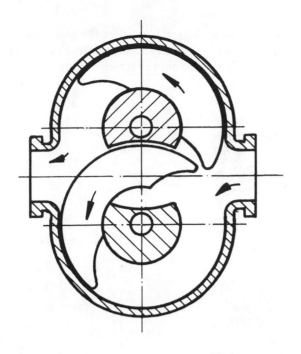

Fig. 4.5 Rotary piston pump designed by Behrens in 1867

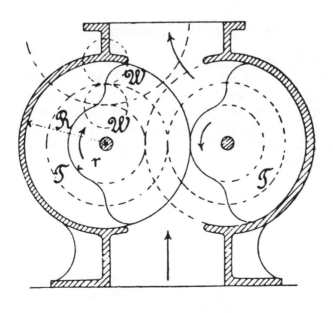

Fig. 4.6 Rotary piston pump designed by Repsold

395

the one designed by <u>Repsold</u> (Fig. 4.6), which , like the pump designed by Behrens, has two single-toothed impellers rotating in one casing, the tooth (piston) points being rounded. <u>Henry</u> (Fig. 4.7) also used single toothed impellers, but with pointed tooth (piston) profiles.

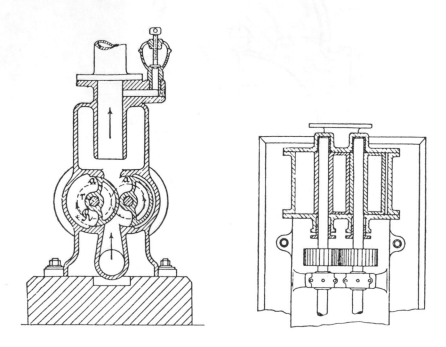

Fig. 4.7 Rotary piston pump designed by Henry

<u>Root</u> (Fig. 4.8) on the other hand, used double vaned wheels as displacers. The tooth profile line is in the shape of an arc circle and designed to correspond to that of the tooth base; alternatively, this theoretical design is made up to several circle sections to facilitate production and the tip of the tooth is not fully formed, solely the summit. <u>Greindl</u> (Fig. 4.9) created yet another tooth (piston) shape using double-toothed (nowadays we say double-vane) displacers with pointed profiles. This construction proved to be eminently functional and soon made its mark.

After practicable models had been developed, industry took an interest in them and refined them until they were suitable for series production.

In the 80s of the previous century, <u>Lederle</u> in Freiburg (Fig. 4.10) adopted Greindl's fluid delivery principle and developed it further. During the same period, similar types of rotary piston pump were manufactured by <u>Enke</u> in Schkeuditz near Leipzig, by the <u>Peninger-Maschinenfabrik</u> in Peningen, and by <u>Jäger</u> in Leipzig-Plagwitz. However, all of these companies used different displacer profiles. From this time onwards, the rotary piston pump became one of the many operating machines for fluid conveyance alongside piston pumps and gear pumps and maintained this position until the turn of the century, the time when the centrifugal pump began to be used in industrial and municipal operations in significant numbers. This was made possible by the simultaneous development of the electric motor, which allowed the use of direct drive systems.

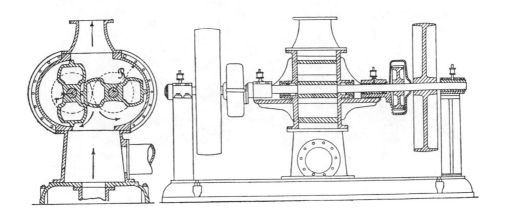

Fig. 4.8 Rotary piston pump designed by Root

The following translation of an excerpt from a catalogue of the company Motoren- und.

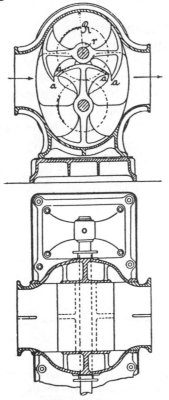

Pumpenfrabrik W. Lederle, Freiburg, Baden, gives some idea of the way in which rotary piston pumps were generally handled at the turn of the century (Fig. 4.11).

Although over the last hundred years numerous other systems of rotary displacement pumps together with centrifugal pumps have covered an increasingly wide range of ever more varied pumping tasks, this rotary piston pump still has its own specific application and optimum areas of use as will be seen in later discussions.

Fig. 4.9 Rotary piston pump designed by Greindl/Enke

397

Fig. 4.10 Rotary piston pump designed by Lederle in 1890

W.Lederle, Motoren-u. Pumpenfarbik, Freiburg, Baden

Installation and handling
of rotary piston pumps

In order to ensure the long life of the pump, it is recommended that the interior of the pump be lubricated perhaps once a month with tallow mixed with a tenth part pulverised graphite. The best way to do this is to fill the pump with hot water until it is warm to the touch. Then drain off the water, melt the appropriate quantity of tallow mixed with the corresponding amount of said graphite. This mixture is then poured in via the top metal screw. First of all, however, the rotary piston of the pump should be set to a position which allows a corresponding amount of the melted tallow to be poured in by turning the pulley. This procedure totally prevents rust formation in the interior of the pump and ensures that the latter will have long service life. This lubrication is especially recommended for pumps which are only operated at intervals: in particular if the pump is to be out of operation for any length of time.

Fig. 4.11   Installation and operating instructions from the year 1900 for rotary piston pumps by the Motoren- und Pumpenfabrik W. Lederle

4.2.2   Rotary piston pumps of new design, operating principle and design of the displacers

4.2.2.1   Operating principle

The twin shaft parallel axis design is common to all rotary piston pumps. The shafts are moved against one another by an external sync gear. The displacers fitted to the shafts roll against one another with low axial and radial clearance in an encapsulating casing (i.e. also in opposite directions to one another and at the same angular velocity) without coming into contact with each other or the casing wall. By this means, they create the pressure required for fluid delivery. The displacers are designed in such a way that they perform a block-sealing function in all positions and thus seal suction and pressure chambers off from each other. At the casing circumference this sealing is known as <u>working sealing point</u> and as <u>block-sealing point</u> at the point at which the displacer acts at the piston hub (Fig 4.12).

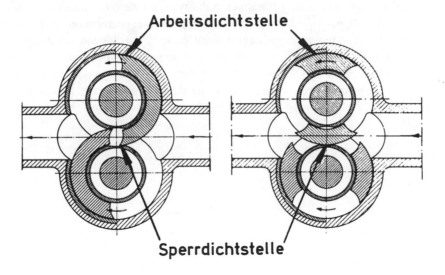

Arbeitdichtstelle   =   working sealing point
Sperrdichtstelle   =   block-sealing point

Fig. 4.12 Working sealing and block-sealing point

During the rolling action of the displacers against one another the profile gaps in the suction chamber are filled with the delivery product due to the atmospheric or tank overpressure acting on the suction fluid level. The fluid is then carried along in a circumferential direction and drained into the discharge line on the discharge side by the immersion of the counter vane (Fig. 4.13).

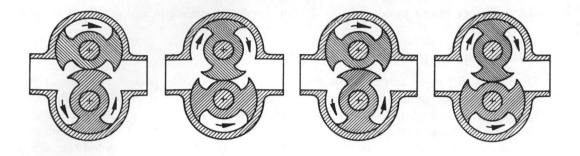

Fig. 4.13 Operating principle of a double vane rotary piston pump

This process creates an even flow from the suction to the pressure line so that surge tanks and valves are not required. The load on the fluid is low as the fluid is transported almost pulsation-free, without acceleration or delay, and at a low shear rate. In many cases, this is the precondition for maintaining the inner stability of the delivery product. The achievable delivery pressures ($p_2 - p_1$) are not a function of the conversion of flow energy into pressure energy as is the case with centrifugal pumps, but depend on the pumping head resulting from the pipeline friction losses and the pressure of the connected plant. In comparison to centrifugal pumps the flow rate of rotary piston pumps is relatively insensitive to fluctuations in pressure. The maximum possible delivery pressure is limited by the mechanical load capacity of the pump parts, the losses of the internal gap flow and the occurring gap cavitation.

4.2.2.2 Design of the displacers

Of the large number of displacer designs which have been developed, only a few have been deemed suitable for series production. Their geometric shape depends mainly on the delivery task in question, e.g. the type of delivery product (mechanically pure or with admixtures, high or low viscosity, fibrous, with gas inclusions, shear sensitivity), the delivery pressure (high or low pumping heads), the suction characteristics (the realisation of NPSHR values which are as low as possible) and the admissible pulsation. A basic distinction must be made between rotary piston pumps with rotating and those with fixed shaft or casing hubs, and the pumps in each group must be classified according to the number of rotor vanes.

4.2.2.3 Rotary piston pumps with rotating hub

4.2.2.3.1 Single vane design with rotating hub (Fig. 4.14)

Two single vane pistons rotate in opposite directions in an encapsulating casing. The working sealing point can be sufficiently dimensioned (surface seal), the control sealing point as a linear seal both in the displacer vane - piston hub position and in the control change position. Rotary piston pumps with these types of displacers are mainly suitable for transporting extremely high viscous fluids at low speeds (max. 350 r.p.m.). The delivery pressure ($p_2 - p_1$) is limited to 6 bar.

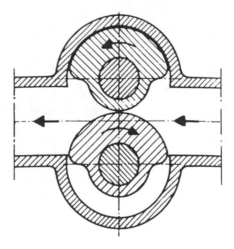

Fig. 4.14 Single vane design with rotating hub

4.2.2.3.2 Double vane design (Fig. 4.15)

Two rotor vanes are situated on the piston hubs symmetrical to the axis. In this case also

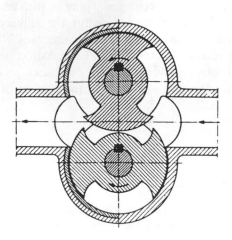

Fig. 4.15 Double vane design with rotating hub

the working sealing point is in the form of a surface seal, and the block-sealing point designed as a line seal. Due to the symmetry of the displacer vanes, however, this pump can be operated at higher speeds than those prescribed. It is suitable for transporting highly and slightly viscous fluids with and without mechanical admixtures, and of small grain size. The maximum delivery pressures are in the region of $(p_2 - p_1) = 8$ bar, the maximum speed is n = 500 r.p.m. The relatively simple design of this machine has made it one of the most commonly used of its kind.

4.2.2.3.3 Multi-vane (gear vane) design (Fig. 4.16)

At first glance this version would seem to be a normal gear pump; in reality, however, it is a multi-vane rotary piston pump, for the tooth-like displacers are in the engagement area but not in contact. The torque to be transferred from the drive unit is taken by the sync gear outside the pressure chamber. This makes it possible to convey non-lubricating fluids as well as to manufacture the tooth piston from austenitic Cr-Ni-Mo steel - a considerable advantage as regards the use of such pumps in the chemical industry. The delivery principle is the same as that of gear pumps. Two or more teeth on

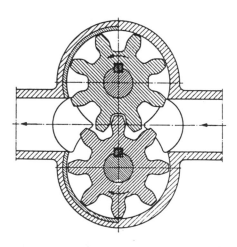

Fig. 4.16 Multi-vane (gear vane) piston

the rotating piston are always in the engagement area and not in contact When the wheel piston is rotated, the teeth in the engagement area emerge from the corresponding gaps in the counter-wheel. An underpressure area is formed at these points into which the delivery product flows through the connected suction line. The filled tooth gaps are moved to the pressure side by rotation and there the teeth of the counter-wheel entering these gaps force the fluid into the discharge line. With the relatively low degree of edge clearance of the vane teeth it is necessary to create compensation pockets in the form of recesses in the casing front surfaces for the "squeezed fluid" enclosed between the tooth head and the tooth base of the counter-wheel to allow the fluid to escape. The creation of such recesses is particularly important where the fluids being transported are highly viscous. Another way of ensuring that the delivery process is "squeeze-free" is to select the tooth shape which ensures that the edge clearance is such that only one pair of edges is in opposition on the profile front at sealing distance, while ensuring greater clearance on the profile rear relative to the counter-wheel. In this version, however, the direction of rotation, and thus also the direction of flow, is fixed. It can only be altered by repositioning the tooth vane pistons (axially by 180°).

4.2.2.4 Rotary piston pumps with fixed hub

4.2.2.4.1 Single vane design (Fig. 4.17)

Two single vane pistons rotate around separate fixed casing hubs equipped with circular hollow sections within an encapsulating casing. This design ensures that there is no engagement of the two vanes at the block-sealing point. Rather, surface sealing along a sufficiently long arc section of the counter hub ensures that no degree of gap loss occurs at this point - a point which is critical in all rotary piston pumps. The smaller gap reverse flow due to the considerably higher sealing quality during the entire rotation provides these machines with a far superior volumetric efficiency and ensures pulsation-free delivery. Moreover, it is possible to achieve greater pumping heads with this system (up to 25 bar). In addition, the NPSH value of the pump is considerably improved. The "squeeze-free" mode of delivery permits the delivery of fluids with high admixture content and of a

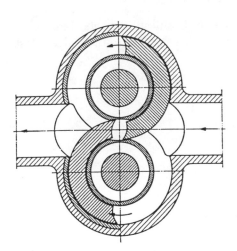

Fig. 4.17 Single vane design with fixed hub

fibrous product (up to 20% abs. dry), as in this case the delivery fluid is not taken in "wedge" fashion as in the case of the versions with rotating hubs. The single vane version of this type is also suitable for extremely high viscosities (over 1000 pa · s).

4.2.2.4.2 Double vane design (Fig. 4.18)

What was written in Section 4.2.2.4.1 basically applies here as well. The advantage of the double vane design over the single vane system lies in the symmetrical arrangement of the piston vanes relative to the rotational axis. This makes it possible to operate these types of pumps at high speeds and thus with lower unit weight (advantage in terms of price).

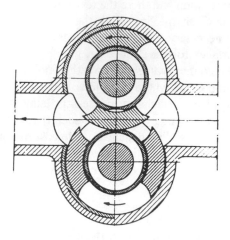

Fig. 4.18 Double vane design with fixed hub

4.2.2.5   Displacer shapes according to delivery characteristics

The following table [4-I.1 to 4-I.6] provides a classification of the best suited and at present most commonly manufactured piston forms according to the delivery task in question.

Suitable displacer form depending on the composition of the fluid in the viscosity range **<u>1 - 200 mm²/s</u>**	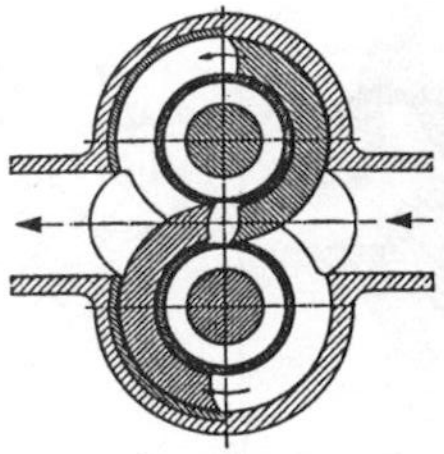	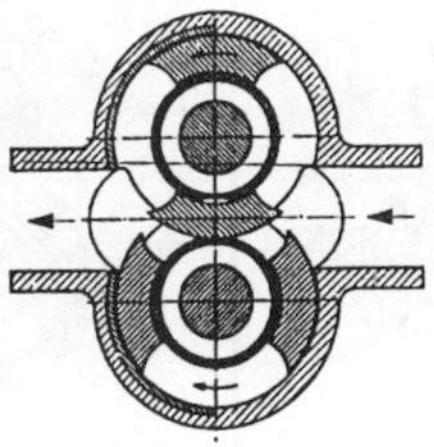	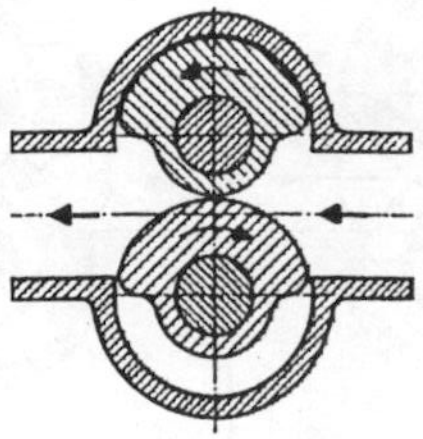	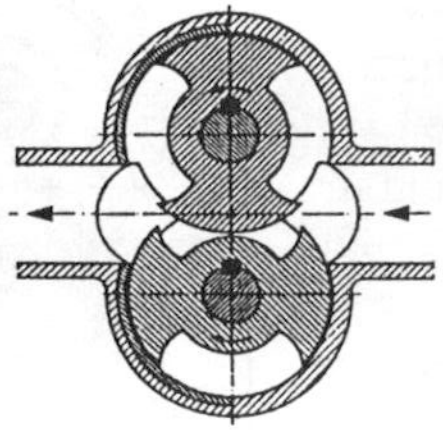	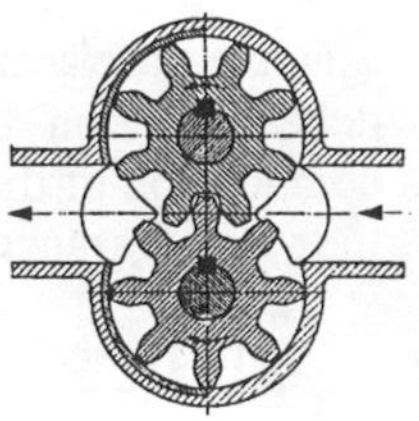
Type of fluid	**single-vane with fixed hub**	**double-vane with fixed hub**	**single-vane with rotating hub**	**double-vane with rotating hub**	**gear piston**
mechanically pure	●	●	●	●	●
with solid particles in the µm range, soft	●	●	●	●	●
with solid particles in the µm range, hard	●	●	●	●	
with solid particles up to serval mm diameter, soft	●	●	●	●	
with solid particles up to serval mm diameter, hard	●	●			
fibrous with low consistency	●	●			
fibrous with low consistency	●	●			
with low gas content	●	●	●	●	●
with high gas content	●	●		●	●

Table 4 - I.1 Displacer shapes and their fields of operation

Suitable displacer form depending on the composition of the fluid in the viscosity range 200 - 10.000 mm²/s

Type of fluid	single-vane with fixed hub	double-vane with fixed hub	single-vane with rotating hub	double-vane with rotating hub	gear piston
mechanically pure	●	●	●	●	●
with solid particles in the μm range, soft	●	●	●	●	
with solid particles in the μm range, hard	●	●	●	●	
with solid particles up to serval mm diameter, soft	●	●	●	●	
with solid particles up to serval mm diameter, hard	●	●			
fibrous with low consistency	●	●			
fibrous with low consistency	●		●		
with low gas content	●	●	●	●	●
with high gas content		●		●	●

Table 4 - I.2   Displacer shapes and their fields of operation

Suitable displacer form depending on the composition of the fluid in the viscosity range <u>10.00 - 100.000 mm²/s</u> / Type of fluid	single-vane with fixed hub	double-vane with fixed hub	single-vane with rotating hub	double-vane with rotating hub	gear piston
mechanically pure	●	●	●	●	●
with solid particles in the μm range, soft	●	●	●	●	●
with solid particles in the μm range, hard	●	●	●	●	●
with solid particles up to serval mm diameter, soft	●	●	●	●	
with solid particles up to serval mm diameter, hard	●	●			
fibrous with low consistency	●	●			
fibrous with low consistency	●				
with low gas content	●	●	●	●	●
with high gas content		●		●	●

Table 4 - I.3 Displacer shapes and their fields of operation

Table 4 - 1.4 Displacer shapes and their fields of operation

Suitable displacer form depending on the composition of the fluid in the viscosity range above 100.000 mm²/s — Type of fluid	single-vane with fixed hub	double-vane with fixed hub	single-vane with rotating hub	double-vane with rotating hub	gear piston
mechanically pure	●	●	●	●	●
with solid particles in the μm range, soft	●	●	●	●	●
with solid particles in the μm range, hard	●	●	●	●	●
with solid particles up to serval mm diameter, soft	●	●	●	●	
with solid particles up to serval mm diameter, hard	●	●	●	●	
fibrous with low consistency	●	●			
fibrous with low consistency	●	●			
with low gas content	●	●	●	●	●
with high gas content	●	●	●	●	●

Suitable displacer form depending on the type of delivery task	single-vane with fixed hub	double-vane with fixed hub	single-vane with rotating hub	double-vane with rotating hub	gear piston
Delivery pressure $p_2 - p_1$ up to 4 bar	●	●	●	●	●
Delivery pressure $p_2 - p_1$ up to 6 bar	●	●		●	●
Delivery pressure $p_2 - p_1$ up to 10 bar	●	●			●
Delivery pressure $p_2 - p_1$ up to 20 bar	●	●			●
Speeds up to 100 rpm	●	●	●	●	●
Speeds from 100 - 400 rpm	●	●	●	●	●
Speeds above 400 rpm		●		●	●
NSPHA (plant) low	●	●			
NSPHA (plant) medium	●	●			●
NSPHA (plant) high	●	●	●	●	●

Table 4 - I.5 Displacer shapes and their fields of operation

409

Table 4 – I.6 Displacer shapes and their fields of operation

Suitable displacer form depending on the specific properties of the delivery fluid / Type of fluid	single-vane with fixed hub	double-vane with fixed hub	single-vane with rotating hub	double-vane with rotating hub	gear piston
thixotropic	•	•			
shear-sensitive	•	•	•	•	
prone to coagulation	•	•			
dispersion	•	•			•
admissible pulsation - low		•			•
admissible pulsation - medium	•	•	•	•	•
admissible pulsation - high	•	•			•

4.2.3 The characteristic curve (Fig. 4.19 and 4.20)

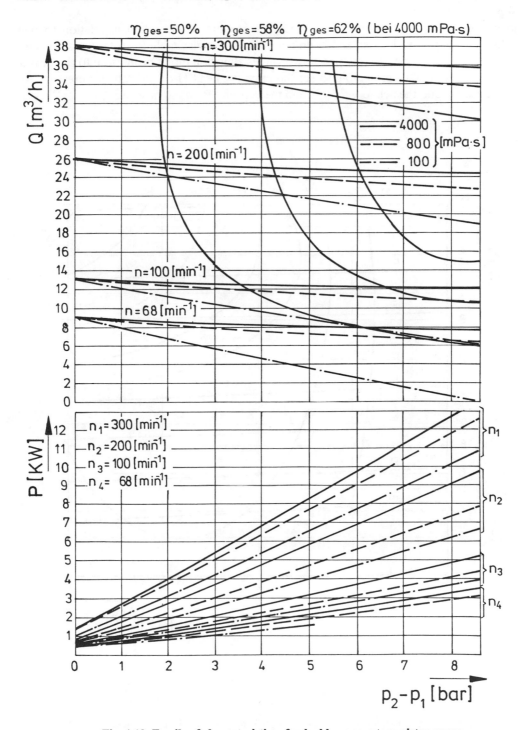

Fig. 4.19 Family of characteristics of a double vane rotary piston pump

In technical flow terms volume flow and pressure difference (p_2 - p_1) determine the operating point of the pump.

The normal method of portraying the main parameters of the pump in relation to one another, i.e. Q on the abscissa and H on the ordinate, is unsuitable for rotary position displacement pumps as the delivery pressure changes almost in direct proportion to the power consumption. It is thus preferable to set the pumping head H or the delivery pressure Δp on the abscissa and the delivery flow Q and the capacity P on the ordinate.

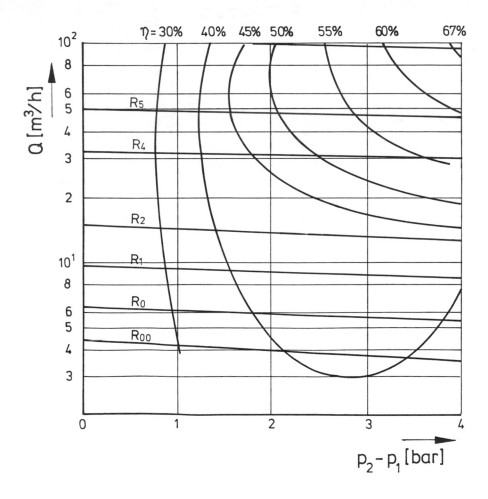

Fig. 4.20 Characteristic curves of a single vane pump (Industriewerk Chemnitz)

4.2.3.1 Delivery flow Q

The theoretical working volume V_u per rotation is the same for all rotary piston pumps and is achieved at a delivery pressure $(p_2 - p_1) = 0$.
It is

$$V_u = \tfrac{\pi}{4} \cdot [D^2 - d^2] \cdot L \qquad\qquad (4 - 1)$$

where   D = displacer diameter
         d = hub diameter
         L = axial length of the displacer

The theoretical delivery flow Q_{th} is then:

$$Q_{th} = V_u \cdot \frac{n}{60} \qquad l \cdot s^{-1} \ \ bei \ n = r.p.m \qquad\qquad (4 - 2)$$

The actual effective delivery flow Q is reduced by the amount of the internal losses caused by the required piston clearance and the possible incomplete filling of the vane cells, and is:

$$Q = Q_{th} - Q_{sp} \qquad\qquad (4 - 3)$$

where Q_{sp} represents the loss flow. It can be expressed by the volumetric efficiency η_v. It is formed from the quotients of the difference $(Q_{th} - Q_{sp})$ and Q_{th}.

$$\eta_v = \frac{Q_{th} \cdot Q_{sp}}{Q_{th}} \qquad\qquad (4 - 4)$$

and the actual effective delivery flow is thus:

$$Q = Q_{th} - \eta_v \qquad\qquad (4 - 5)$$

This requires that the suction side is in a cavitation-free operating state.
The gap losses impair the overall efficiency of the pump, which is made up of the hydraulic η_h, mechanical η_m and volumetric η_v efficiencies.

It is therefore:

$$\eta_{ges} = \eta_h \cdot \eta_m \cdot \eta_v \qquad\qquad (4 - 6)$$

The hydraulic efficiency η_h depends particularly on the viscosity of the fluid. In contrast to centrifugal pumps there is no intentional conversion of velocity to pressure energy so that

413

no pressure conversion losses occur. The mechanical efficiency is expressed in terms of friction losses from bearings and shaft seals.

The flow rates of today's rotary piston pumps range from 1 to 6000 l/min.

4.2.3.2 Pumping head H or delivery pressure p

Hydrostatic operating machines, of which the rotary piston pumps is one, possess the property of always generating the pressure required from them by the plant at the respective working volume. Theoretically this would mean that the H(Q) line is a line parallel to the abscissa in the H(Q) diagram, Q changing in proportion to the speed. Due to the internal losses, however, Q is in a way dependent on H, since with an increase in pumping head the effective delivery flow Q falls in accordance with the volumetric efficiency η_v. η_v drops with increasing pumping head and falling viscosity and speed. It can reach the value of zero, thus causing the total delivery flow volume to flow back to the suction side via the gap. Increasing viscosity and speed in the permissible range thus improve the volumetric efficiency η_v. In order to prevent uncontrolled pressure increase in rotary piston pumps therefore, safety valves, rupture disks or automatic pressure shut-off devices should be fitted in the pressure lines. Torque limited couplings can also be used for overload protection. There are also cases in which overpressure valves are used to meter the delivery flow in order to achieve a specific take-off quantity. This is permissible for short-term operation by effecting a return flow to the suction side. In the case of permanent operation, however, the consequent heating-up of the delivery product must be taken into account and it is advisable to route the return flow to the suction tank.

4.2.3.3 Power P

In the characteristic curve, the specifications are always plotted over the delivery head as coupling power. The required drive power P is a function of the flow volume Q and thus also of the speed n, the pumping head H and the overall efficiency rate η_{ges}. This in turn results from the mechanical, hydraulic and volumetric efficiencies.

The hydraulic efficiency η_h also includes the losses from the power transfer caused by the lateral friction and shear work of the vane pistons on the fluid. The delivery capacity P_Q is calculated as follows:

$$P_Q = \rho \cdot g \cdot Q \cdot H \qquad (4 - 7)$$

where

$$\begin{aligned}
\rho &= \text{density} \\
g &= \text{acceleration due to gravity } (9.81 \text{ m/s}^2) \\
Q &= \text{delivery flow} \\
H &= \text{pumping head}
\end{aligned}$$

The power requirements of the pump are thus:

$$P = \frac{\rho \cdot g \cdot Q \cdot H}{\eta_{ges}} \tag{4 - 8}$$

With increasing pumping head H, the volumetric efficiency η_v falls whilst the overall rate of efficiency increases (Fig. 4.19 and 4.20). The latter is explained by the fact that the shear momentum and friction losses are independent of the delivery pressure. In addition, the mechanical losses and friction power of the shaft seal do not increase to the same extent as the pumping head.

With increasing speed and viscosity there is the danger of working chamber cavitation because the working chamber is no longer filled with the fluid (for further details refer to Chapter 4.2.7 "Suction capacity, NPSHR and cavitation").

4.2.3.4 Piston clearances

As has already been mentioned, the rotors rotate in the casing in contact-free mode; in other words they have a certain amount of clearance. There is clearance between rotor and casing circumferences, between casing cover and rotor face ends and between rotor vane and piston hub. The amount of clearance depends on various factors as well as on operating conditions. Normally the aim is to keep the tolerances between the moving parts and between the moving and stationary parts of the pump to a minimum in order to achieve a high delivery rate. Low piston clearances also improve the suction characteristics of the pump; something which should be taken into account in the case of pumps with a positive suction head or those which deliver from a vacuum system.

If pumps are made of stainless steel with an austenitic structure, the inner tolerances have to be increased in order reliably to prevent contact of the rotating parts with each other or with the casing wall due to shaft deflection caused by the pressure load on the rotors. Otherwise the poor friction properties of the material lead to seizing and blocking of the pump.

It is also necessary to increase the rotor clearance if the pump has to be operated at high temperatures. The differing thermal expansion characteristics of casing and rotor vanes cause a reduction of the clear cross-section between rotor and casing as the heat is dissipated from the casing and the latter does not become as hot as the rotors rotating in the fluid. As a result there is a danger that the rotors will laterally and radially (working sealing point) contact the casing covers. There is also danger of contact on the corresponding roll-off line between rotor vane and rotor hub (block-sealing point) as both rotor vane and rotor hub expand as the temperature increases.

If the rotary pump is to convey a suspension which must not be damaged or whose particles are harder than the pump material, the piston clearance must also be increased insofar as this is permissible; in such cases the piston clearance is set to correspond to the maximum grain size of the suspension.

This measure has an extremely favourable effect on the wearing characteristics of the pump (Fig. 4.21). The increase in flow loss can mostly be compensated for by increasing the speed.

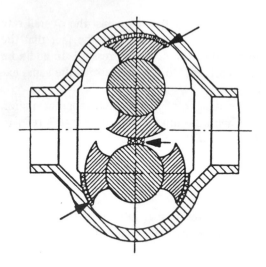

Fig. 4.21 Increased piston clearance for pumping abrasive substances

The tolerance should also be increased with rising viscosity of the fluid to be transported. Part of the power required by the pump is used to shear off the fluid film between the fixed and the moved surfaces. The larger the distance between these surfaces within certain limits, the lower is the power required for shearing off of the fluid film.

If two surfaces (A), (Fig. 4.22) separated by fluid are moved against one another at distance (y) and speed (v), the force required to overcome the internal friction of the fluid is equal to F. Friction increases proportional to the surface A and the displacement speed (v) and viscosity (η), and decreases proportional to the distance (y).

F is then calculated from the equation:

$$F = \eta \cdot A \cdot \frac{dv}{dy}$$

(4 - 9)

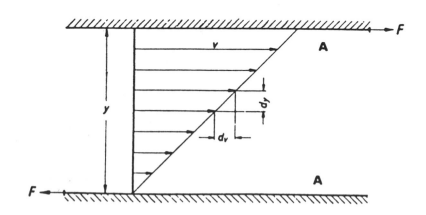

Fig. 4.22 Shear force at the fluid

416

Although the volumetric efficiency is reduced due to the increased piston clearance and the improvement in the overall efficiency rate is thus partly offset, the tendency of the fluid to flow back from the pressure to the suction side decreases with increased viscosity.

In double vane rotors with rotating hubs, an increase in rotor clearance with increasing viscosity is necessary for an additional reason. When the rotors move against one another the return flow column Q_R must be able to flow off without being "squeezed" (Fig. 4.23). In some cases, therefore, the edges of the rotor vanes are reduced a little at "a", so that the vane edges are < 90°; which however leads to a reduction in flow capacity (Fig. 4.24).

The increase in the piston clearance and chamfering of the rotor edges to less than 90° relative to the vane circumference increases the reverse flow of the fluid from the pressure to the suction side. The greater part of the returning flow strikes the sealing point between the rotor vane and rotor hub of the opposing rotor because

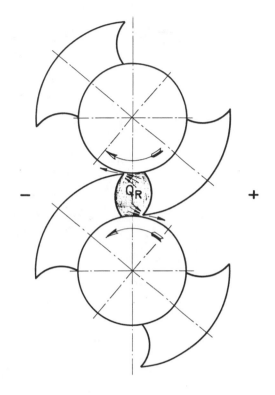

Fig. 4.23 Clearance flow at mesh changeover between the rotors

there is only a line seal at this point, but acts on the surface seal between the rotor vane and casing, both on the circumference and on both front faces. Also, at the outer sealing faces (casing circumference and front face) the pistons move in the opposite direction to the reverse flow with the casing wall stationary, whilst at the sealing point at the rotor hub both the hub and the rotor vanes move in the direction of the reverse flow and therefore support the "slip" of the pump (Fig. 4.25).

The flow loss therefore depends on the internal dimensions of the pump and also on the pressure difference p_2 - p_1 between the suction and discharge sides, which increases approximately linearly, and also on the viscosity of the fluid.

The slip decreases with increase in viscosity because increasing viscosity reduces the reverse flow tendency. The relative flow loss increases with reducing speed and reduces with increasing speed. This is explained by the fact that the dwell time at the sealing gap is longer with reducing speed and shorter with increasing speed.

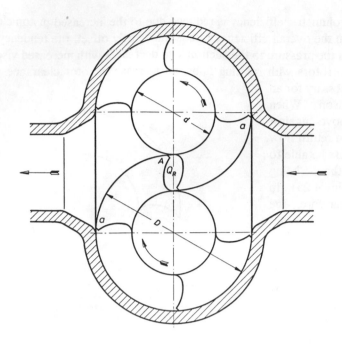

Fig. 4.24 Correction of the vane edges

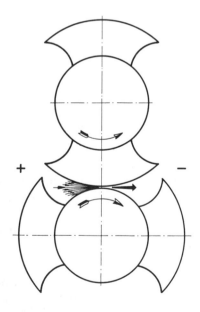

Fig. 4.25 Gap flow between the rotor vanes and shaft hub of the opposing rotor

The amount of the pressure difference between the suction and pressure sides ($p_2 - p_1$) is a measure of the deflection of the rotor. It must not exceed the size of the gap between the rotor vane and pump casing if striking the casing wall is to be safely avoided.

418

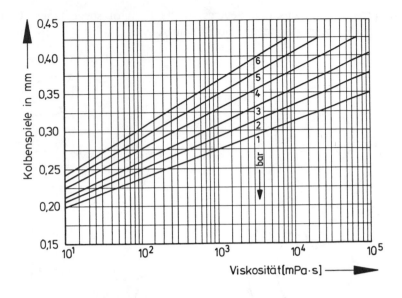

Kolbenspiele = Piston clearances
Viskositat = Viscosity

Fig. 4.26 Piston clearances of a rotary piston pump in stainless steel depending on delivery pressure and viscosity of the fluid

There is therefore an optimum for the efficiency of a rotary piston pump at which the piston clearance has a maximum value for the existing viscosity, temperature and delivery pressure of the fluid. Fig. 4.26 gives the values of the piston clearances of a stainless steel pump with an 80 mm suction connection and bearings at both ends, relative to the pressure difference between the suction and pressure connections and the viscosity of the liquid at normal temperature.

4.2.4 Vane profiles and delivery flow fluctuations

The vane profiles of single and double vane rotors are of such design that in theory there are no fluctuations in delivery volume above a rotation angle of 360 degrees (Fig. 4.27 and 4.28), therefore dv/dφ = constant.

That this is not the case in practice has been shown in a study [4-2] carried out using a rotary pump with deliberately increased piston clearance. The pressure characteristic was measured with a scan recorder over the entire angle of rotation. Fig. 4.29 shows the results of these measurements. A particularly large degree of slip, and thus pressure loss, always occurs in a certain rotor position at the moment of transition of the engagement point (control change position) from one shaft hub to the other. The beginning of the pressure drop is at 30, 120, 210 and 300° angles of rotation respectively and the pressure drop reaches its lowest level at 45, 135, 225 and 315°. This pressure fluctuation, due to the rotation angle, is mostly of no significance for day-to-day operation as long as it is not necessary to perform quantity measurements using nozzles or orifices. In such cases it is

419

advisable to provide compensation devices in the form of surge tanks. Unlike piston pumps, rotary pumps otherwise require no damping devices.

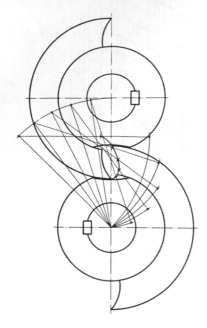

Fig. 4.27 Roll-off line of a single vane rotor pair with fixed hub

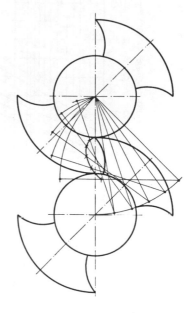

Fig. 4.28 Roll-off lines of a double vane rotor pair with rotating hub

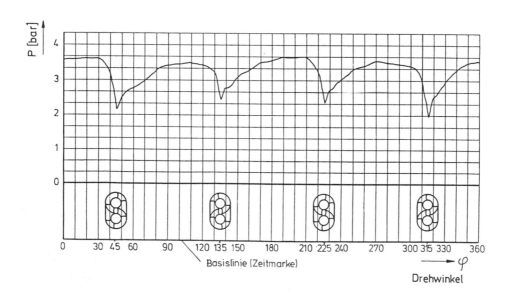

Basisline (Zeitmarke) = Base line (Time mark)

Drehwinkel = Rotation angle

Fig. 4.29 Measured pressure characteristics with large clearance at point of engagement

420

4.2.5 Speed

For reasons of economy, the speed of a rotary piston pump should be as high as possible in order to obtain a low unit weight and thus a reasonably priced machine. In this respect rotary pumps, with the exception of the single vane profile models, have the advantage over oscillating positive-displacement pumps that their speed is not limited by inertia forces. The maximum possible speed depend on:

1. the viscosity,
2. the size of the machine,
3. the rotor vane type and the number of vanes,
4. the pressure at the inlet (suction nozzle of pump, NPSHA).

The viscosity of a fluid limits the speed of the pump by virtue of its flow characteristics. The fluid moves more slowly to the extent that the inner flow losses increase due to rising viscosity. There is therefore a pump speed above which the fluid is not able to flow fast enough to fill the empty spaces of the vane cells, although total vacuum prevails at these points. If it is not possible to achieve complete filling of the suction chambers, the volumetric efficiency rate begins to fall as the ratio of actual delivery capacity to the normal capacity achieved when the vane cells are full. The pump then enters the cavitation range. The result is noise and uneven operation (see Section 4.2.7 "Suction characteristics, NPSHR and cavitation" for more details on cavitation). The more viscous the fluid the slower the speed that should be selected. Fig. 4.30, 4.30.1 and 4.30.2 shows good values for maximum speeds depending on machine size, viscosity and rotor shape, taking the suction pressure into account. The suction line connected to the pump should be such a size that satisfactory flow characteristics are ensured for the viscosity of the fluid in question.

The flow speed v_s is calculated from the volume flow Q and the pipe cross-section is:

$$v_s = \frac{4 \cdot Q}{\pi \cdot d_2^2} \qquad (4 - 10)$$

Q = volume flow
d_2 = diameter of pipe

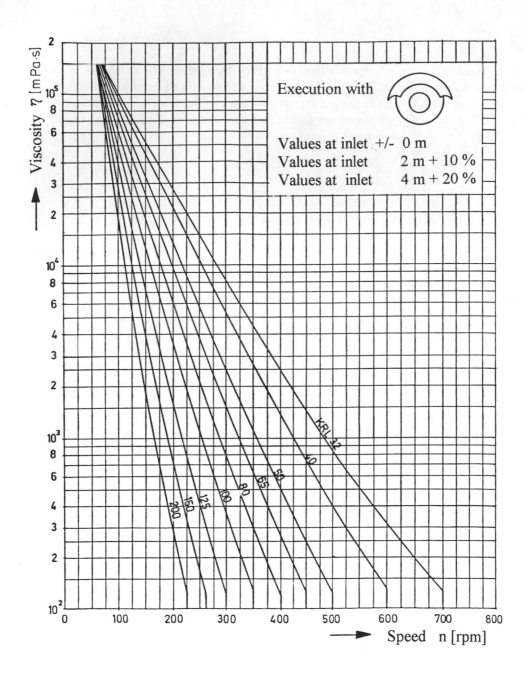

Fig. 4.30 Maximum speeds relative to size, vane shape and number, and viscosity

422

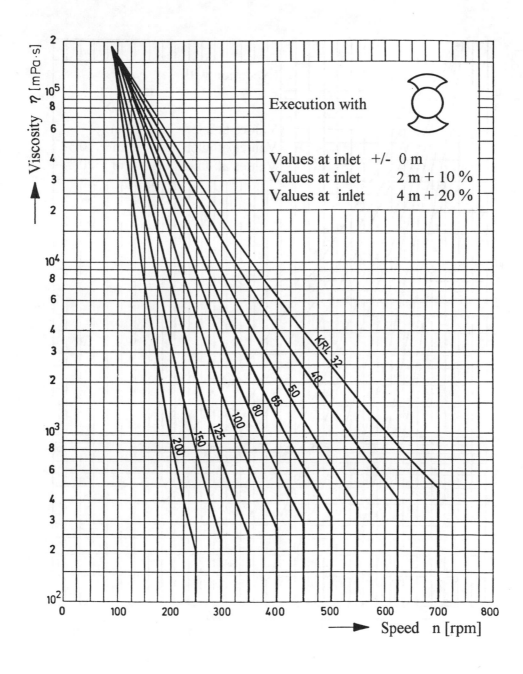

Fig. 4.30.1 Maximum speeds relative to size, vane shape and number, and viscosity

423

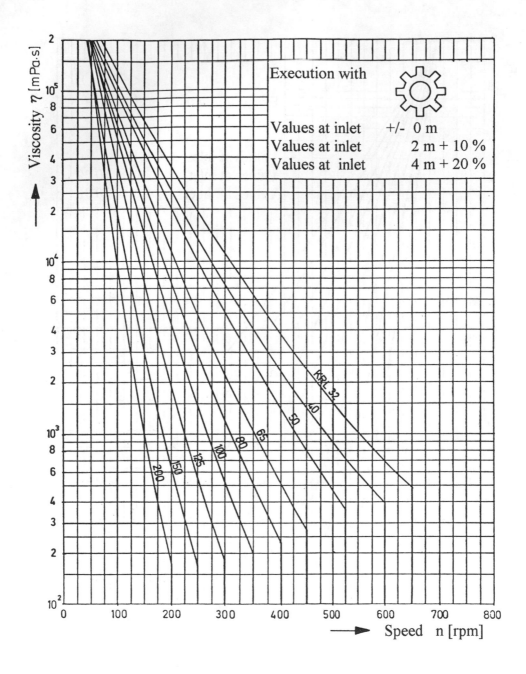

Fig. 4.30.2 Maximum speeds relative to size, vane shape and number, and viscosity

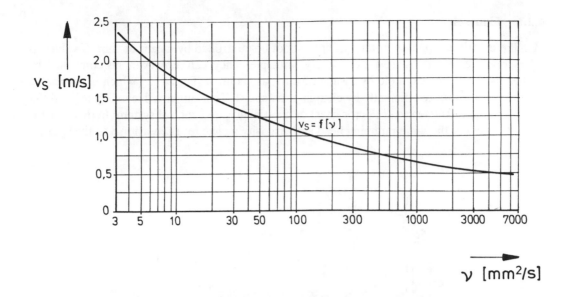

Fig. 4.31 Speed in the suction nozzle V_s as f (v)

Fig. 4.31 shows realistic values for V_s relative to viscosity.

The larger a pump the lower the speed which should be selected in order to give the vane cells sufficient time to fill up.

The rotor shape and the number of rotor vanes also affects the maximum possible speed. The filling time increases as the profile gaps become larger so that the speed has to be reduced. Symmetrically arranged rotor vanes in double or multi-vane rotors allow higher speeds compared to single vane systems (unbalance).

It is possible to explain why the filling level is influenced by rising or falling pressure at the intake of the pump (suction nozzle). The greater the inlet pressure the faster the cells fill and the higher the speed which can be selected.

4.2.6 Control

Unlike centrifugal pumps, rotary pumps cannot be controlled by means of shutoff valves as they operate on the hydrostatic delivery principle. They can, however, be controlled by other methods; for example by means of a bypass control (Fig. 4.32) which is associated with losses, or more efficiently by changing the speed. This can be achieved by using different fixed speeds (stepped selector gear) or a variable speed control system. In this way the volume flow at the same pumping head and viscosity a changes proportional to the speed.

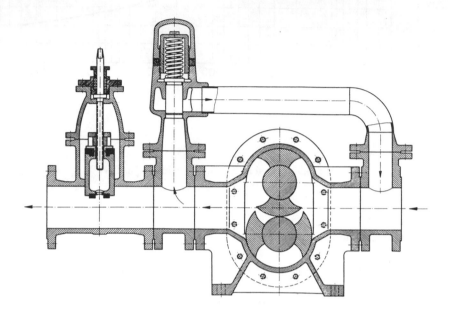

Fig. 4.32 Bypass control

The relatively low speeds at which rotary pumps are operated (between 1 and 750 r.p.m.) generally require the use of a reduction gear between the drive motor and pump. The pumps are mostly equipped with gear motors with direct flange connection of motor and gear unit. If the control process is to entail a minimum of performance loss it is advisable to select a steplessly variable drive. If it is necessary continuously to adapt the volume flow to a particular process, the control gear unit is fitted with a pneumatic or electric remote adjustment facility (servomotor). The necessary adjustment pulses can be given either manually, by means of level monitor, or flowmeter. Using the variable speed control method it is possible to employ the rotary pump for metering tasks in the range beyond that catered for by oscillating piston metering pumps.

4.2.7 Suction characteristics, NPSHR and cavitation

Due to their hydrostatic delivery characteristics, rotary pumps are self-priming. Their suction capacity depends on their NPSHR response or on the maximum permissible degree of cavitation. This in turn depends on the pressure conditions on the suction side, i.e. absolute

426

pressure on the surface of the inlet fluid, geodetic suction head, pipeline losses and inlet conditions, which are formed by the geometric configuration of the rotary pistons.

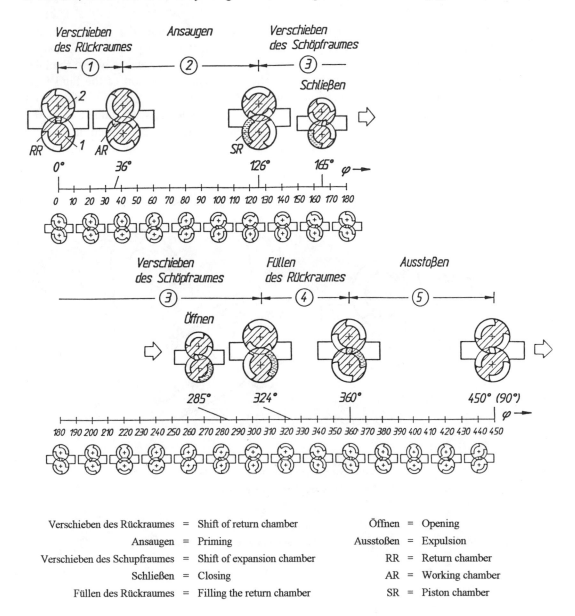

Verschieben des Rückraumes	=	Shift of return chamber		Öffnen	=	Opening
Ansaugen	=	Priming		Ausstoßen	=	Expulsion
Verschieben des Schupfraumes	=	Shift of expansion chamber		RR	=	Return chamber
Schließen	=	Closing		AR	=	Working chamber
Füllen des Rückraumes	=	Filling the return chamber		SR	=	Piston chamber

Fig. 4.33 Displacement position during delivery of lower rotary piston

Furthermore, in contrast to centrifugal pumps, the pressure drop P_2-P_1, at least in the area of low to medium viscosities of the liquid, is responsible for the NPSHR. A co-operative study by the VDMA and the Forschungskuratorium Maschinenbau eV [Mechanical Engineering Research Committee] as well as research work supported by AIF, Vetter and Zimmermann [4-3] [4-4] have described the qualitative and quantitative conditions which enable $NPSHR_S$ = suction head to be determined.

The <u>suction side cavitation</u> can be described as the piston chamber and expansion chamber cavitation.

The <u>suction chamber cavitation</u> begins when the pressure is less than the vapour pressure at the point at which the suction chamber is just opening with the piston position at an angle of rotation of $\varphi = 36°$ as shown in Fig. 4.33 and 4.34. This point is where the inlet cross section of the lower rotary piston is smallest. A strong diversion of the fluid flow takes place at this point. The expansion chamber of the upper rotary piston is filled and when shifted further in the direction of the pressure chamber has no influence on the process of cavitation. With an increasing angle of rotation the inlet cross section A_E (Fig. 4.34) opens so that the inlet velocity V_E (Fig. 4.35), which is a measure for the start of cavitation, is reduced.

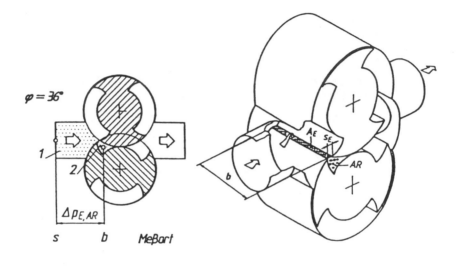

1 - Piston chamber, 2 - Suction chamber

Meßart = measuring point

Fig. 4.34 Geometry at the point where the suction process is just beginning ($\varphi = 36°$)

When the expansion chamber is filling it first carries vapour bubbles with it which formed at rotary piston position $\varphi = 36°$. These disappear with the increasing size of the cross section A_E and associated pressure rise so that they do not affect the volume flow. Nevertheless this can lead to cavitation erosion (Fig. 4.37, 4.34, and 4.44).

<u>Expansion chamber cavitation</u> occurs when the pressure drop ΔP_{ESR} in the suction chamber drops below the vapour pressure and vapour bubbles can no longer be reformed. The expansion chamber is therefore partly filled with vapour and partly with liquid (Fig. 4.36 and 4.37). As the rotary piston rotates further the expansion chamber moves to the pressure side at angular position $\varphi = 285°$ so that from there liquid flows at higher velocity into the expansion chamber and the vapour bubbles collapse.

428

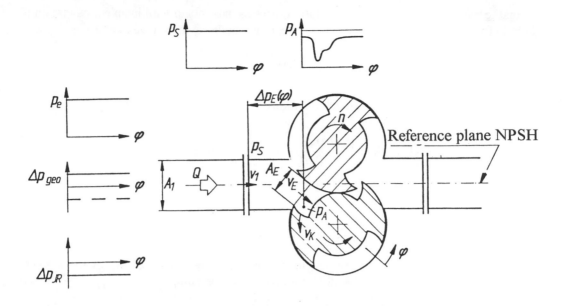

Fig. 4.35 Pressure in the working chamber of a rotary piston pump due to the pressure conditions on the suction side and the inlet pressure loss

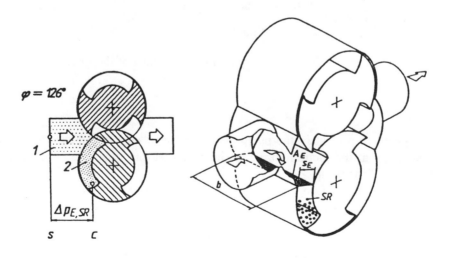

1 = piston chamber, 2 = expansion chamber

Fig. 4.36 Geometry for the φ = 126° position.

A mathematical treatment of this cavitation process must begin with incipient cavitation if bubble formation is to be avoided. Therefore, to preclude cavitation we must have:

$$NPSHA \geq NPSHR_s \qquad (4-11)$$

NPSHA is obtained from the system conditions:

$$NPSHA = \frac{p_e + p_{amb} - p_v}{\rho \cdot g} + \frac{v_e^2}{2g} \pm h_s \pm Z_s - H_{vs} \qquad (4-12)$$

g	Normal gravity acceleration
h_s	Geodetic height between the inlet cross section of the system and the inlet cross section of the pump
H_{vs}	Inlet losses
p_e	Absolute pressure in the inlet of the system
p_{amb}	Atmospheric pressure
p_v	Absolute vapour pressure
Z_s	Difference in level between the liquid on the inlet side and the entrance plane
ρ	Density of fluid
v_e	Mean velocity of flow in the inlet of the system

The inlet pressure present in the pump at the suction flange p_s reduces due to the mean flow velocity v_1 at the suction nozzle to the mean velocity v_K at the piston sides and the internal losses of the pump

The pressure in the working chamber, which is influenced by the piston configuration, depends on the angle of rotation, and is therefore:

$$p_a(\varphi) = p_s + \frac{\rho \cdot v_1^2}{2} - \frac{\rho \cdot v_k^2}{2} - \Delta p_E(\varphi) \qquad (4-13)$$

$$v_k = \frac{D+d}{4} \cdot \omega \qquad (4 - 14)$$

v_K Mean velocity of flow at the piston flank
D   External diameter of rotary piston
d   Hub diameter of rotary piston

The NPSHR due to condition on the suction side is obtained from the inlet pressure loss of the pump p_E

$$NPSHR_s = \frac{\Delta p_E}{\rho \cdot g} \qquad (4 - 15)$$

To be able to determine Δp_E it is necessary to find the pressure resistance coefficient ξ_E. This is obtained from the Reynolds number formed with the hydraulic diameter of the inlet geometry

$$\zeta_E(\varphi) = \frac{2 \cdot \Delta p_E(\varphi)}{\rho \cdot v_E(\varphi)^2} \qquad (4 - 16)$$

$$Re_E(\varphi) = \frac{\rho \cdot v_E(\varphi) \cdot d_{Eh}(\varphi)}{\eta} \qquad (4 - 17)$$

$$d_{Eh} = \frac{4 A_E}{U_E} = \frac{2 S_E \cdot b}{S_E + b} \qquad (4 - 18)$$

$$v_E = \frac{Q}{A_E(\varphi)} \qquad (4 - 19)$$

A_E Cross-sectional area of the inlet geometry determinant for cavitation (Fig. 4.34)

d_E Hydraulic diameter of the inlet geometry determinant for cavitation

b   Axial length of rotary piston

S_E Width of the inlet geometry determinant for cavitation

U   Circumferential velocity of rotary piston

η Dynamic viscosity

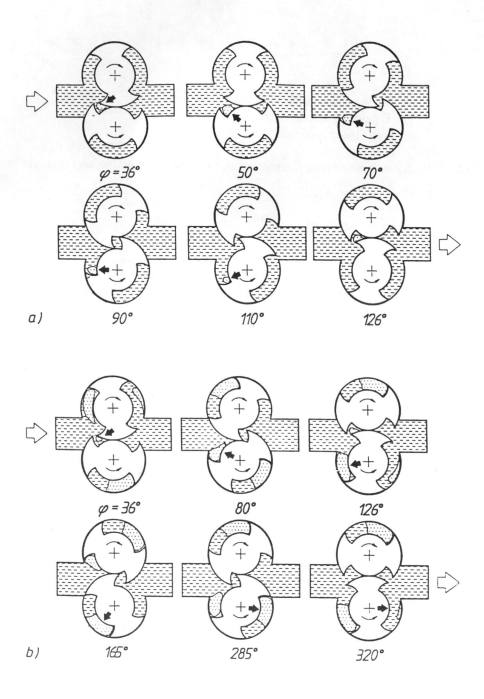

φ = 36° 50° 70°

a) 90° 110° 126°

φ = 36° 80° 126°

b) 165° 285° 320°

Fig. 4.37 Cavitation in a rotary piston pump (schematic)
a) pronounced suction chamber cavitation
b) pronounced expansion chamber cavitation

we therefore get:

$$\Delta p_E(\varphi) = \zeta_E[Re_E(\varphi)] \cdot \frac{\rho}{2} \cdot v_e(\varphi)^2 \qquad (4 - 20)$$

The resistance coefficients ξ_E can be taken from Fig. 4.38 as a function of the Reynolds number Re_E for piston positions 36°, 40° and 45°.

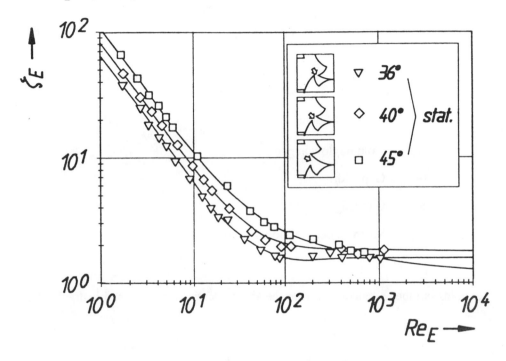

Fig. 4.38 Resistance coefficients for the piston positions at the start of the suction process (stationary) of a test pump
(φ =36°: S_E= 3,65 mm; φ = 40°: S_E= 6,4 mm; φ = 45°: S_E = 10,7 mm)

Gap cavitation occurs due to leakage flows between the pressure and suction chambers due to the pressure drop P_2 - P_1. It depends on the geometry of the clearance and the Reynolds number. To determine the $NPSHR_{SP}$ required for this, it is necessary to consider the point at which the largest reverse flow losses occur, which, for example, arise on the sealing gap between the piston vanes and on the piston hub in the case of twin vane pumps with a rotating hub (Fig. 4.33 to 4.37). Only a line seal exists at this point between the suction and pressure chambers. Furthermore, the upper and lower limiting bodies rotate in the same direction, in the direction of the flow at the same angular velocity. In the case of single or twin vane rotary pumps with a fixed hub and rotating pistons the point which is important for the reverse flow is the relatively small surface seal at the block-sealing point Fig. 4.12). This type of seal provides better $NPSHR_{SP}$ values.

Mathematical treatment of gap cavitation on the pressure side

The gap flow occurring at the block-sealing point (Fig. 4.39) results from the pressure drop

$$\Delta p = p_2 - p_1 \qquad (4-21)$$

According to Surek [4-5] this would be:

$$\Delta p = \zeta_D \cdot \frac{\rho}{2} v_{sp}^2 = \zeta_D \cdot \rho \cdot \frac{Q_{sp}^2}{b^2 \cdot s^2} \qquad (4-22)$$

ξ_D Pressure loss coefficient as a function of the Reynolds number of the gap flow

ρ Density of fluid

Q_{sp} Volume flow in gap

b Gap width

S Gap height

v_{sp} Mean velocity of flow in gap

The Reynolds number is calculated from the kinematic viscosity of v and hydraulic diameter $d_h = 2 \cdot S$

$$Re = \frac{v_{sp} \cdot d_h}{v} = \frac{v_{sp} \cdot 2 \cdot s}{v} \qquad (4-23)$$

The hydraulic diameter of the gap in rotary piston machines d_{dh} is calculated in the same way as Equation (4-18)

$$d_{dh} = \frac{2 \cdot b \cdot s}{b + s} \qquad (4-24)$$

For very small gap dimensions $S \ll b$, we can use:

$$d_{dh} = 2 \cdot s \qquad (4-25)$$

For a given gap geometry Re depends on the gap velocity v_{sp} and the viscosity of the fluid.

434

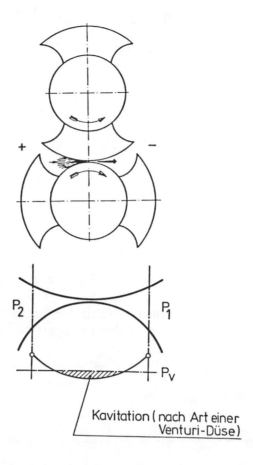

Kavitation (nach Art einer Venturi-Düse) = Cavitation (similar to a Venturi tube)

Fig. 4.39 Pressure drop at the block-sealing point of a twin vane rotary piston pump with a rotating hub

The pressure loss coefficient ξ_{sp} in the gap at the block-sealing point of rotary piston pumps when pumping viscose liquids can be regarded as proportional to the kinematic viscosity v, so that it is relative to the viscosity of water [4-5]:

$$\zeta_{sp} = \zeta_w \cdot \frac{\upsilon}{\upsilon_w} = \zeta_w \cdot \frac{\eta}{\eta_w} \cdot \frac{\rho_w}{\rho} \qquad (4-26)$$

η = dynamic viscosity Pa $\cdot$ s

where $\quad \eta = \rho \cdot \upsilon \qquad\qquad\qquad\qquad (4-27)$

becomes $\quad \Delta p = \zeta_w \cdot \frac{\upsilon}{\upsilon_w} \cdot \frac{\rho}{2} \cdot v_{sp}^2 = \zeta_w \cdot \frac{\upsilon}{\upsilon_w} \cdot \frac{\rho}{2} \cdot \frac{Q_{sp}^2}{b^2 \cdot s^2} \qquad (4-28)$

From this we get a gap volume flow of

$$Q_{sp} = b \cdot s \cdot \sqrt{\frac{2}{\rho} \cdot \frac{\Delta p \cdot \upsilon_w}{\zeta_w \cdot \upsilon}} \qquad (4 - 29)$$

From Q_{sp} and the gap cross section A_{sp} the gap velocity can be determined

$$v_{sp} = \frac{Q_{sp}}{A_{sp}} = \frac{Q_{sp}}{b \cdot s} = \sqrt{\frac{2}{\rho} \cdot \frac{\Delta p \cdot \upsilon_w}{\zeta_w \cdot \upsilon}} \qquad (4 - 30)$$

and from this we get $NPSHR_{SP}$ at the block-sealing point from the gap flow for rotary piston pumps.

$$NPSHR_{SP} = \frac{v_{sp}^2}{2g} = \frac{[p_2 - p_1] \cdot \upsilon_w}{\zeta_w \cdot \upsilon \cdot g \cdot \rho} \qquad (4 - 31)$$

The ξ_w-values required for this equation relative to Δp can be taken from Fig. 4.40. Fig. 4.41 shows the $NPSHR_{SP}$ for water relative to Δp and the gap configuration (line or surface sealing).

The NPSHR required for a particular application must be obtained after estimation of the two partial $NPSHR_S$ and $NPSHR_{SP}$ levels which are independent of each other, with the higher numerical value being adopted.

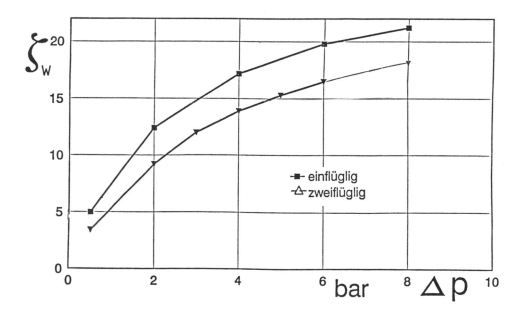

einflüglig = single vane ;   zweiflüglig = twin vane

Fig. 4.40 Pressure loss coefficients in the gap of single and twin vane rotary piston pumps for water at 20 °C

Gap cavitation occurs, as already stated, in the area of low to medium fluid viscosity. This can also been seen from Fig. 4.42 which shows the relationship between pump efficiency η_p and the viscosity of the fluid. In the low viscosity range (1 to 200 mPa·s), the pump efficiency is particularly dependent on the amount of gap reverse flow, whilst with increasing viscosity the gap flows reduce from the culmination point 200 mPas and the increasing friction in the casing gaps to the side of the rotor vanes as well as the working sealing point are increasingly responsible for reducing the pump efficiency. Above 5000 mPa·s the gap flows are negligible and gap cavitation moves towards zero.

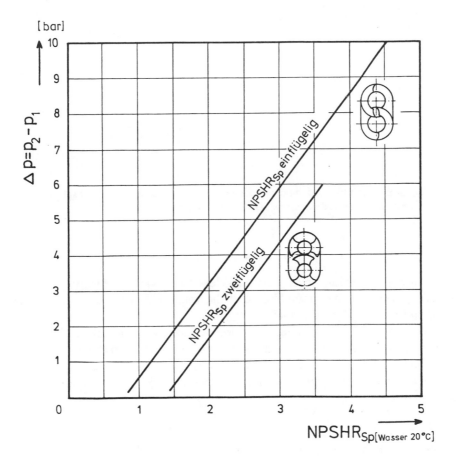

einflügelig = single vane ; zweiflügelig = twin vane ; Wasser = Water

Fig. 4.41 NPSHR$_{sp}$ values for single vane rotary piston pumps with a fixed hub and twin vane rotary piston pumps with a rotating hub for fluids with a viscosity of 1 mPa·s

However, the suction cavitation increases simultaneously by the same amount as the reduction in gap flow due to the increase in viscosity because the increasing velocity reduces the degree of admission of the working chamber and therefore the NPSHR$_S$ value increases if cavitation is to be avoided.

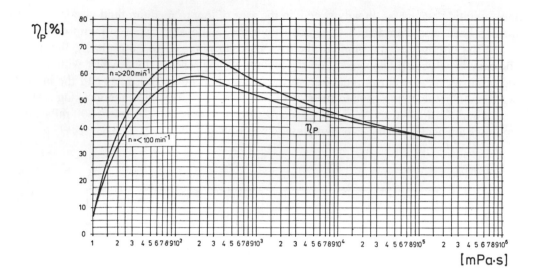

Fig. 4.42 Total efficiency η_P of rotary piston pumps relative to viscosity and speed (in this case the measured mechanical efficiency was 83%)

4.2.7.1 Cavitation erosion due to suction and gap cavitation

Suction cavitation acts mainly on the shaft hub close to the vane root (Fig. 4.43 and 4.44). Gap cavitation occurs at the circumference of the vanes and at the point of the shaft hub corresponding to it (Fig. 4.43 and 4.45).

The cast iron rotor vane shown in Fig. 4.45 already shows considerable cavitation erosion, whilst the steel fixing bolt shows only slight damage. This is due to the material properties of cast iron and steel. Cast iron has an amorphous structure whilst steel is a tough material. This is also an example of how a suitable choice of material can counteract cavitation erosion, at least for a time.

Where there is low or zero head $NPSH_{SP} = 0$. $NPSHR_{SP}$ increases linearly with increase in the pressure difference $p_2 - p_1$, regardless of the supply conditions at the pump inlet. In other words the pump can enter the cavitation state with rising backpressure, without going below the $NPSHR_S$ value on the inlet side.

4.2.7.2 Vapor-gas cavitation

The majority of delivery applications where rotary piston pumps are used are for delivering highly viscous fluids with different consistencies, with the pumped product to some extent fractionated, which depending on its composition has diverging vapor pressures. Also it is not uncommon for it to contain dissolved gases. These can dissolve out of the fluid where there is a pressure drop or high flow velocities and associated pressure reduction, thus resulting in gas bubbles similar to the formation of gas bubbles when the vapor pressure is

438

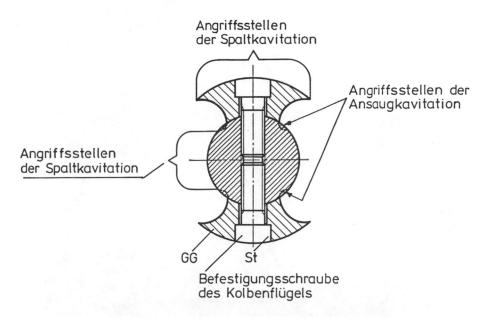

Angriffsstellen
der Spaltkavitation

Angriffsstellen der
Ansaugkavitation

Angriffsstellen
der Spaltkavitation

GG   St

Befestigungsschraube
des Kolbenflügels

Angriffsstellen der Spaltkavitation	=	Points which are attacked by gap cavitation
Angriffsstellen der Ansaugkavitation	=	Points what are attacked by suction cavitation
Befestigungsschraube des Kolbenflügels	=	Rotor vane attachment bolts

Fig. 4.43 **Points susceptible to cavitation on a twin vane rotary piston pump with rotating hub**

Fig. 4.44 **Cavitation erosion caused by suction cavitation on the rotor hub of a twin vane rotary piston pump as shown in Fig. 4.43**

not maintained. This is referred to as gas cavitation, a not quite correct expression which has however become established. 'Gas cavitation' is commonly observed where oils are being pumped. After the gas bubble enters an area of higher pressure it then goes into solution, without causing cavitation damage to the pump material.

Fig. 4.45 Cavitation erosion due to gap cavitation on the rotor vanes and hub of a rotary piston pump as shown in Fig. 4.43

This account of a test with a rotary piston pump pumping vaseline oil is provided to show how gap cavitation works. To enable the gap flow to be seen, the pump was fitted with a glass cover and operated under a stroboscopic light at n = 250 r.p.m. The cavitation threshold was determined by increasing the suction and pressure heads [4-6], [4-7]. Fig. 4.46 clearly shows the gap flow triggered by the gap cavitation, as it passes from the pressure to the suction side. The left (suction) side is filled with a liquid-gas-vapor mixture which gives the oil a milky color, whilst the oil on the right (pressure) side is bubble free. The onset of gap cavitation is accompanied by a knocking noise which occur in proportion to the rotational frequency and number of rotor vanes. Increasing pressure difference $(p_2 - p_1)$ amplifies the intensity of the noises and increases the sound to an unpleasant level. This effect also confirms the effect of gap cavitation and therefore of the pressure drop $(p_2 - p_1)$ on the suction capacity of the rotary piston pump.

It was said that the delivery flow of rotary piston pumps was theoretically independent of backpressure, but that due to unavoidable gaps between the various pressure chambers a dependency between Q and H exists. This is all the greater the wider the gap. For test purposes (Fig. 4.47) [4-7] the aforementioned pump was subjected to various tests with increased piston clearance. The vaseline oil used for the test was pumped at a mean temperature of 22; 25,8; 33,7 °C and four different characteristic curves were plotted. The individual operating temperatures correspond to viscosities of 47; 40; 36 and 33,7 mm²/s.

The mean density was 0.858 kg/dm³ and the speed was held steady at n = 200 r.p.m. during all tests. The increase in the piston clearance enabled the pump to be throttled to Q = 0 without a damaging pressure rise. The curve clearly shows the dependency of an achie-

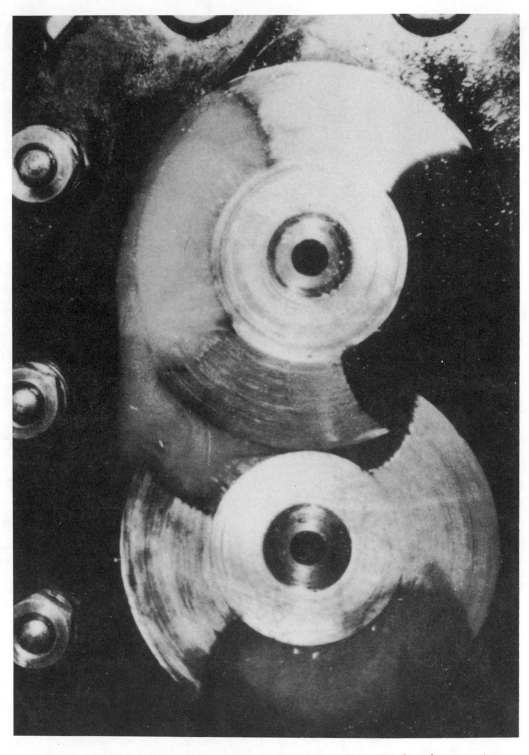

Fig. 4.46 Vapor-gas cavitation and gap flow at the block-sealing point of a twin vane rotary piston pump with a rotating hub

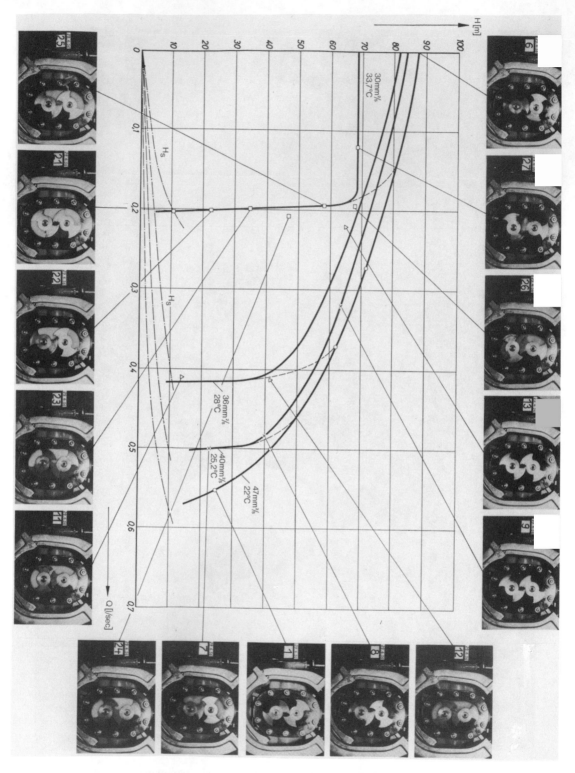

Fig. 4.47 H (Q) curve of a twin vane rotary piston pump relative to viscosity showing the cavitation-free and cavitation operating states

442

vable head or delivery rate on the viscosity of the fluid. The flow condition of the pump was photographed in individual operating states. This enabled the point at which cavitation began and its pattern to be established. For example, photograph 27 shows cavitation-free operation, whereas there is severe cavitation in photograph 25. The H (Q) values of the throttling curves is assigned in each case to a corresponding suction head on the H_s (Q) curve in the lower part of the diagram. As long as no cavitation is present the pumped fluid (vaseline oil) on the suction and discharge side has a dark coloration. When cavitation occurs (where the curve breaks off), the oil becomes a milky white color due to the inclusion of gas and vapor bubbles.

4.2.7.3 Avoidance of cavitation without changing the static suction head, the pipeline area and backpressure

There are cases in which rotary piston pumps have to operate in the cavitation range, as the operator cannot provide a lower suction head or a larger static head. The result is uneven pump running with loud knocking noises which are transferred to the entire pipeline system, causing the latter to vibrate. This vibration can be so great that the pipeline is torn out of its anchorage or that pipe bursts occur. Quite apart from the fact that such operating conditions can destroy the pump itself, the noise may become so excessive that operating personnel cannot be expected to work in the same room. Insofar as it is permitted by the delivery task, it is possible to prevent cavitation or a least reduce it to an acceptable level by relatively simple means. If the pump operates with suction head, a blow valve should be fitted a short distance in front of or directly on the suction nozzle. This ensures that a small quantity of air is carried along with the flow, but only as much as is necessary to provide for smooth pump operation. If the pump operates with a static head, an appropriate amount of compressed air is added from the compressed air network. The air content of the fluid has a damping effect on the cavitation process. Use is made of this property in turbines, for example, by injecting compressed air at points where there is danger of cavitation. The addition of air naturally also affects the usable delivery volume, as the pump must now deliver a fluid-air mixture, something it is able to do due to its hydrostatic delivery characteristics. If the pump is being used purely for fluid delivery tasks, the resulting capacity loss can mostly be accepted. Before an attempt is made to remedy cavitation by the addition of air, however, it must be established that this will not have a negative effect on the delivery product.

If rotary piston pumps are used to convey gaseous or effervescent media, they can be operated at speeds considerably higher than the normal levels, since although the high speed of the gap flow caused by the gas content leads to gas bubble formation it does not result in actual cavitation, i.e. vapor formation.

4.2.8 Types of rotary piston pumps

It is basically possible to install the displacer designs listed in Section 4.2.2 in all the pump versions described here. In many cases it is also possible to interchange existing piston shapes. All these pumps are designed in inline version and are suitable for reversible operation. For this reason it is expedient to provide both flanges with the same nominal diameter.

When the direction of rotation changes, so does the direction of fluid delivery. This is particularly advantageous in cases where the suction and pressure lines have to be drained following operation to prevent solidification of the delivery product and blockage of the pressure line. One common feature of all the rotary piston pumps presented here is that the bearings are outside the pressure-fed casing; i.e. the delivery product does not come into contact with the bearings, something which is extremely important in the case of non-lubricating fluids and fluids with admixtures. In addition, all these pumps can be designed with or without a heating jacket. It is also possible to fit the drive shaft either at the top or bottom of the bearing seat. This is generally of no significance as regards the operation of the pump, but may have an effect on the type of drive to be installed. In some cases, the shaft bearings are to be positioned next to one another rather than one on top of the other; for example if the aim is to deliver an extremely viscous product from a tank directly through the tank drain into the displacers (Fig. 4.48 and 4.49). This type of displacer feed is, however, unsuitable for conveying tasks in the vacuum range or for fluids with gas inclusions, as the gas bubbles escaping from the delivery product rise upwards and can thus interfere with the delivery process.

4.2.8.1   Bearings

4.2.8.1.1 Pumps with bearings on both sides for the low pressure ranges (Fig. 4.50 and 4.51)

Pumps of this type are used for simple delivery tasks with pumping heads up to 6 bar. The relatively large bearing spacing leads to correspondingly great deflection loads. This is a design which is used relatively successfully under rough operating conditions, e.g. in the bitumen processing industry for high temperatures (up to 300 °C). The disadvantage of this design is the fact that four shaft seals are required. Fig. 4.52 shows this type of pump used for producing latex.

4.2.8.1.2 Pumps with single-side (overhung) bearing for the medium pressure ranges (Fig. 4.53 and 4.54)

These types of pumps are used for pressures up to 10 bar. The one-sided bearing arrangement allows fast cleaning of the displacer chamber without dismantling the pump. In addition, all the wearing parts (displacer, any wearing bowls used as casing inserts, mechanical seals) can be removed and replaced or cleaned after the cover has been taken off (process design). The sync gear unit operating in an oil bath is practically non-wearing, as are the roller bearing which are overdimensioned due to the large shaft necessary to take up the deflection loads which occur.

444

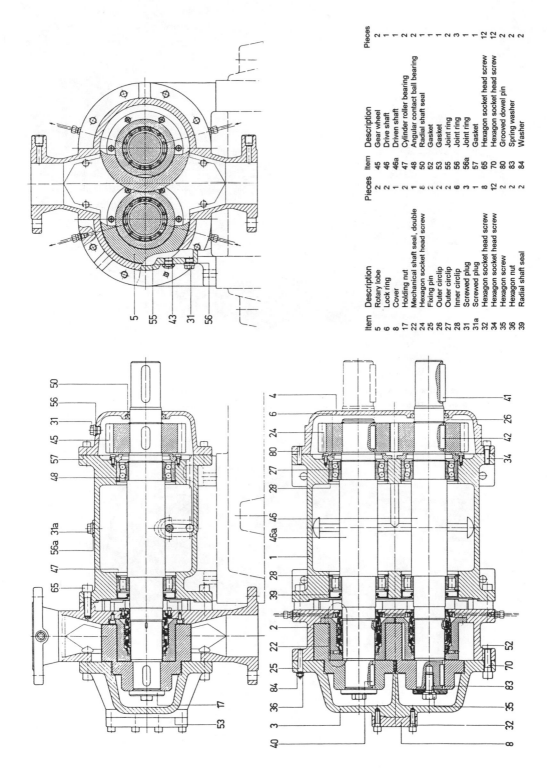

Item	Description	Pieces
5	Rotary lobe	2
6	Lock ring	2
8	Cover	1
17	Holding nut	2
22	Mechanical shaft seal, double	1
24	Hexagon socket head screw	8
25	Fixing pin	2
26	Outer circlip	2
27	Outer circlip	6
28	Inner circlip	2
31	Screwed plug	3
31a	Screwed plug	1
32	Hexagon socket head screw	8
34	Hexagon socket head screw	12
35	Hexagon screw	2
36	Hexagon nut	2
39	Radial shaft seal	2

Item	Description	Pieces
45	Gear wheel	2
46	Drive shaft	1
46a	Driven shaft	1
47	Cylinder roller bearing	2
48	Angular contact ball bearing	1
50	Radial shaft seal	1
52	Gasket	2
53	Gasket	1
55	Joint ring	2
56	Joint ring	3
56a	Joint ring	1
57	Gasket	1
65	Hexagon socket head screw	12
70	Hexagon socket head screw	12
80	Grooved dowel pin	2
83	Spring washer	2
84	Washer	2

Fig. 4.48 Single-vane, rotary piston pump with vertical flange connection (Lederle)

445

Fig. 4.49 View of a rotary piston pump with a vertical flange connection as shown Fig. 4.48 and a cardan shaft for swivel mounting of feed tank (Lederle)

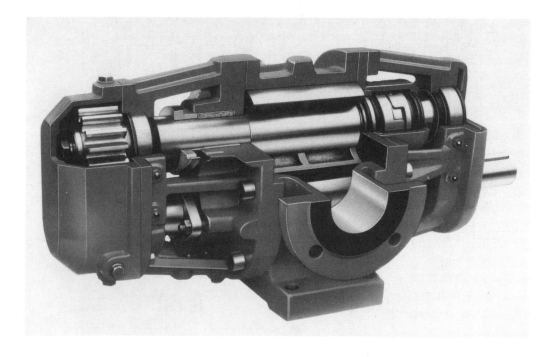

Fig. 4.50 Sectioned model of a rotary piston pump as shown in Fig. 4.51, stuffing box or mechanical sealing option (Lederle)

446

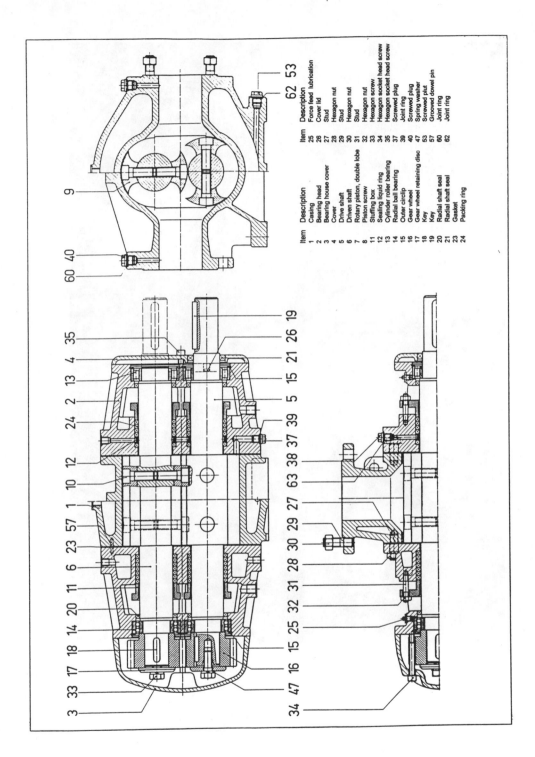

Item	Description	Item	Description
1	Casing	25	Force feed lubrication
2	Bearing head	26	Cover lid
3	Bearing house cover	27	Stud
4	Cover	28	Hexagon nut
5	Drive shaft	29	Stud
6	Driven shaft	30	Hexagon nut
7	Rotary piston, double lobe	31	Stud
8	Piston screw	32	Hexagon nut
11	Stuffing box	33	Hexagon screw
12	Sealing liquid ring	34	Hexagon socket head screw
13	Cylinder roller bearing	35	Hexagon socket head screw
14	Radial ball bearing	37	Screwed plug
15	Outer circlip	39	Joint ring
16	Gear wheel	40	Screwed plug
17	Gear wheel retaining disc	47	Spring washer
18	Key	53	Screwed plut
19	Key	57	Grooved dowel pin
20	Radial shaft seal	60	Joint ring
21	Radial shaft seal	62	Joint ring
23	Gasket		
24	Packing ring		

Fig. 4.51 Rotary piston pump with bearing at both ends, with or without a heating jacket as option (Lederle)

447

Fig. 4.52 Rotary piston pumps as shown in Fig. 4.51 for handling latex (Lederle)

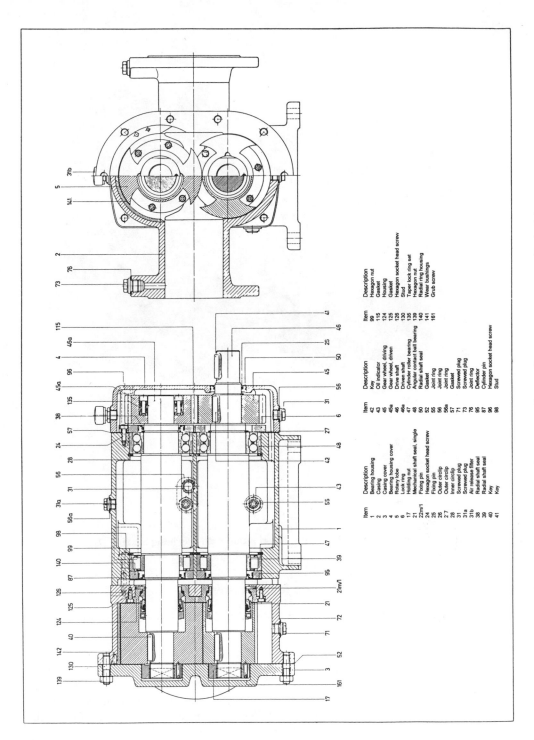

Fig. 4.53 Process-type rotary piston pump for medium pressure ranges (Lederle)

Item	Description
1	Bearing housing
2	Casing
3	Casing cover
4	Bearing housing cover
5	Rotary lobe
6	Lock ring
17	Holding nut
21	Mechanical shaft seal, single
22m/1	Fixing pin
24	Hexagon socket head screw
25	Outer circlip
26	Outer circlip
27	Inner circlip
28	Gasket
31	Screwed plug
31a	Screwed plug
31b	Air release filter
38	Radial shaft seal
39	Radial shaft seal
40	Key
41	Key

Item	Description
42	Key
43	Oil indicator
45	Gear wheel, driving
45a	Gear wheel, driven
46	Drive shaft
46a	Driven shaft
47	Cylinder roller bearing
48	Angular contact ball bearing
50	Radial shaft seal
52	Gasket
55	Joint ring
56	Joint ring
56a	Joint ring
71	Gasket
73	Screwed plug
76	Screwed plug
95	Joint ring
87	Deflector
96	Cylinder pin
98	Hexagon socket head screw
	Stud

Item	Description
99	Hexagon nut
115	Gasket
124	Housing
125	Gasket
126	Hexagon socket head screw
130	Stud
135	Taper lock ring set
139	Hexagon nut
140	Radial ring housing
141	Wear bushings
161	Grub screw

449

Fig. 4.54 Sectioned model of a rotary piston pump with overhung bearing (Lederle)

4.2.8.1.3 Pumps with single-side bearing for high pressures

In the case of pressure differences $(p_2 - p_1)$ from 10 to 20 bar, types as shown in Fig. 4.55 and 4.56 are used. In principle these designs correspond to those shown in Fig. 4.53, except that they have suitably strengthened components and a surface seal at the block-sealing point. To save space the pumps can also be provided with an angular gear drive motor (Fig. 4.57). For large pump power outputs above approximately 50 kW the provision of hydrodynamic torque converters, electric smooth start devices or frequency converters for a smooth start is recommended (Fig. 4.58). They have the advantage of providing a stepless, controllable drive system. Fig. 4.59 shows pumps of this kind installed in a large scale chemical plant.

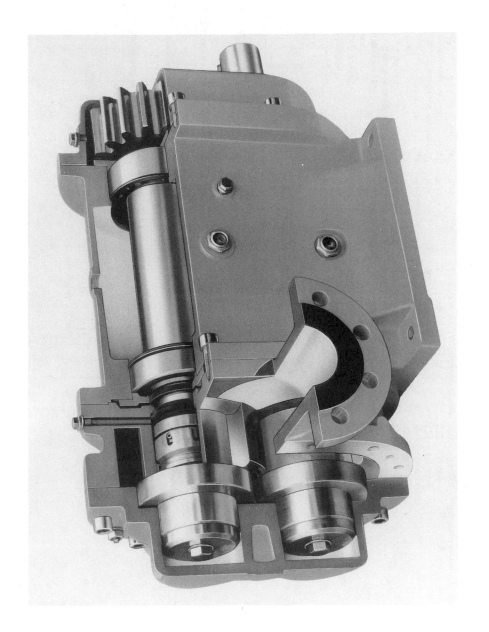

Fig. 4.55 Sectioned model of a rotary piston pump (Fig. 4.56) (Lederle)

451

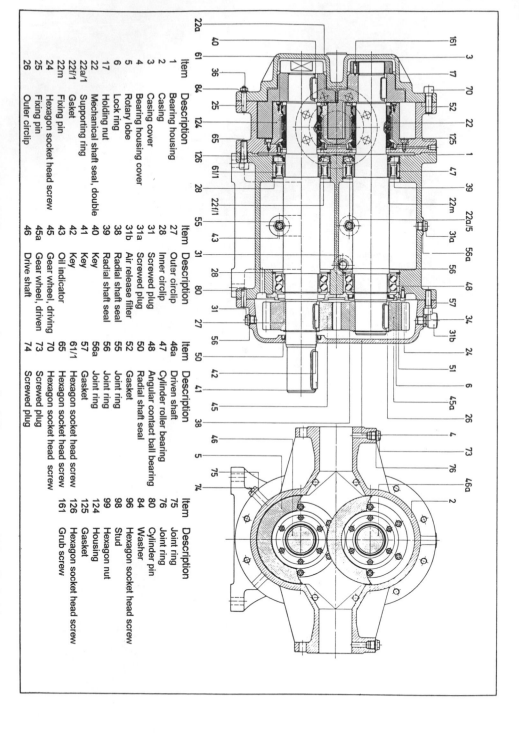

Item	Description
1	Bearing housing
2	Casing
3	Casing cover
4	Bearing housing cover
5	Rotary lobe
6	Lock ring
17	Holding nut
22	Mechanical shaft seal, double
22a/1	Supporting ring
22f/1	Gsket
22m	Fixing pin
24	Hexagon socket head screw
25	Fixing pin
26	Outer circlip

Item	Description
27	Outer circlip
28	Inner circlip
31	Screwed plug
31a	Screwed plug
31b	Air release filter
38	Radial shaft seal
39	Radial shaft seal
40	Key
41	Key
42	Key
43	Key
45	Gear wheel, driving
45a	Gear wheel, driven
46	Drive shaft

Item	Description
46a	Driven shaft
47	Cylinder roller bearing
48	Angular contact ball bearing
50	Radial shaft seal
52	Gasket
55	Joint ring
56	Joint ring
56a	Joint ring
57	Gasket
61/1	Hexagon socket head screw
65	Hexagon socket head screw
70	Hexagon socket head screw
73	Screwed plug
74	Screwed plug

Item	Description
75	Joint ring
76	Joint ring
80	Cylinder pin
84	Washer
96	Hexagon socket head screw
98	Stud
99	Hexagon nut
124	Housing
125	Gasket
126	Hexagon socket head screw
161	Grub screw

Fig. 4.56 Process-type rotary piston pump for delivery pressures up to 20 bar (Lederle)

452

Fig. 4.57 Rotary piston pump with an overpressure release valve and angular gear drive
motor as shown in Fig. 4.56 (Lederle)

Fig. 4.58 Rotary piston pump with hydrodynamic torque converter for smooth starting and
speed control (Lederle)

453

Fig. 4.59 Rotary piston pumps as shown in Fig. 4.58 installed in a large scale chemical plant
(Lederle)

4.2.8.1.4 Pumps with single-side bearing for high system pressures (Fig. 4.60 and 4.61)

In delivery tasks in circuits under high internal (system) pressures (e.g. up to 100 bar), axial thrust problems arise at the shaft supports. These thrust forces cannot be taken up by normal double-row roller bearings so that axial bearing is required in addition to structural strengthening of the casing.

4.2.8.1.5 Pumps with single-side bearing with plastic rotors (Fig. 4.62 and 4.63)

Rotary piston pumps with plastic rotors, the vane form of which is similar to a three-vane gear piston, have been developed for particularly good suction conditions (polygon profile). The engagement (block-sealing) point is equipped with surface sealing. The required sealing gap can be kept to a minimum and the gap flow and gap cavitation thus also reduced to a minimum. The low fit tolerances are made possible by plastic coating of the stainless steel displacer cores. The ring nuts on the cover enable quick dismantling. Fig. 4.64 shows the polygon displacer profile and the ease of replaceability of the displacer.

Parallel to the industrial model (Fig. 4.62 and 4.63), there is a development which takes account of the requirements in the foodstuffs industry (Fig. 4.65 and 4.66).

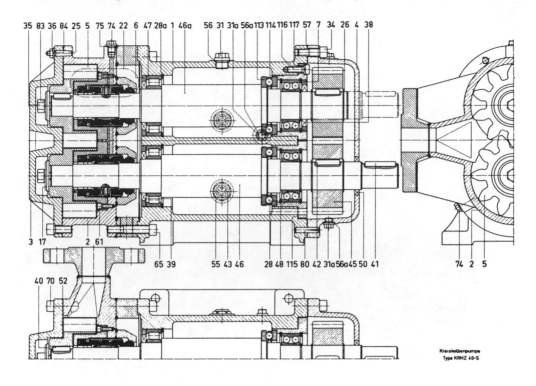

Fig. 4.60 Rotating piston pump for high system pressures (Lederle)

Fig. 4.61 Rotating piston pump for high system pressures with vertical flanges (Lederle)

Fig. 4.62 View of a rotary piston pump according Fig. 4.63 (Lederle)

The most important aspects in this model are the possibility of rapid cleaning and that of antiseptic delivery during operation. Ring nuts are used for fastening the casing cover in order to facilitate rapid cleaning. To ensure antiseptic operation, steam is injected at 0.5 bar overpressure through the borehole closed by bolt no. 156 in Fig. 4.65. The displacers are loosely attached to a square section of the shaft journal with the result that no fastening bolts are required. These pumps can be used for delivery pressures up to 10 bar. Fig. 4.67 is an exploded view of this pump. The pump is designed in such a way that the hydraulic section and the sync gear unit are separated from one another. The advantages of this design are optionally rapid manual cleaning or fully automatic CIP cleaning as well as easy replacement of the pump parts which come into contact with the product.

This series complies with the 3A hygiene regulations for rotary piston pumps used for the delivery of milk and dairy products in the USA and has been awarded the 3A hygiene standard symbol for dairy design by the competent authorities.

A rapid cleaning pump, as shown in Fig. 4.68, has been developed for delivery tasks in the low pressure range (up to 6 bar). In this case the sync gear unit is arranged between the roller bearings to make the pump shorter.

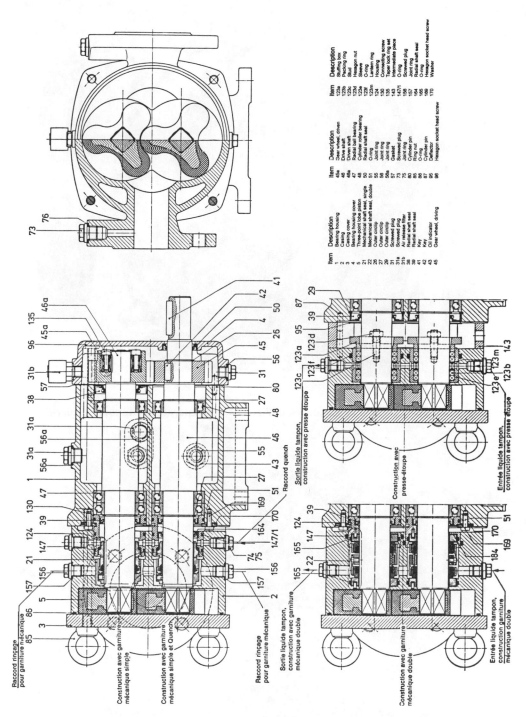

Fig. 4.63 Rotary piston pump for rapid cleaning with poygon profile plastic rotors (Lederle)

Item	Description
1	Bearing housing
2	Casing
3	Casing cover
4	Bearing housing cover
5	Three-point lobe piston
21	Mechanical shaft seal, single
22	Mechanical shaft seal, double
26	Outer circlip
27	Outer circlip
29	Outer circlip
31	Screwed plug
31a	Screwed plug
31b	Air release filter
38	Radial shaft seal
39	Radial shaft seal
41	Key
42	Key
43	Oil indicator
45	Gear wheel, driving

Item	Description
45a	Gear wheel, driven
46	Drive shaft
46a	Driven shaft
47	Radial ball bearing
48	Cylinder roller bearing
50	Radial shaft seal
55	O-ring
56	Joint ring
56a	Joint ring
57	Joint ring
73	Screwed plug
75	Joint ring
80	Cylinder pin
85	Ring nut
86	O-ring
87	Cylinder pin
95	Deflector
96	Hexagon socket head screw

Item	Description
123a	Stuffing box
123b	Packing ring
123c	Stud
123d	Hexagon nut
123e	Sleeve
123m	O-ring
124	Lantern ring
130	Housing
135	Connecting screw
143	Taper lock ring set
147/1	Intermediate piece
156	O-ring
157	Screwed plug
164	Radial shaft seal
165	O-ring
169	Joint ring
170	Hexagon socket head screw
184	Washer

Raccord rinçage pour garniture mécanique simple

Construction avec garniture mécanique simple

Construction avec garniture mécanique simple et Quench

Raccord quench

Raccord rinçage pour garniture mécanique

Sortie liquide tampon, construction avec garniture mécanique double

Sortie liquide tampon, construction avec presse étoupe

Construction avec presse-étoupe

Entrée liquide tampon, construction avec presse étoupe

Construction avec garniture mécanique double

Entrée liquide tampon, construction garniture mécanique double

457

Fig. 4.64 Rotary piston pump as shown in Fig. 4.63 with separate casing cover (Lederle)

Fig. 4.66 View of a rotary piston pump as per Fig. 4.65 (Lederle)

458

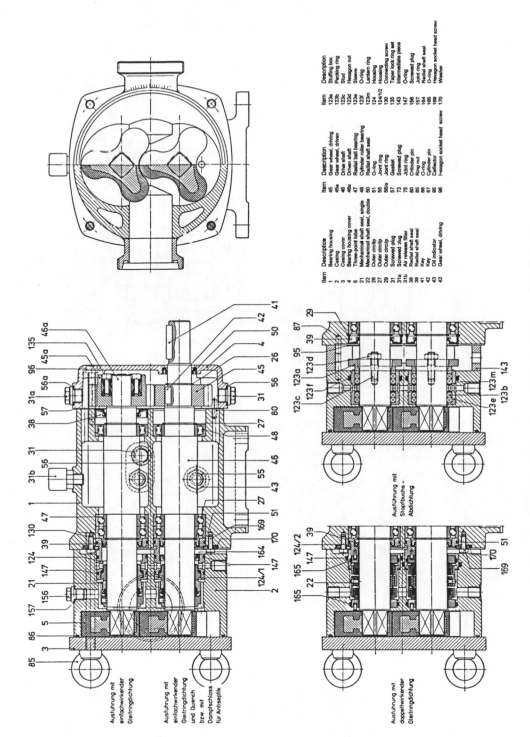

Item	Description	Item	Description
1	Bearing housing	45	Gear wheel, driving
2	Casing	45a	Gear wheel, driven
3	Casing cover	46	Driven shaft
4	Bearing housing cover	46a	Driven shaft
5	Three-point lobe	47	Radial ball bearing
21	Mechanical shaft seal, single	48	Cylinder roller bearing
22	Mechanical shaft seal, double	50	Radial shaft seal
26	Outer circlip	51	O-ring
27	Outer circlip	56a	Joint ring
29	Outer circlip	56	Joint ring
31	Screwed plug	57	Gasket
31a	Screwed plug	73	Screwed plug
31b	Air release filter	75	Cylinder pin
38	Radial shaft seal	80	Ring nut
39	Radial shaft seal	85	O-ring
41	Key	86	Cylinder pin
42	Key	95	Deflector
43	Gear wheel, driving	96	Hexagon socket head screw

Item	Description
123a	Stuffing box
123b	Packing ring
123c	Stud
123d	Sleeve
123e	O-ring
123f	Lantern ring
123m	Housing
124	Housing
124/1/2	Connecting screw
130	Connecting screw
135	Taper lock ring set
143	Intermediate piece
147	O-ring
156	Screwed plug
157	Joint ring
164	Radial shaft seal
165	O-ring
169	Oil indicator
170	Washer

Ausführung mit einfachwirkender Gleitringdichtung

Ausführung mit einfachwirkender Gleitringdichtung und Quench bzw. mit Dampfschloss für Antiseptik

Ausführung mit Stopfbuchs-Abdichtung

Ausführung mit doppeltwirkender Gleitringdichtung

Fig. 4.65 Rotary piston pump for delivery of foodstuffs (e.g. milk) (Lederle)

459

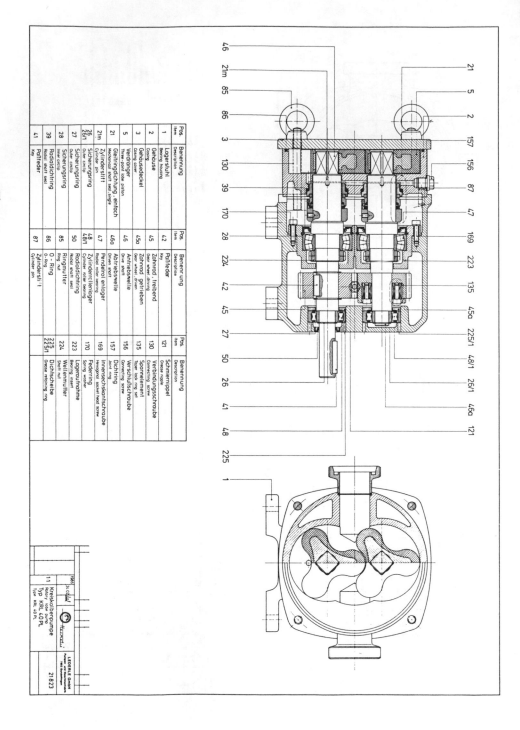

Fig. 4.68 Rotary piston pump for food industry and small delivery capacity (Lederle)

Pos. Item	Benennung Description	Pos. Item	Benennung Description	Pos. Item	Benennung Description
1	Lagerstuhl Bearing housing	42	Paßfeder Key	121	Schmiernippel Grease nipple
2	Gehäuse Casing	45	Zahnrad, treibend Gear wheel driving	130	Verbindungsschraube Connecting screw
3	Gehäusedeckel Casing cover	45a	Zahnrad getrieben Gear wheel driven	135	Spannelement Taper lock ring set
5	Verdränger Three-point lobe piston	46	Antriebswelle Drive shaft	156	Verschlußschraube Connecting screw
21	Gleitringdichtung, einfach Mechanical shaft seal, single	46a	Abtriebswelle Driven shaft	157	Dichtring Joint ring
21m	Zylinderstift Cylinder pin	47	Pendelrollenlager Radial roller bearing	169	Innensechskantschraube Hexagonal socket head screw
26 26/1	Sicherungsring Outer circlip	48 48/1	Zylinderrollenlager Cylinder roller bearing	170	Federring Spring washer
27	Sicherungsring Outer circlip	50	Radialdichtring Radial shaft seal	223	Lagernaufnahme Bearing insert
28	Sicherungsring Inner circlip	85	Ringmutter Ring nut	224	Wellenmutter Shaft nut
39	Radialdichtring Radial shaft seal	86	O - Ring O-Ring	225 225/1	Dichtscheibe Grease retaining ring
41	Paßfeder Key	87	Zylinderstift Cylinder pin		

460

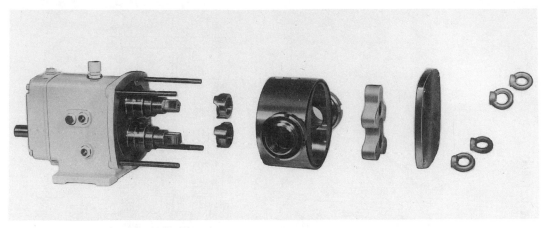

Fig. 4.67 Exploded view of a rotary piston pump as shown in Figs. 4.65 and 4.66 (Lederle)

4.2.9. Pumps of monoblock construction for the foodstuffs industry

To avoid the need for base plates or foundation frames and to have a particularly space-saving pump unit, the pump series shown in Fig. 4.69 and 4.70 was developed, and this also meets all hygiene requirements.

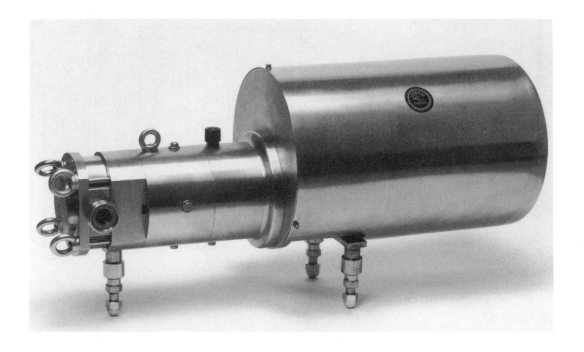

Fig. 4.69 View of a rotary piston pump as shown in Fig 4.70 (Lederle)

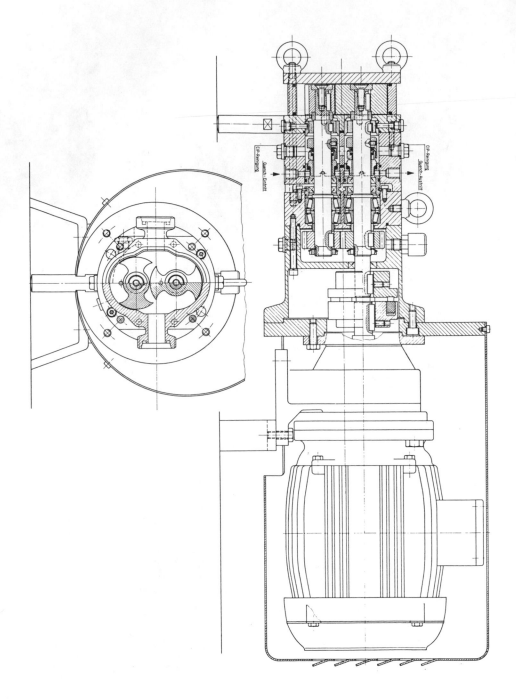

Fig. 4.70 Monobloc construction of a rotary piston pump for handling foodstuffs (Lederle)

462

4.2.10 Materials used in rotary piston pumps

The physical and chemical properties of the delivery product determine the materials used for the parts which come into contact with the fluid. The following table lists the materials most commonly used at the present time.

Table [4.II]

Casing:	Grey cast iron	GG 20
	Spheroidal graphite cast iron	GGG 40
	Bronze	GBZ 14
	Stainless steel	1.4571
		1.4581
	Hastelloy	B/C
Casing cover:	Grey cast iron	GG 20
	Spheroidal graphite cast iron	GGG 40
	Bronze	GBZ 14
	Stainless steel	1.4571
		1.4581
	Hastelloy	B/C
Bearing seat:	Grey cast iron	GG 20
Shafts:	Steel	St 60-2
	Stainless steel	1.4571
	Hastelloy	B/C
Rotor vanes:	Grey cast iron	GG 20
	Spheroidal graphite cast iron	GGG 40
	Bronze	GBZ 14
	Stainless steel	1.4571
		1.4581
	Hastelloy	B/C
	Plastic/Tefzel	

4.2.11 Rotary piston pumps for fluids with abrasive ingredients

All pumps which are used to convey fluids with abrasive solid ingredients are subject to wear. Taking this fact into consideration, the manufacturer and the operator must reach agreement on the degree of wear to be anticipated. Rotary piston pumps with relatively large gap widths at both the working sealing point and the block-sealing point as well as at the piston face ends are particularly suitable for delivery of abrasive products as long as the following conditions are fulfilled:

1. Low speeds
2. Gap widths as large as possible
3. Wear-resistant materials

463

4.2.11.1 Low speed ranges

The circumferential speeds of the rotary pistons provide a direct indication of the degree of wear. They should therefore be kept as low as possible. This is achieved by selecting a larger pump with the same volume flow. Speeds in the lower range, in other words lower than 100 r.p.m., have an extremely favorable effect on the service life of the pump.

4.2.11.2 Large gap widths

As long as the gap widths are larger than the grain diameter of the hard particles contained in the delivery flow, extremely favorable results as regards wear resistance can be anticipated. Wear is reduced to the extent to which the viscosity increases. This is due to the resulting reduced flow speed in the gaps. In this case as well, the flow speed is a direct measure of the degree of wear.

4.2.11.3 Wear-resistant materials

The hardness of the parts in contact with the fluid is of decisive significance as regards the operating life of a pump used to deliver fluids with abrasive ingredients. It should be noted however that, due to the danger of corrosion, it is not always possible to select the material which would be most suitable for the abrasive delivery product. Cast iron casings and rotor vanes made of hardened GGG 40.1 have proven their suitability for the delivery of products in cases where the corrosion resistance of the pump materials is of no significance. The depth of hardness is between 0.2 and 0.3 mm. If stainless steel is required due to its chemical resistance, type 1.4591 is used. Pumps subjected to a high degree of wear are best fitted with replaceable casing inserts (Fig. 4.12, left half).

There is a whole number of delivery tasks which can only be successfully and properly performed using rotary piston pumps while still ensuring a low degree of wear and a long pump service life.
Delivery of bitumen with abrasive ingredients and similar application in the sugar, paper and chemical industries are just a few examples.

4.2.12 Shaft sealing

Particular attention must be paid to shaft sealing not only for ecological reasons but also for efficiency and safety.
This problem can be solved in various ways, such as:

> - *Stuffing box seals with and without a sealing chamber ring*
> - *Single mechanical seal*
> - *Double mechanical or tandem seals*

Fig. 4.50 shows a cutaway view of a rotary piston pump with bearing at both ends and alternative shaft sealing (stuffing box - single mechanical seal)

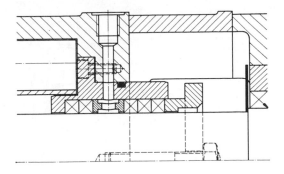

Fig. 4.71 Stuffing box packing with or without sealing or rinsing liquid chamber.
For high, low-viscous, hardening, glutinous, crystalline or soiled products at
high temperatures (max. = 180 °C) and system pressures up to p max. 10 bar.

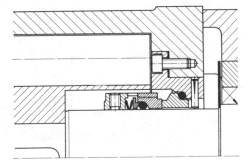

Fig. 4.72 Single, internal mechanical seal
for normal condition; i.e. products with viscosities up to max. 10000 cP with good
lubricating properties, which are not prone to hardening, coagulation or polymerization. The delivery product must be pure and free of crystals and solid matter.
Temperatures up to t = 150 °C. System pressures up to p max. 15 bar

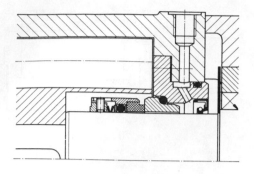

Fig. 4.73 Single, internal mechanical seal with quench connection.
The rinsing fluid feed between contra-ring and lead seal prevents the direct escape of leakage. Area of application as under 4.72. For products which are delivered under vacuum. For toxic or malodorous media and media with a tendency to harden and conglutinate in the atmosphere.
For cooling or heating.
Temperatures t max. = 150 °C. System pressures up to p max. 15 bar.

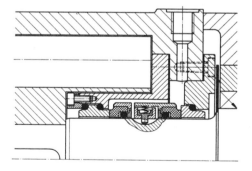

Fig. 4.74 Double mechanical seal
with sealing fluid for soiled, highly viscous, toxic, malodorous, crystalline, polymerizing or conglutination-prone delivery products and media which have to be delivered under vacuum.
Temperatures to max. - 180 °C.
System pressures up to p max. 20 bar.
For delivery or system pressures above 20 bar, balanced double mechanical seals with sealing liquid devices are used.

466

4.2.13 Example of applications of rotary piston pumps

Several examples from various industrial sectors help to show just how versatile rotary piston pumps are. Their application ranges from fluids with a viscosity below 1 mPas right through to doughy, still deformable textures at temperatures up to 300 °C.

Table [4 - III]

Chemical and Dyestuff Industries:

Acetic ether
Acetic acid
Acids
Acurol
Aluminium hydroxide
Aluminium sulphate
Ammonium carbonate
Ammonium chloride
Ammonium nitrate
Ammonium phosphate
Ammonium sulphate
Anhydrite acid
Antistatin
Araldite resin
Asphalt

Barium chloride
Barium hydroxide
Barium nitrate
Barium sulphate
Bitumen
Bunker fuel
Butyl acetate

Calcium chlorate
Calcium chloride
Calcium hydrate lye
Caprolactam
Carbamide
Carbon bisulphide
Carbon disulphide
Caurite glue
Caustic soda hydrated
Caustic soda lye
Cellulose
Chlorinated lime
Chlorobenzole
Chloroparaffin
Clay sludge
Cold glue
Colophonic acid
Colophony
Colour suspensions
Concentrates
Copper nitrate
Copper sulphate
Creams
Crude oil
Crude tar
Cyclohexane

Dichlorobutane
Diethylene glycol
Dispersions
Dodecyl benzole sulphonate
Dyestuffs

Ethyl acetate
Ethylene
Emulsifying agent

Emulsions

Fat/grease
Fatty acid
Fatty alcohol sulphonate
Formaldehyde
Fuel oil (heavy and light)
Fufurol

Glazing
Glucose
Glue
Gum

Hide glue

Impregnating varnish
Isocyanate

Lacquer paste
Lacquers
Latex
Lead sludge
Lime milk
Lithopene sludge
Lubricating oils
Luviskol
Lye

Magnesium chloride
Magnesium sulphate
Marlon
Mercury
Mercurous chloride
Monylite dispersions
Moviol
Mowilith

Naphtha

Oils
Oil spirit

Paraffin
Pastes
Perchloroethylene
Petroleum spirit
Plastic adhesives
Plastic dispersions
Plastic suspensions
Plastics
Plastinol
Polycondensate
Polyester resin
Polymerisates
Polystyrene
Polyvinyl acetate
Potassium carbonate
Potassium chloride
Potassium cyanide

Potassium hydroxide
Potassium nitrate
Potassium phosphate
Potassium sulphate
Printing ink
PVC dispersions
PVC paste

Residues from evaporators
Resins
Retexturing agent

Saltwater
Salt sludge
Schellac
Sludge lubricating grease
Soda lye
Sodium aluminate
Sodium bisulphite
Sodium carbonate
Sodium chloride
Sodium hydroxide
Sodium nitrate
Sodium silicate
Sodium suplhate
Softener
Standard bonding agents
Starch glue
Stearic acid
Styrene
Sulphite lye
Sulphonate paste
Sulphonic acid
Sulphuric acid
Sulphur (liquid)
Synthetic resin
Synthetic rubber

Tannin extract
Texapone
Thixotropic lqiuids
Titanium chloride
Titanium oxide sludge
Toluol
Trichloroethylene
Trisodium phosphate

Varnish
Vinyl acetate
Viscose
Vitriol

Water
Water glass

Zinc chloride
Zinc chloride lye
Zinc nitrate
Zinc sulphate

Tabelle [4 - III]

Soap, Cosmetics and Pharmaceutical Industries:

Alkyl sulphonate
Castor oil
Cream
Detergent paste
Detergents
Disitillates
Egg shampoo
Emulsifyer
Fatty acid
Fatty alcohol
Foam bath liquids

Glycerine
Hair cream
Herb extracts
Laundry soap
Marlopon
Monoglycerine mixture
Oils
Ointments
Olein
Pastes
Plaster adhesive

Raw pine soap
Skin cream
Slurry
Soaps
Soft soap
Starch
Sulphonate paste
Toothpaste
Vegetable extract
Waxes

Foodstuffs Industry, Breweries, Refridgeration Industry and Sugar Factories:

Alcohol
Apple pulp
Apple puree
Beer
Bread dough
Cetenine
Cheese
Chocolate
Claircs
Cocoa
Cocoa butter
Coconut fat
Coconut oil
Curds
Edible oil
Egg flip
Egg yolk
Essence
Evaporated milk
Fats
Fatty acid
Filling compound
Fish mush
Fish oil
Fruit juice

Fruit wash
Gelatine
Gelatine paste
Glucose
Glycerine
Grape juice
Herb extracts
Honey
Ice cream
Jam
Juice concentrate
Juices
Lactose solution
Lecithin
Lecithin sludge
Liquor
Liver paste
Malt extract
Mayonnaise
Meat extract
Molasses
Milk
Mineral water
Must
Mustard

Oils

Pectine
Plum jam
Pomaceous fruit wash
Preserves
Runnings
Sausage mixture
Spice
Spinach
Sugar solution
Sugar syrup
Syrup
Tomato pulp
Vegetable mush
Vinegar
Whisky
Wine
Wine yeast
Wort
Yeast
Yeast milk

Paper and Cellulose Industries, Spinning Mills and Tanneries:

Air water mixture
Alum
Bleaching lye
Cellulose
Cellulose acetate
Cellulose nitrate
Chemical pulp

Coating paste
Finishing agents
Kaolin sludge
Mechanical wood pulp
  up to 20% abs. dry
Paper fabric
Precipitants

Spinning solutions
Sulphate lye
Sulphite lye
Viscose

Fig. 4.75 Rotary piston pump with double mechanical sealing and sealing liquid device (Lederle)

Fig. 4.76 Mounting of a rotary piston pump (Lederle)

Fig. 4.77 Rotary piston pump in a large scale chemical plant (Lederle)

Fig. 4.78 Rotary piston pump for delivery of high viscosity rubber (Lederle)

4.3 Gear pumps

As one of the most commonly manufactured rotary positive-displacement pumps, the gear pump has conquered a wide market thanks to its simple design and the possibility of economical production. It is used in large numbers in hydraulic systems as well as for conveying tasks in the chemical, petrochemical, pharmaceutical and bitumen-processing industries.

4.3.1 Operating principle

Two or more engaged gearwheels rotate in an encapsulating casing: during this process power is transferred from one gear wheel to the other. This creates and extremely good seal between suction and discharge side at the block-sealing point - tooth profile against tooth profile. At the working sealing points between the tooth heads and the casing there is a gradual reduction in pressure which enable gear pumps to achieve good volumetric rates of efficiency. The precondition for this, however, is that the speed is appropriate for the viscosity of the fluid, thus avoiding working chamber cavitation. The tooth gaps created when the gearwheels rotate into the suction chamber are filled with the fluid by virtue of the atmospheric or system pressure acting on the fluid. Upon further rotation, the fluid is conveyed in the direction of the circumference and then pressed into the pressure chamber as the counter-tooth on the discharge side enters this tooth gap (Fig. 4.79).

When the wheels rotate in the direction of the transition from the suction to the discharge side, a chamber (squeeze chamber) is formed in the tooth gaps, the content of which is reduced to a minimum during rotation of the wheels (Fig. 4.79). In order to prevent the enclosed residual fluid from being excessively compressed - this would lead to a pressure increase which could damage the pump - measures must be taken to ensure that the squeeze fluid can be guided into the pressure chamber of the pump.

This is achieved by means of cutouts in the side casing walls of the pump opposite the face ends of the gearwheels (Fig. 4.80). These cutouts should not extend as far as the suction chamber - otherwise they would lead to an internal "short circuit" which would impair the suction capacity of the pump. This squeezing process is accompanied by a loss of energy. The production accuracy of the gearwheels, which should have as hard a surface as possible, is a measure of the performance of the pump; it should be ensured that the contact ratio factor of the gearing is greater than 1. The axial and radial clearance between casing and gearwheels should be kept to a minimum, particularly in pumps which are used to deliver low-viscosity fluids, in order to obtain good volumetric rates of efficiency. Care must be taken to ensure, however, that the heat expansion and the deflection due to the pressure load on the wheels are taken into account. In the case of high-viscosity fluids, larger clearances are admissible, and although this leads to a reduction in the volumetric efficiency it also considerably reduces lateral friction of the wheels (see also Section 4.2.3.4 "Piston clearances"). The low degree of clearance at the side and head and the non-positive power transmission at the gearwheels render these pumps unsuitable for fluids with solid, abrasive inclusions. They may only be used to convey fluids which posses a certain lubricating effect and which are free of solid, hard components. If used to deliver particularly viscous fluids, the pumps are equipped with a heating jacket in order to reduce the viscosity.

471

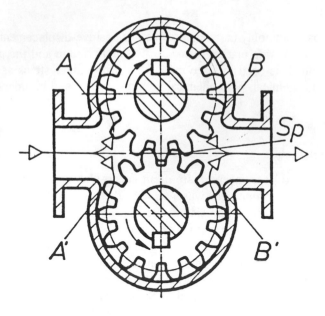

Fig. 4.79 Gear pump, working sealing point from A to B and A' to B', block-sealing at Sp

Cutout on the side walls for the flow-off of the squeeze fluid into the pressure chamber

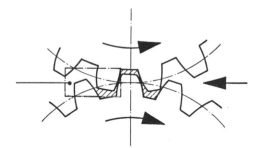

Fig. 4.80 Cutout on the side walls for the flow-off of the squeeze fluid

4.3.2 Toothing

These pumps are equipped with a specific type of toothing appropriate to the delivery tasks they are to perform. Most gear pumps possess straight-toothed spur wheels with an involute profile. If quiet running properties are desired, helical and double helical gear wheels are used. In the case of helical gear wheels, the problem of axial thrust must be dealt with, if necessary by fitting special bearings. In double helical gear wheels, the thrust forces cancel each other out.

472

4.3.3 Irregularity

The volume flow delivered by gear pumps is not free of pulsation. It is subject to a degree of irregularity which depends on the number of teeth, the type of toothing and the discharge of the squeeze fluid. The reason for this is that the fluid is subject to considerable fluctuation in speed as well as strong shearing forces as it flows through the pump. First of all, the fluid must fill the rotating tooth gaps, a process for which only a short time is available. The fluid is then accelerated in the direction of the circumference to the circumferential velocity of the gearwheel, and finally slowed down again to the flow speed in the pressure chamber after it has left the tooth gaps. In addition, the squeeze fluid has to escape from the centre of the wheel to the lateral cutouts of the housing covers during this process in order to flow off into the pressure chamber. Return flows are created above and to the side of the gap and at the tip circle, and these make an additional contribution to irregularity. An increase in the number of teeth and an increasing engagement angle lead to a reduction of delivery flow fluctuation. The average degree of delivery flow fluctuation to be expected in gear pumps is in the order of 2 - 5%.

4.3.4 Delivery flow Q

According to Molly [4-8], the theoretical delivery volume per revolution of a gear pump with two external-toothed wheels with DIN involute spur toothing is:

$$V_u = b \cdot \frac{\pi}{2} \cdot \left[d_k^2 - c^2 - d_g^2 \cdot \frac{\pi^2}{3 \cdot z^2} \right] \qquad (4 - 32)$$

b	=	tooth width
d_k	=	addendum diameter
d_g	=	circle dedendum diameter, not to be confused with base diameter
c	=	axis spacing
z	=	number of teeth on the driving wheel
n	=	speed r.p.m. of the driving wheel, then the theoretical delivery flow is

$$Q_{th} = V_u \cdot n \; l/min \qquad (4 - 33)$$

The actual effective delivery flow Q is reduced by the amount of the internal losses and possible non-filling of the tooth gaps (working chamber cavitation) thus:

$$Q = Q_{th} - Q_v \qquad (4 - 34)$$

where Q_v results from the gap flow losses Q_{sp}, expressed in the degree of sealing

$$\lambda_1 = \frac{Q}{Q + Q_{sp}} \qquad (4 - 35)$$

and the degree of filling

$$\lambda_2 = \frac{Q + Q_{sp}}{Q_{th}} \qquad (4 - 36)$$

resultiert.

The quotient from the difference between $(Q_{th} - Q_v)$ and Q_{th} is called the volumetric efficiency λ_{ges}.

$$\lambda_{ges} = \frac{Q_{th} - Q_v}{Q_{th}} = \frac{Q}{Q_{th}} = \lambda_1 \cdot \lambda_2 \qquad (4 - 37)$$

$$Q = \lambda_{ges} \cdot Q_{th} \qquad (4 - 38)$$

A simplified formula can be used in place of the Molly formula and is generally sufficiently accurate:

$$Q_{th} = 2\left[\frac{\pi}{4} \cdot d_k^2 - \frac{\pi}{4} \cdot d_g^2\right] \cdot b \cdot n \qquad (4 - 39)$$

$$Q_{th} = \frac{\pi}{2} \cdot b \cdot n[d_k^2 - d_g^2]$$

4.3.5 The total head H or delivery pressure p

(temperature and viscosity range)
What was shown to apply to rotary piston pumps in Section 4.2.3 ("The characteristic field") also applies to gear pumps as hydrostatic operating machines with regard to the mode of delivery and their H (Q) characteristics. The applications of these pumps range from the low pressure sector with pumping heads of only a few metres right through to the high pressure range up to delivery pressures of 300 bar. The fluid temperatures for which gear pumps can be used range from -60 to +450 °C, the viscosities from 1 mPas to 2×10^6 mPa·s, and these pumps can therefore be used over a wide range of applications.

474

4.3.6 Power P

The formula derived for rotary piston pumps also applies here, since the preconditions for the delivery process are the same. Thus the power requirement P of the pump is:

$$P = \frac{\rho \cdot g \cdot Q \cdot H}{\eta_{ges}} \qquad (4 - 40)$$

where

ρ	=	density
g	=	acceleration due to gravity
Q	=	capacity
H	=	total head
η_{ges}	=	$\eta_h \cdot \eta_m \cdot \lambda_{ges}$
η_h	=	hydraulic efficiency
η_m	=	mechanical efficiency
λ_{ges}	=	volumetric efficiency

4.3.7 Materials used in gear pumps

The choice of pump materials depends on the task to be performed; according to the aggressiveness of the fluid, as well as its temperature and internal pressure (delivery pressure and feed pressure), a wide spectrum of materials is used, ranging from cast iron, nodulised cast iron, bronze, steel, various types of stainless steel and alloys with high nickel content right through to plastics. The choice of material for the engaged gearwheels is of particular importance. If stainless steels or alloys with high nickel content are required to protect the pump from corrosion, the material combination selected should be such as to prevent seizing up of the tooth profiles and in the housing. The materials listed in the following table [4-IV] have proved suitable for these tasks.

Table [4-IV]

Drive wheel	Driven wheel
Teflon, glass-fiber reinforced	Teflon, glass-fiber reinforced
Stainless steel - high nickel	Stainless steel - high nickel
Hastelloy C	Teflon, glass-fiber reinforced
Hastelloy C	Hastelloy C

4.3.8 Shaft seals

The types of seals listed under Section 4.2.12 ("Shaft sealing") for rotary piston pumps also offer design solutions for gear pumps.

> - *Stuffing box packing with and without a sealing chamber ring*
> - *Single mechanical seal*
> - *Double mechanical seals*

The relevant technical and company publications give details of the versions and shapes of the materials and their specific areas of application.

4.3.9 Gear pump design types

There are many different designs of gear pumps to deal with the wide range of fluid delivery tasks. At this point we would like to take a closer look at the pumps used in the area of plant engineering.

4.3.9.1 Gear pumps with spur-toothed wheels

Fig. 4.81 shows the most elementary gear pump design. A pair of gearwheels rotates in an encapsulating casing. The bearings are journal bearings made mainly of carbon or SiC. Fig. 4.82 shows two series-connected gear pumps. With this method, it is possible to double the delivery pressure. It is also of course possible to arrange the pumps in parallel, thus doubling the delivery flow.

The gear pump in Fig. 4.83 was developed for the high pressure range (100 bar). The precondition for this is that the wheels have a high number of teeth and that the stuffing box is connected downstream of a long throttle section with hydraulic balance towards the suction side. An overpressure safety valve is fitted to the pump.
Fig. 4.84 show a gear pump which achieves delivery pressures up to 20 bar and is used in large numbers in technical process plants.

4.3.9.2 Gear pumps with internal toothing (Fig. 4.85, 4.86 and 4.87)

Internal-toothed gear pumps have proven themselves suitable for the delivery of highly viscous fluids. A spur wheel is driven by a planet wheel, and a sickle-shaped casing section located opposite the tooth engagement point seals the suction chamber off from the pressure chamber. In contrast to other, similar pump designs, the intake of fluid does not take place against the centrifugal force into the working chambers, but is forwarded to the latter axially, and this results in more favorable $NPSHR_S$ values.
This construction has the advantage that the pump runs smoothly even when conveying viscous fluids and can be of relatively compact design. As the drive and driven gearwheels rotate in the same direction, external-toothed gear pumps are not subject to shear forces. In addition, the long tooth engagement phase in internal gear pumps reduces irregularity, thus resulting in lower pressure fluctuations. The air/gas suction capacity of the internal

gear pump is considerably better than that of the external gear pump. The pinion driven by the planet wheel normally has a steel or stainless steel core which is plastic-coated. The wear is thus transferred to an easily replaceable machine part. Compared to external gear pumps, internal gear pumps have only a few teeth of simple geometric design (see German Federal Patent DE 2951952 C2). Fig. 4.88 shows characteristic curve sheet of an internal-toothed gear pump.

The volume flow reduces linearly with increasing delivery pressure, the output increases linearly.

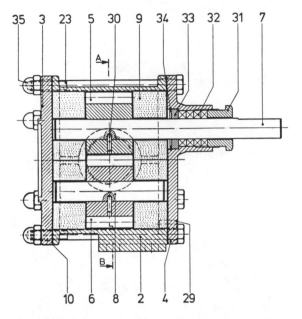
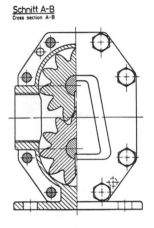

Pos. Item	Benennung Description	Pos. Item	Benennung Description
2	Gehäuse Casing	23	Sechskantschraube Hexagon screw
3	Gehäusedeckel Casing cover	29	Zylinderstift Cylinder pin
4	Gehäusedeckel, motorseitig Casing cover, motorside	30	Zylinderstift Cylinder pin
5	Zahnrad, treibend Gear wheel, driving	31	Stopfbuchse Stuffing box
6	Zahnrad, getrieben Gear wheel, driven	32	Packungsring Packing ring
7	Antriebswelle Drive shaft	33	Scheibe Washer
8	Abtriebswelle Driven shaft	34	Sicherungsring Inner circlip
9	Lager Bearing	35	Hutmutter Domed nut
10	Gehäusedichtung Gasket for casing		

Fig. 4.81 Gear pump with spur-toothed wheels (Lederle)

477

Fig. 4.82 Row/series arranged gear pumps with three-phase motor drive (Lederle)

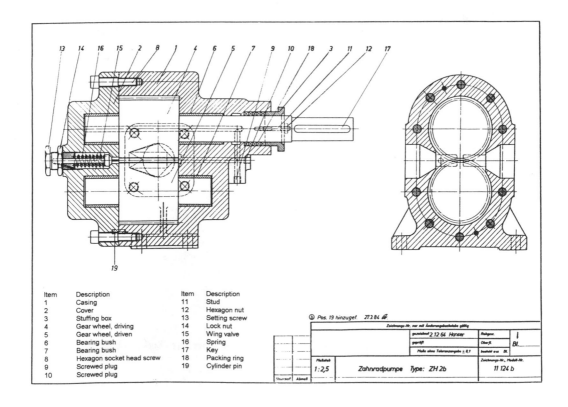

Item	Description	Item	Description
1	Casing	11	Stud
2	Cover	12	Hexagon nut
3	Stuffing box	13	Setting screw
4	Gear wheel, driving	14	Lock nut
5	Gear wheel, driven	15	Wing valve
6	Bearing bush	16	Spring
7	Bearing bush	17	Key
8	Hexagon socket head screw	18	Packing ring
9	Screwed plug	19	Cylinder pin
10	Screwed plug		

Fig. 4.83 Spur-toothed gear pump for 100 bar delivery pressure (Lederle)

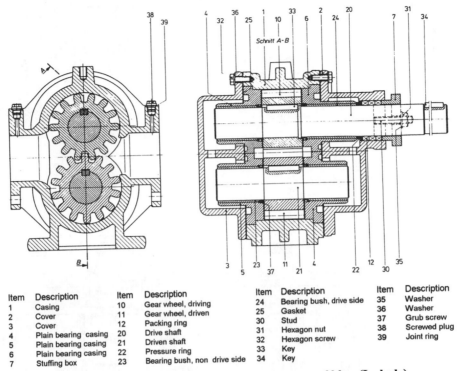

Item	Description	Item	Description	Item	Description	Item	Description
1	Casing	10	Gear wheel, driving	24	Bearing bush, drive side	35	Washer
2	Cover	11	Gear wheel, driven	25	Gasket	36	Washer
3	Cover	12	Packing ring	30	Stud	37	Grub screw
4	Plain bearing casing	20	Drive shaft	31	Hexagon nut	38	Screwed plug
5	Plain bearing casing	21	Driven shaft	32	Hexagon screw	39	Joint ring
6	Plain bearing casing	22	Pressure ring	33	Key		
7	Stuffing box	23	Bearing bush, non drive side	34	Key		

Fig. 4.84 Spur-toothed gear pump for delivery pressure up to 20 bar (Lederle)

Fig. 4.85 Internal gear pumps in a production plant (Lederle)

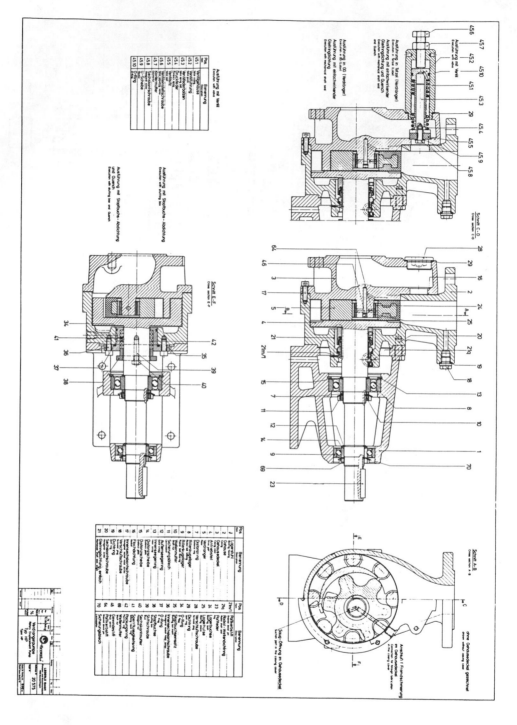

Fig. 4.86 Internal gear pump (Lederle)

480

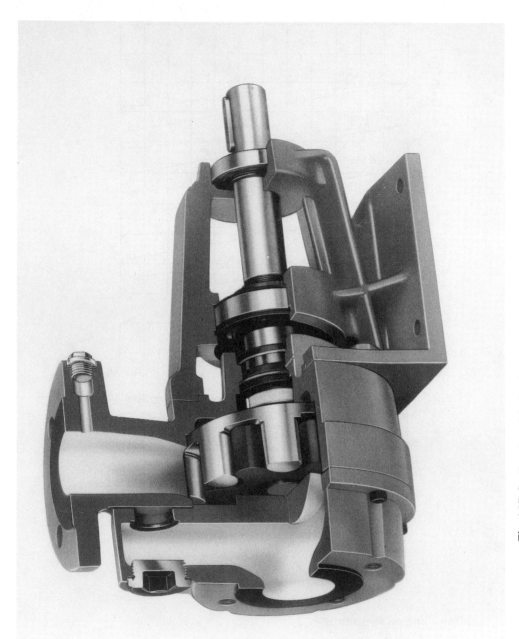

Fig. 4.87 Sectioned model of an internal gear pump

481

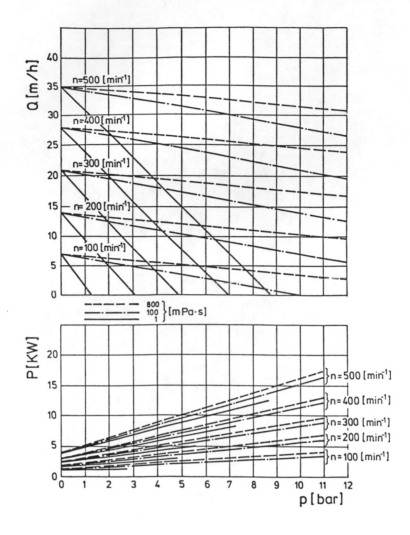

Fig. 4.88 Family of characteristics of an internal gear pump

4.4 Sliding-vane screw pumps

The principle of sliding-vane screw pumps is new to the field of positive-displacement pumps. Their outer appearance is similar to that of the well-known helical screw pump but their mode of operation is completely different. Their parameters are wide ranging and the pumps can deal with viscosities from 1 to 1000 Pa·s, pressures up to 20 bar, delivery flows up to 60 m³/h and temperatures of up to 150 °C at a speed from 1 to 1450 r.p.m.. These figures show that sliding-vane screw pumps are in a position to take on a large proportion of the delivery tasks which have up to now been performed by rotary -positive displacement pumps, and this with an astonishingly low unit-weight and relatively compact design. They represent a real alternative to conventional systems, not least due to their low purchase price and repair costs.

482

4.4.1 Delivery principle of the sliding-vane screw pump

The basic idea of the sliding-vane screw pump is to combine the principles of the cell vane pump with that of an Archimedean screw. In this way it is possible to effect continuous volumetric delivery. If we take a closer look at the cross section of a sliding-vane screw pump at any point in the delivery chamber we can see that the eccentrically rotating screw shaft is located eccentrically to the respective casing sections. A sickle-shaped chamber filled with product is created which is interrupted by the drive or block vane sliding in the shaft (Fig 4.89). If we imagine the finite length of a vane cell to be infinitely small, and if such cells are arranged in a spiral form one after the other, a spiral casing is created as a single-flighted screw with two opposite sealing vane contact lines which separate the suction pressure chambers from one another both during standstill and operation. (Fig. 4.90)

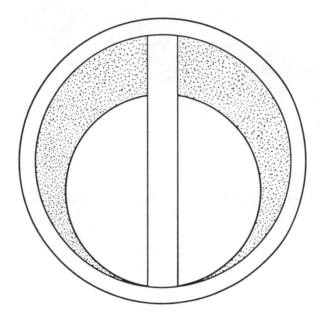

Fig 4.89 Cross-section of the transport chamber of the sliding-vane screw pump.

If the rotor passing through the spiral chamber casing is moving centrical, the sealing elements perform a sliding motion at their outer ends relative to the casing which encapsulates them. At the same time, a to-and-fro motion of the sealing elements is generated when the rotor turns. During this process the sealing elements are forcibly driven by the casing and shifted with regard to one another by the value Δs (Fig 4.91). A fluid situated between casing and shaft is thus transported in spiral fashion in axial direction. A continuous delivery flow without pulsation is created, thus ensuring smooth running of the pump at both high and low speeds. The flow direction can be reversed by changing the rotational direction of the motor. Due to its hydrostatic delivery method, the pump is self-priming and also capable of conveying fluids with gas inclusions. No valves are required for this.

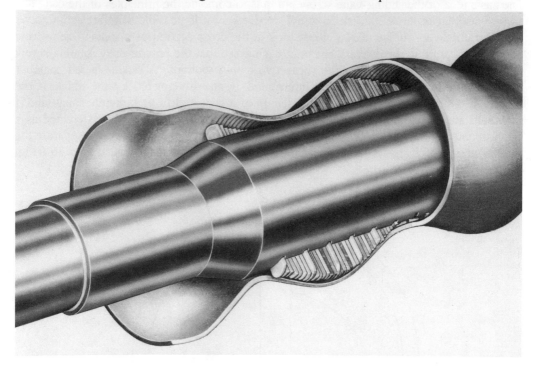

Fig 4.90 Stator (casing cross-section) and rotor of a sliding-vane screw pump

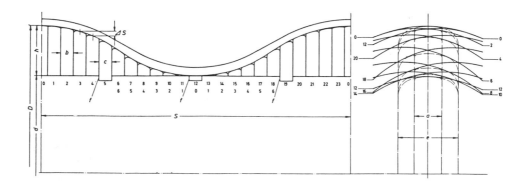

Fig 4.91 Motion characteristic of the vanes

4.4.2. Geometry of the delivery chamber and kinetics of the vane motion

The starting point for the representation of the geometry of the displacer chamber is the circular cross-section of the shaft at any desired point in the casing (Fig 4.92). The shafts rotates centrically about its central point A. A sickle-shaped area in the respective section plane about the shaft is required for constant volume change to be performed by the blocking vane which slides in the shaft. This area is formed by a circle (construction circle) which is eccentric to the centre point A of the shaft and has its centre at B. The point of contact 12 is formed in the direction of the spiral line by mean of a systematic adjacent arrangement of cross sections offset with regard to one another. At 12, the sickle-shaped area begins to widen to the culmination point 0, then narrows again to the point of origin 12. If we observe the connecting line 6 to 18 which travels across the shaft central point A, this line represents the theoretical vane length L. If this line is rotated by 90° in the direction from 0 to 12, L does not contact the construction circle 0. The theoretical vane length L only extends as far as 0'. If we assume that the arc of a circle is formed from 6 to 18 via 12 with the centre B, then the generating curve of an envelope corresponding to the theoretical length L of the blocking vane must be created starting at 18, rising to point 0 and falling from there to point 6.

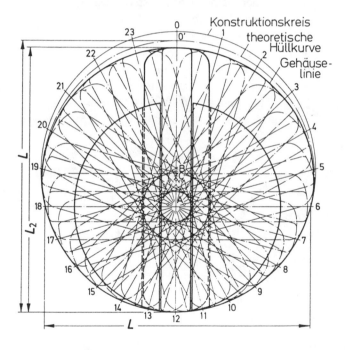

Konstruktionskreis	=	Construction circle
theoretische Hülkurve	=	Theoretical generating curve
Gehäuselinie	=	Casing line

Fig 4.92 Geometry of the displacer chamber

485

4.4.3 Capacity and total head

The theoretical working volume V_{thH} per revolution is computed from the respective sickle area A_1 (Fig 4.93).

$$A_1 = \frac{\pi \cdot [D^2 - d^2]}{4} \qquad (4 - 41)$$

multiplied by the axial pitch s less the volume of the part of the vane protruding in to the delivery chamber and that of any supporting vanes.

Therefore

$$V_{thH} = \frac{\pi \cdot [D^2 - d^2]}{4} \cdot s - [[a \cdot h \cdot \Sigma b] + [e \cdot h \cdot \Sigma c]] \qquad (4 - 42)$$

where

V_{thH}	theoretical working volume
s	axial pitch
D	inner diameter of the spiral casing
d	outer diameter of the shaft
a	width of the vane
h	height of the vane (D-d)
b	thickness of the vane
e	width of the supporting vane
c	thickness of the supporting vane

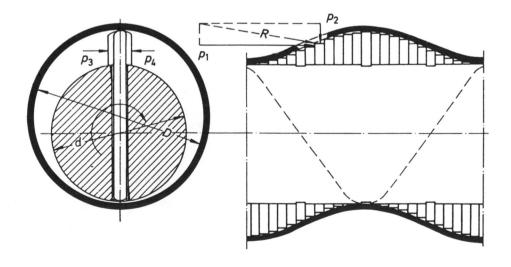

Fig 4.93 Forces at the blocking vane

The summation of the vanes refers to the axial pitch s in each particular case; the theoretical delivery volume Q_{th} per second is thus

$$Q_{th} = V_{thH} \cdot \frac{n}{60} \quad 1 \cdot s^{-1} \tag{4 - 43}$$

$$V_{thH} = dm^3 \; ; \; n = r.p.m.$$

The finite widths a of the blocking vanes do not permit the realisation of the theoretical length L as described above. If the blocking vanes are plotted in their constructional widths a necessary to achieve the corresponding delivery pressures with their centre lined through the shaft point A leading to different angle positions 0-12 with regard to the vertical plane (in analogy to Fig 4.92 from 0-23), a new curve of an envelope is created which corresponds to the actual path of the blocking vanes. This means a reduced delivery flow relative to the described surfaces of the theoretical and actual envelope curve path A_2/A_1. A_1 is the planimetered theoretical sickle area, and A_2 the planimetered actual sickle area. Sliding-vane pumps possess a hydrostatic delivery characteristic. The delivery flow should therefore be independent of the pumping head. Gaps are, however, inevitable with rotating parts and parts moving with relation to one another, and this results in reverse flows, the so-called gap losses Q_v. The effective delivery volume Q is thus:

$$Q = Q_{th} \cdot \frac{A_2}{A_1} - Q_v \tag{4 - 44}$$

$$Q_v = Q_{v1} + Q_{v2} + Q_{v3} \tag{4 - 45}$$

Q_{v1}	=	gap loss shaft-casing
Q_{v2}	=	gap loss vane-vane
Q_{v3}	=	gap loss vane-casing

The movement pattern of the blocking vanes is shown in Fig 4.91. The vanes always follow the same course on their envelope curve path; thus they never leave their path of orbit at the casing wall in the direction of the axial pitch. In order to obtain the best possible sealing characteristics between casing wall and blocking vane head, the heads are roof-shaped at their run-in ends. the run-off edge thus creates the sealing line between the two pressure chambers p_3 and p_4 (Fig 4.93). The blocking vanes of a chamber ; e.g. 13, 14, 15, 16, 17, 18 and 19 are shifted in the shaft, oscillating as an overall packet. During this process, there is only a small degree of movement of the vanes relative to one another, correspondingly to the axial pitch Δs (Fig 4.91). The spiral thread motion described above, with which the shaft is in linear contact, forms individual chambers together with the blocking vane sets. In one revolution of the shaft the working volume of two chambers is displaced.

It is evident that the axial pitch of the spiral casing is directly related to the achievable pumping head of the pump. Small axial pitches provide more chambers over the same casing length of the axial pitches. A greater number of chambers, however, means a smaller pressure difference from chamber to chamber at a given pumping head, and thus a reduced reverse flow, in other words reduced losses.

It is therefore possible to achieve different pumping heads with one and the same pump by using different axial pitches. In the reverse case, it is possible to deliver a correspondingly larger volume flow by using a large axial pitch of the spiral casing and a lower pumping head, since the sealing chambers have greater volumes (Fig. 4.94)

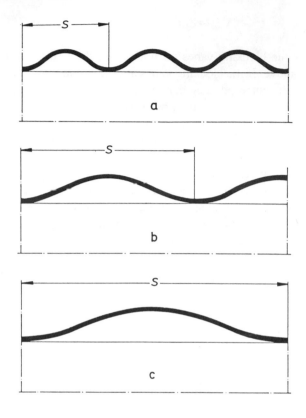

Fig 4.94 Stator designs with various axial pitches

4.4.4   Design of the sliding-vane screw pump

The parts which make up the sliding-vane screw pump are relatively simple. The casing (stator) is a stainless steel pipe in the form of a spiral in which a shaft (rotor) moves centrically. Sliding-vanes are arranged opposite each one another vertical to the axis of the shaft and moveable on it. (Fig 4.90). In order to keep reverse flow loses at the outer sealing line to a minimum, these sliding-vanes should theoretically be infinitesimally slim. In practice, these vanes are up to 3 mm thick, with the result that a certain degree of reverse flow must be accepted. Their size is dependent on the pressure difference between the front and rear sides of the vane, the viscosity of the delivery product and the speed. The spiral casing is connected to a suction and pressure nozzle. Shaft sealing can take the form of a stuffing box or single or double mechanical seal. For double seals, the two single seals are used "back to back". The seal used varies in accordance with the task to be performed. It is advisable to choose the transport direction of the pump so that the shaft seal is against the suction side in order to keep leakage to a minimum. Sliding-vane pumps are equipped with a heating jacket as standard. It should be noted that the ideal heat transfer from thermal medium to the delivery fluid via the thin-walled stainless steel stator is an advantage which basically distinguishes this machine form the double helical screw pump.

The standard version of the sliding-vane screw pump is fitted with a bearing seat (Fig. 4.95, 4.96 and 4.97). The block type is an extremely short design (Fig. 4.98). An upright version has been designed for applications where there is a minimum of available space (Fig. 4.99). The sliding-vane system is also used as an immersion and sample-taking pump (Fig. 4.100, 4.101 and 4.102). In all cases, either direct drive systems (n = 750, 900, 1450 r.p.m), gear motors or variable drive units can be used, depending on the viscosity of the delivery product or the degree of regulation of the volume flow.

Fig. 4.95 View of sliding-vane screw pump as shown in Fig 4.97 (Lederle)

Fig. 4.96 Sliding-vane screw pumps with V-belt drive (Lederle)

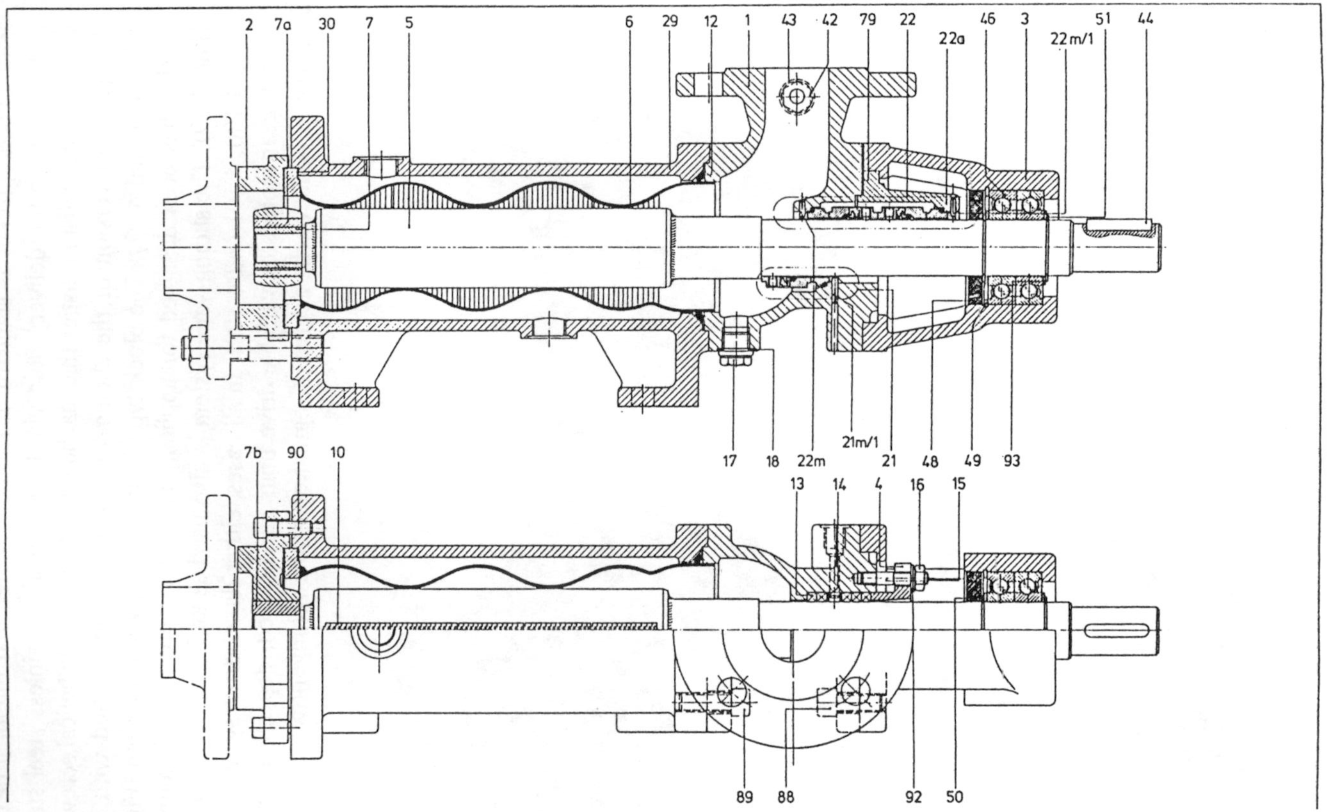

Fig. 4.97 Sliding-vane pump (Lederle)

Item	Description
1	Pressure casing
2	Suction flange
3	Bearing housing
4	Stuffing box gland
5	Shaft
6	Stator casing
7	Shaft sleeve
7a	Bearing bush

Item	Description
7b	Bearing bush
10	Vane
12	O-ring
13	Packing ring
14	Sealing liquid ring
15	Stud
16	Hexagon nut
17	Screwed plug

Item	Description
18	Joint ring
21	Mechanical shaft seal, single
21m/1	Grooved dowel pin
22	Mechanical shaft seal, double
22a	Casing for mechanical seal
22m	Casing for mechanical seal
22m/1	Grooved dowel pin
29	Pump casing

Item	Description
30	Gasket
42	Screwed plug
43	Joint ring
44	Key
45	Grooved ball bearing
46	Grease regulator
48	Inner circlip
49	Outer circlip
50	Disc

Item	Description
51	Outer circlip
79	Gasket
88	Hexagon socket head screw
89	Hexagon socket head screw
90	Hexagon socket head screw
92	Washer
93	Disc

490

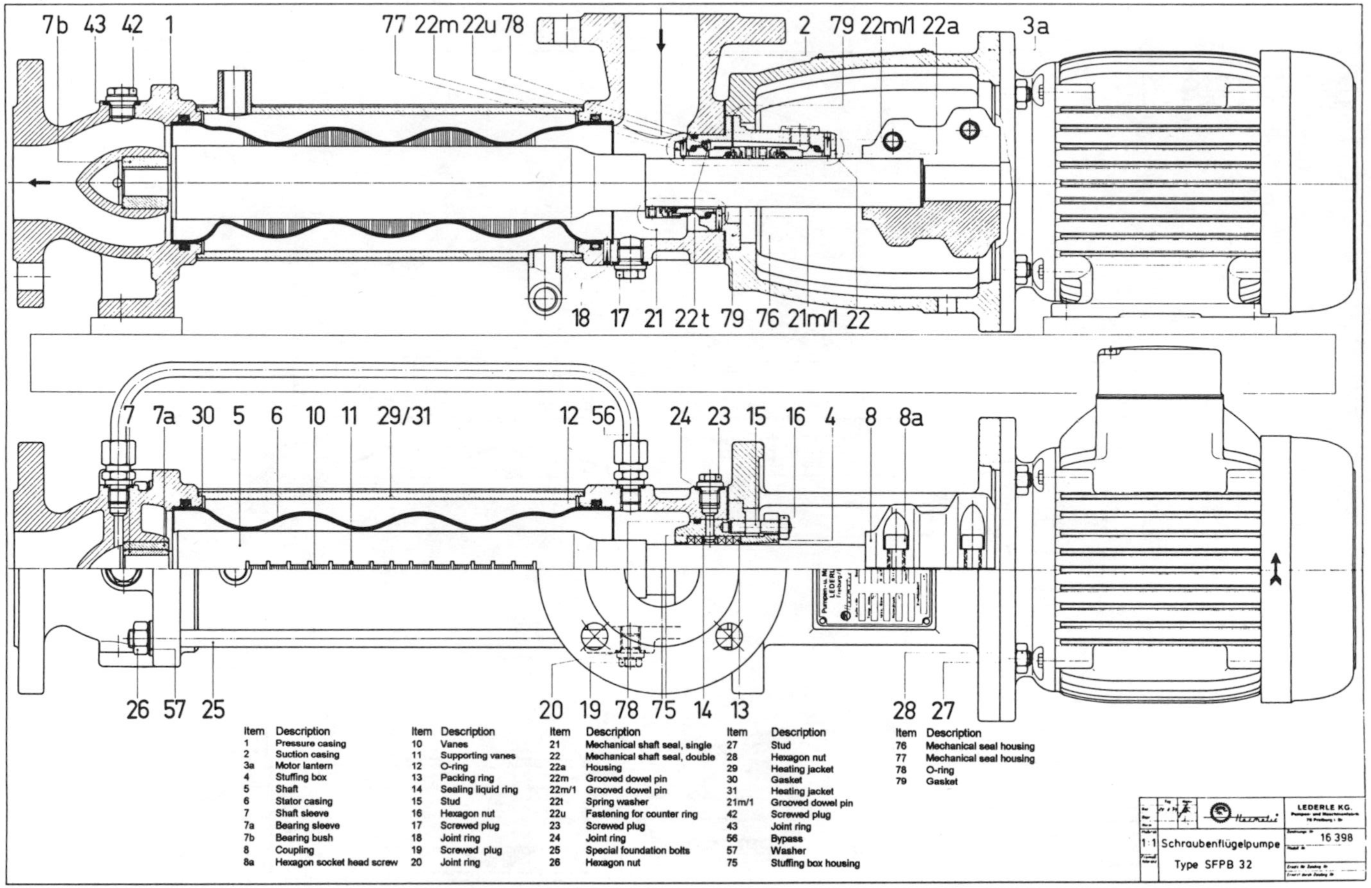

Item	Description
1	Pressure casing
2	Suction casing
3a	Motor lantern
4	Stuffing box
5	Shaft
6	Stator casing
7	Shaft sleeve
7a	Bearing sleeve
7b	Bearing bush
8	Coupling
8a	Hexagon socket head screw

Item	Description
10	Vanes
11	Supporting vanes
12	O-ring
13	Packing ring
14	Sealing liquid ring
15	Stud
16	Hexagon nut
17	Screwed plug
18	Joint ring
19	Screwed plug
20	Joint ring

Item	Description
21	Mechanical shaft seal, single
22	Mechanical shaft seal, double
22a	Housing
22m	Grooved dowel pin
22m/1	Grooved dowel pin
22t	Spring washer
22u	Fastening for counter ring
23	Screwed plug
24	Joint ring
25	Special foundation bolts
26	Hexagon nut

Item	Description
27	Stud
28	Hexagon nut
29	Heating jacket
30	Gasket
31	Heating jacket
21m/1	Grooved dowel pin
42	Screwed plug
43	Joint ring
56	Bypass
57	Washer
75	Stuffing box housing

Item	Description
76	Mechanical seal housing
77	Mechanical seal housing
78	O-ring
79	Gasket

Fig. 4.98 Sliding-vane pump of monoblock construction (Lederle)

491

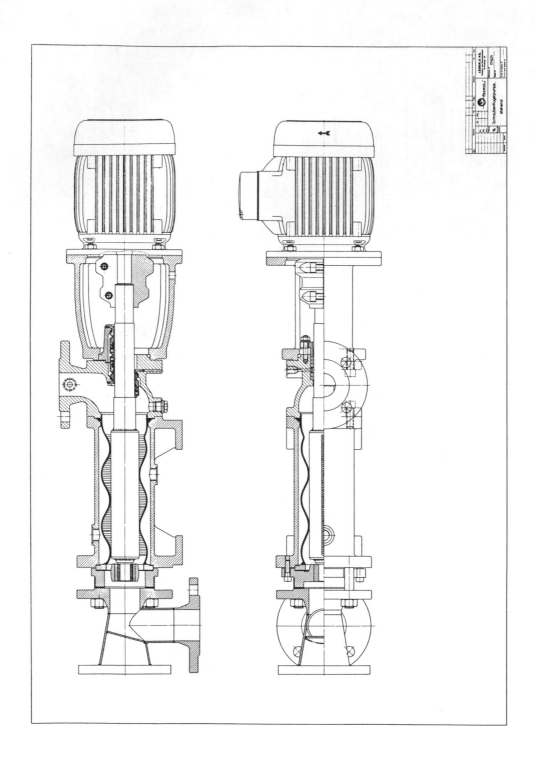

Fig. 4.99 Sliding-vane screw pump in vertical design (Lederle)

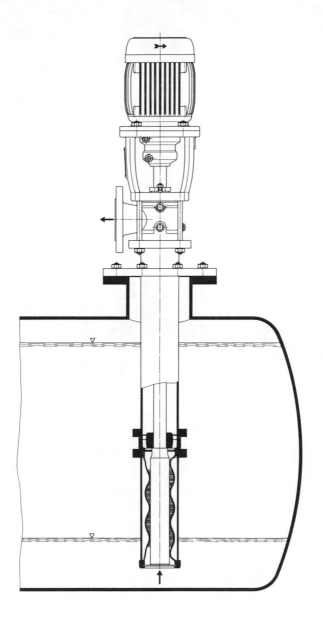

Fig. 4.100 Sliding-vane screw pump mounted in tank (Lederle)

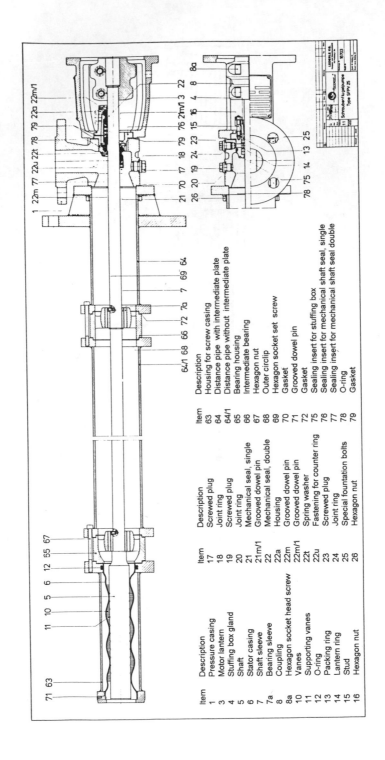

Item	Description	Item	Description	Item	Description
1	Pressure casing	17	Screwed plug	63	Housing for screw casing
3	Motor lantern	18	Joint ring	64	Distance pipe with intermediate plate
4	Stuffing box gland	19	Screwed plug	64/1	Distance pipe without intermediate plate
5	Shaft	20	Joint ring	65	Bearing housing
6	Stator casing	21	Mechanical seal, single	66	Intermediate bearing
7	Shaft sleeve	21m/1	Grooved dowel pin	67	Hexagon nut
7a	Bearing sleeve	22	Mechanical seal, double	68	Outer circlip
8	Coupling	22a	Housing	69	Hexagon socket set screw
8a	Hexagon socket head screw	22m	Grooved dowel pin	70	Gasket
10	Vanes	22m/1	Grooved dowel pin	71	Grooved dowel pin
11	Supporting vanes	22t	Spring washer	72	Gasket
12	O-ring	22u	Fastening for counter ring	75	Sealing insert for stuffing box
13	Packing ring	23	Screwed plug	76	Sealing insert for mechanical shaft seal, single
14	Lantern ring	24	Joint ring	77	Sealing insert for mechanical shaft seal double
15	Stud	25	Special fountation bolts	78	O-ring
16	Hexagon nut	26	Hexagon nut	79	Gasket

Fig. 4.101 Sliding-vane screw pump for tank emptying, as shown in Fig. 4.100 (Lederle)

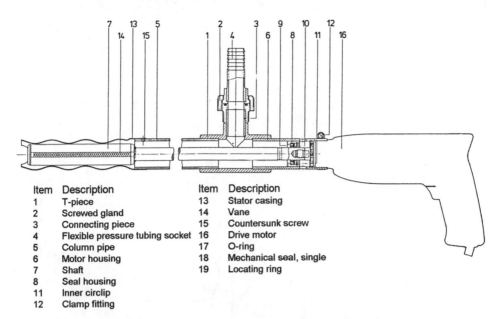

Item	Description	Item	Description
1	T-piece	13	Stator casing
2	Screwed gland	14	Vane
3	Connecting piece	15	Countersunk screw
4	Flexible pressure tubing socket	16	Drive motor
5	Column pipe	17	O-ring
6	Motor housing	18	Mechanical seal, single
7	Shaft	19	Locating ring
8	Seal housing		
11	Inner circlip		
12	Clamp fitting		

Fig. 4.102 Sliding-vane screw pump used for sampling (Lederle)

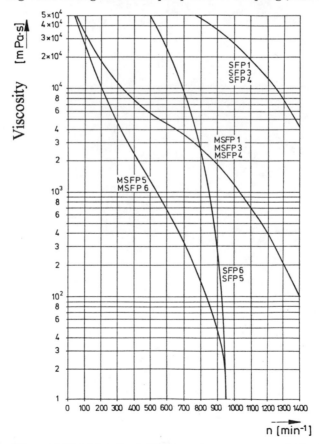

Fig. 4.103 Maximum rotational speeds of sliding-vane screw pumps relative to viscosity and size of pump (Lederle)

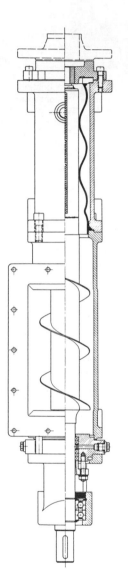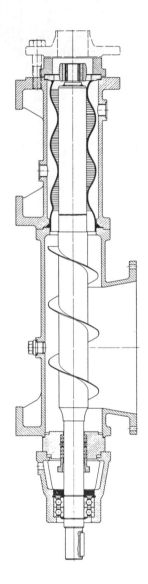

Fig. 4.104 Sliding-vane screw pump with inducing screw for very viscous fluids (Lederle)

496

4.4.5 Technical aspects of operation

The multi-chamber system and the subsequent division of the delivery pressure into partial pressures (Fig 4.93) make it possible to keep the deflection load on the blocking vanes to a minimum. It is thus possible to use plastic vanes for high pressures. In sliding-vane screw pumps, the low radial load on the rotor is of particular advantage. The delivery pressure generated in the pump surrounds the shaft in an axial direction, rising according to the lead of the spiral casing. A sinusoidal deflection load thus occurs along the shaft and only a small residual thrust remains in the radial direction. These deflection loads are of no significance for the shaft, as design considerations require that its cross-section is many times that of a permissible deflection cross-section. The guide bearing located in the delivery chamber is subjected to only a small load and is not required at all in certain pump versions. If shaft sealing is effected on the suction side, an axial load on the shaft face surface is created which corresponds to the delivery pressure. In pumps with bearing seat devices this axial pressure is taken up by axial bearings; in block pumps compensatory boreholes are drilled to effect thrust balance . The continuous delivery mode of sliding-vane screw pumps has a favourable effect even with high viscosities (Fig 4.103). For extremely high viscosities, the pump is fitted with a supply device in the form of a helical inducer mounted on the suction section. With the help of such devices, it is possible to helical inducer with viscosities of up to 1.000.000 mPa·s (Fig 4.104). Sliding-vane screw pumps may be operated without fluid supply for a certain of time without the pump being damaged. This is one of the main advantages of these types of pumps. In day-to-day operation it is not always possible to avoid dry-running. This can be caused by supply or suction tanks which have become empty, inadequate or interrupted fluid delivery from vacuum vaporisers or imprudent switch-on.

Sliding-vane pumps ensure that the unpleasant consequences of such malfunctions are kept to a minimum, since the shaft is centrically guided, and its radial load low during fluid delivery and zero during dry-running. In the absence of a pressure-generating fluid there are no hydraulic axial pressures on the vanes, and the degree of the vane-to-vane friction is thus minimal. The small amounts of heat generated by the movement of the vanes relative to one another and at the spiral casing can easily be dissipated via the stainless steel stator and the shaft. Sliding-vane pumps thus have a number of considerable advantages over other systems.

4.5 Screw pumps

The screw pump is a working machine, related to the gear pump, with an almost constant delivery over the angle of rotation. With this type of pump, the fluid is not pumped in a circumferential   direction as with the gear pump but is instead moved axially from the suction to the pressure nozzle, squeeze-free and eddy-free and without the centrifuging. This is particularly satisfactory for fluids with a viscose structure which are susceptible to shearing. Depending on their design principle, screw pumps are divided into single or multiscrew types with inner or outer bearing systems. Multiscrew pumps are constructed with two, three or five screws.

The delivery parameters determine the number of spindles and the bearing arrangement. The main area of application of screw pumps is in hydraulic systems for the transmission of energy, as burner pumps in boiler systems, for filling/discharging tanker ships, in the chemical, petrochemical and paint and varnish industries as well as for food processing.

4.5.1 Single-screw design

Single screw pumps are used mainly for very viscous fluids. Their basic operating principle (Fig. 4.105) is an externally-threaded rotor enclosed by a tubular casing in which a drag effect is generated in a viscous substance by the internal forces, thus transporting the fluid from the suction to the pressure nozzle. This also produces reverse flows both within the rotor channel and also in the gaps between the screw heads and casing wall. There is no block-sealing point, rather a working sealing point (on the outer diameter of the screw) between the pressure and suction chambers, so that the pump can also operate safely within a closed system, in contrast to the multiscrew type. The principle of the friction pump is that the delivery flow Q is heavily dependent on the delivery pressure Δp and the viscosity of the fluid. For this reason, pumps of this kind are normally only used for viscosities above 1000 mm²/s. These pumps are used mainly as single-screw extruders for pumping molten plastics.

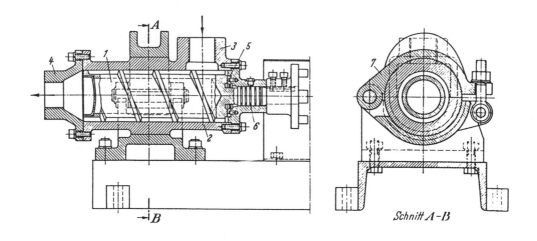

Schnitt A-B = Section A-B

Fig 4.105 Single-screw pump (Walbersdorf)

4.5.2. Two-screw design

Pumps of this kind are provided with either external or internal bearing systems depending on the application.

4.5.2.1. Two-screw pump with an external bearing system

This type is normally used as a transport pump for large rates of flow up to 2000 m³/h and pressures up to 50 bar (Fig 4.106). The screws rotating in a stator casing are driven in opposite directions from an external synchronous gear unit similar to that used for rotary piston pumps. The screw faces and screw tips rotate without contact. The running clearance required is variable and depends on the viscosity and temperature of the fluid.

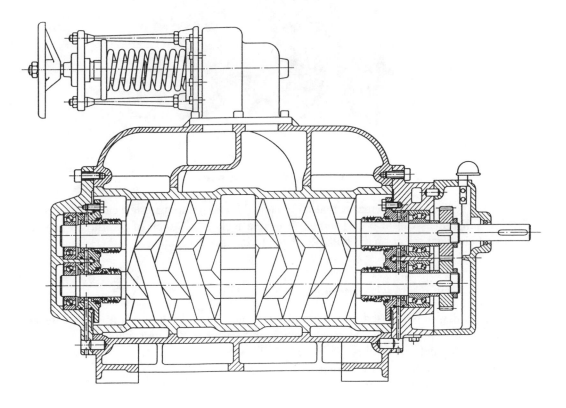

Fig 4.106 Two-screw pump with external bearing system (Allweiler)

Such pumps are normally of a double suction design which provides almost complete axial thrust balance. Each screw has a left and right hand thread for this purpose. The thread profile is flat trapezoidal (Fig 4.107). Delivery parallel to the axis is achieved by shifting the chambers in the profile gaps thus produces a gentle, eddy-free movement. The transverse forces and the resulting bending stresses are minimised by maintaining suitable clearances between the screw tips and the casing wall. Appropriate running clearances enable screw pumps to also be used for fluids up to 400 °C where the expansion of the casing and screws is different. The gap clearances enable a good venting of the pump and it can therefore handle liquid, air/gas mixtures without difficulty.

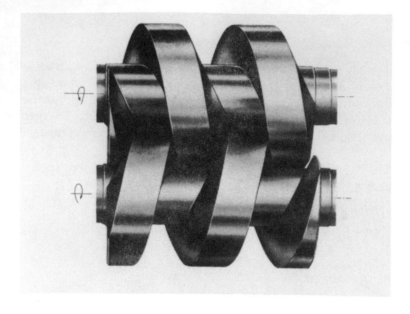

Fig 4.107 Thread profile of a two-screw pump

The positive screw guidance means that all the usual metal pump materials can be used and corrosive substances can also be handled. The relatively small pitch of the thread profile also has a beneficial effect on the NPSHR characteristics of the pump.

4.5.2.2. Two-screw pump with internal bearing system

These pumps can only be used with a liquid which also provides lubrication. Two screws mesh in the enclosing casing with the main screw with two flights driving the auxiliary screw (Fig 4.108 and Fig 4.109).

The screws are supported in the casing so that the complete length of the delivery chamber acts as a plain bearing. The references circle diameters of the main and auxiliary screws are different. This design is however subject to axial thrust, which is compensated for by suitable dimensioning of the bearing journal.

Fig 4.108 Cross-section of the main and auxiliary screws of a two-screw pump

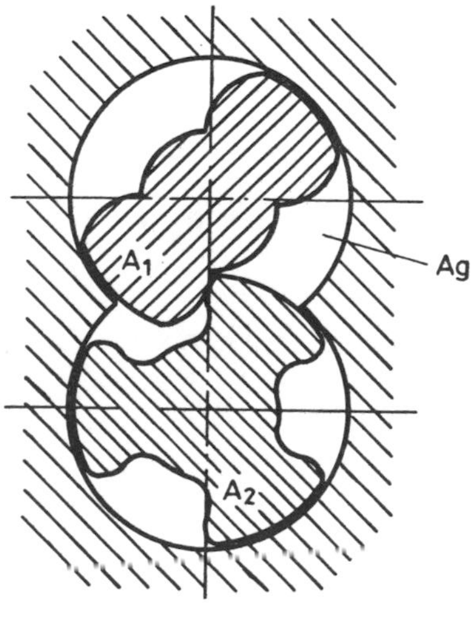

This produces a hydraulic connection between the outside of the bearing position on the suction end and the pressure chamber.

500

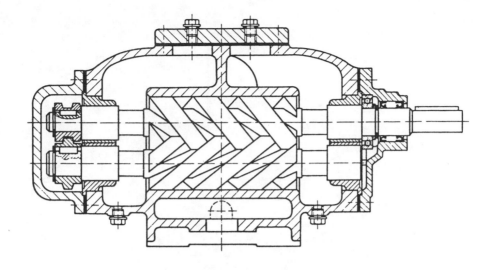

Fig 4.109. Single-flow, two-screw pump with internal bearing system (Leistritz)

4.5.3. Three-screw pump with internal bearing system (Fig 4.110 and Fig 4.111)

The transmission of energy using hydraulic oil demands particularly quiet running in additi-on to high pressures. The screw pump is superior to the gear pump for this purpose even though it is limited to a maximum delivery pressure of 250 bar. The delivery principle is similar to that of the two-screw type.

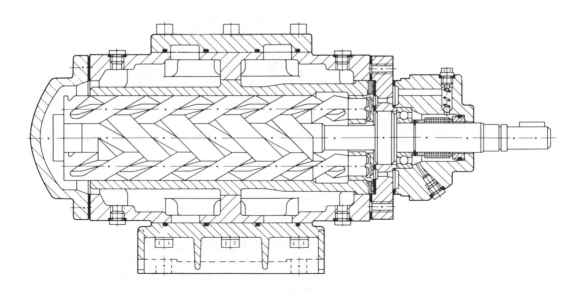

Fig 4.110. Three-screw pump (Allweiler)

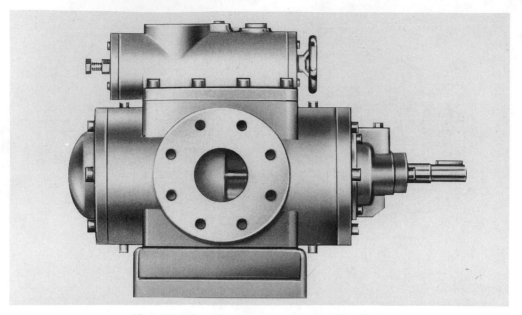

Fig 4.111. View of a three-screw pump (Allweiler)

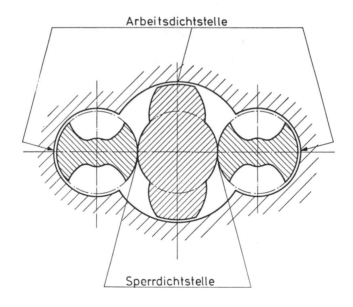

Arbeitsdichtstelle = Working sealing point
Sperrdichtstelle   = Block-sealing point

Fig. 4.112 Profile cross section of a three-screw pump

Fig. 4.112 shows the profile section of a three-screw pump. The design of the thread profile is explained in Fig 4.113 [4-9]. The main and auxiliary screws have the same reference circle diameter d (i.e. the same rotational speed). The reference circles are rolling circles for the cycloidal paths which are followed from the points during mutual roll off. The pointed episcycloid a-a passes through vertex A and in doing so describes the profile face of the main screw. The extended (looped) epicycloids b-b and c-c pass through the vertices BC of the main screw and therefore determine the faces of the auxiliary screws.

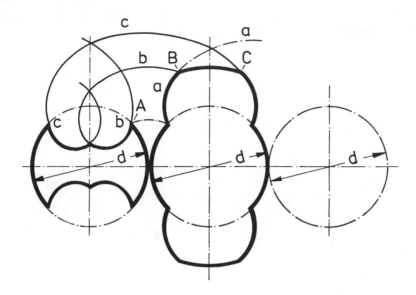

Fig 4.113 Design of the profile of a three-screw pump

The profile is designed such that only the driving screw delivers and the two block-sealing screws are at the same time driven by the pressure of the fluid. There is no direct, only a hydraulic, force transmission between the main and auxiliary screws. The displacement volume per rotation of the main screw is calculated by:

$$V_u = [A - \Sigma A'] h \qquad dm^3/U \qquad\qquad (4 - 46)$$

where

$A =$ of the cross-section configuration cut out from the tip circles of the screws (equals cross-sectional area of casing minus gap area)

$A' =$ sum of the screw cross-sectional areas

$h \;=$ screw pitch of main screw

$U \;=$ rotations

we therefore get

$$Q_{th} = V_u \cdot \frac{n}{60} \quad l \cdot s^{-1} \qquad (4 - 47)$$

$$n = r.p.m.$$

Because of the pressure drop $p_2 - p$, reverse flows occur in the gaps in the direction of the suction side, with the number of chambers determining the pressure drop from chamber to chamber. The flow at discharge Q is reduced by the losses caused by the reverse flow and any incomplete filling of chambers (suction cavitation).
It is therefore,

$$Q = Q_{th} - Q_V \qquad (4 - 48)$$

whereby Q_v from the gap flow losses Q_{sp} is expressed by

the tightness:
$$\lambda_1 = \frac{Q}{Q + Q_{sp}} \qquad (4 - 49)$$

and the

degree of admission:
$$\lambda_2 = \frac{Q + Q_{sp}}{Q_{th}} \qquad (4 - 50)$$

results.

The quotient of the difference between $(Q_{th} - Q_{sp})$ and Q_{th} is called the volumetric efficiency λ_{ges}

$$\lambda_{ges} = \frac{Q_{th} - Q_{sp}}{Q_{th}} = \frac{Q}{Q_{th}} = \lambda_1 \cdot \lambda_2 \qquad (4 - 51)$$

The actual rate of flow at discharge is therefore:

$$Q = \lambda_{ges} \cdot Q_{th} \qquad (4 - 52)$$

Three-screw pumps with an internal bearing system have extremely low pulsation compared with other displacement pumps. According to Laiber [4-10], at a delivery pressure of 100 bar an unevenness (pulsation) of 2,5% is found at a reference speed of 1500 r.p.m. This result can be further improved in many cases at possible speed of 3000 r.p.m.

4.5.4. Suction capacity and cavitation

4.5.4.1. Suction cavitation

The drop in dynamic pressure as the fluid enters into the displacer profile is what mainly determines the suction capacity of the pump (Fig 4.114). The entry velocity v_R results from the axial and circumferential velocity of the delivering screw [4-11].

$$v_R = \sqrt{v_a^2 + v_u^2} \qquad (4 - 53)$$

The dynamic energy $v_R^2/2g$ is equal to the head on the suction side $\Delta h = NPSHR_s$.

$$\Delta h = \frac{v_R^2}{2g} = NPSHR_S \qquad (4 - 54)$$

If the screw pitch S is related to the outer diameter of the driving screw D_a and if the pitch is taken to be $S = K \cdot D_a$,

we then get:
$$NPSHR_s = \left[\frac{D_a \cdot n}{60}\right]^2 \cdot \frac{K^2 + \pi^2}{2g} \qquad (4 - 55)$$

where K = number of delivery chambers; D_a = m; n = r.p.m; g = m/s^2

similar to a value already determined for rotary piston pumps

4.5.4.2 Gap cavitation NPSHRsp

The pressure distance $p_2 - p_1$ generates reverse flows in the gaps of the working and block-sealing points and therefore gap cavitation has to be taken into account. In contrast to rotary piston pumps where the pressure gradient $p_2 - p_1$ is present between the suction

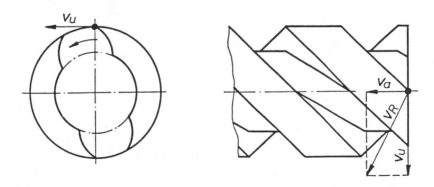

Fig 4.114 Illustration of the entry velocities into the displacer chamber

505

and discharge sides, with the screw pumps this is spread over the number of chambers so that a reduced reverse flow must be expected at the working sealing point.

The amount of the NPSHR induced by the gap flows is $NPSHR_{sp}$ but is, however, strongly dependent on the viscosity of the fluid so that there is a reciprocal interaction between $NPSHR_{sp}$ and viscosity.

Gap cavitation at the working sealing point is obtained from the pressure gradient $p_2 - p_1$ divided by the number of working chambers (K) so that the differential pressure Δp required for determining the gap flow is as follows:

$$\Delta p = \frac{p_2 - p_1}{K} \qquad (4 - 56)$$

The determination of $NPSHR_{sp}$ for screw pumps is based on the two sealing gaps in question at the working and block-sealing point (Fig 4.112). The gap flow at the working sealing point can be described as a flow in concentric gaps with a laminar character and a relative velocity w = 0, in the gap surfaces forming the boundary of the gap. According to Pfund [4-12] the following should be used for this gap flow Q_{spl}.

$$Q_{spl} = \frac{\Delta p \cdot s^3 \cdot d_m \cdot \pi}{12 \cdot \eta \cdot L} \qquad (4 - 57)$$

Δp	= effective pressure difference at gap
η	= dynamic viscosity
d_m	= mean diameter
L	= axial length of gap
s	= radial width of gap

The transverse force vertical to the rotational axis, due to the delivery pressure, acting on the screws deflects the shafts and results in an eccentric annular gap thus causing a larger gap flow than previously determined. According to Chaimowitsch [4-13], this can be calculated as follows:

$$Q_{spexz} = Q_{konz} \cdot [1 + 1,5 \cdot e^2] \qquad (4 - 58)$$

$$e = \text{Eccentricity whereby } e_{max} = 1$$

which means that at maximum eccentricity of the ring gap the flow loss is 2.5 times that for a concentric gap.

Because of the bend line of the shaft the eccentricity at the cross section which determines the gap cavitation (start of the screw) is low so that the gap loss flow can be determined with sufficient accuracy using equation 4-57.

The flow velocity v_{SPI} in the gap at the working sealing point can be determined from Q_{sp1} and the gap cross section as follows:

$$v_{sp1} = \frac{Q_{sp1}}{A} = \frac{Q_{sp1}}{d_m \cdot \pi \cdot s} \qquad (4 - 59)$$

A = Gap cross section = $d_m \cdot \pi \cdot s$ at the working sealing point
s = Gap width

which produces the static pressure drop and therefore the NPSHR for this gap:

$$NPSHR_{sp1} = \frac{v_{sp1}^2}{2g} = \frac{1}{\xi_1} \cdot \frac{\Delta p}{\rho \cdot g} \qquad (4 - 60)$$

ξ_1 = pressure loss coefficient (working sealing point)

Hamelberg [4-14] states that the gap flow at the block-sealing point increases by a factor of 3 because at this point there is only a line seal between the suction and pressure spaces of the first chamber. Increased gap flow, however, also means an increased velocity of the reverse flow particularly as the line sealing gap has smaller dimensions than the annular gap at the circumference of the screw and extends along the entire length of the screw.
A rough calculation therefore shows that the head at the block-sealing point $NPSHR_{SP2}$ must be three times that of $NPSHR_{SP1}$.

$$NPSHR_{sp2} = \frac{v_{sp2}^2}{2g} = \frac{1}{\zeta_2} \cdot \frac{\Delta p}{\rho \cdot g} \qquad (4 - 61)$$

ξ_2 = Pressure loss coefficient (block-sealing point)

$$NPSHR_{sp2} \approx 3 \cdot NPSHR_{sp1} \qquad (4 - 62)$$

The largest of the two partial heads determined must be used as a basis for determining the NPSH of the pump.
Fig. 4.115 and 4.116 show families of characteristics for screw pumps with internal and external bearing systems and Fig. 4.117 shows the recommended speeds relative to the viscosity of the fluid.

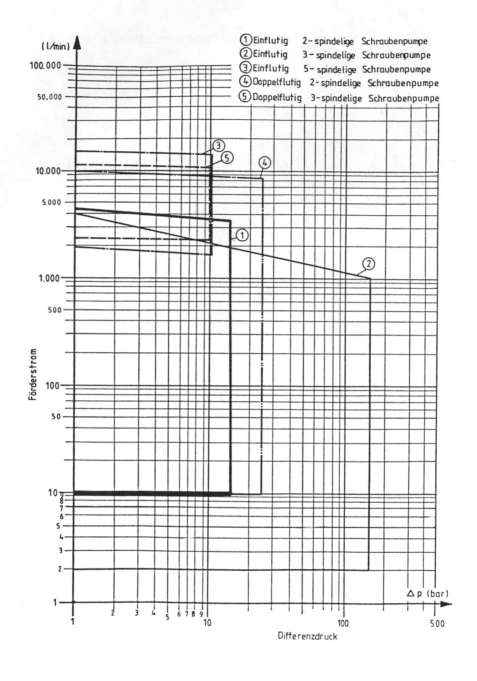

| Förderstrom | = | Delivery flow |
| Differenzdruck | = | Differential pressure |

1 = Single suction, two-screw pump
2 = Single suction, three-screw pump
3 = Single suction, five-screw pump
4 = Double suction, two-screw pump
5 = Double suction, three-screw pump

Fig. 4.115 Family of characteristics of screw pumps with an internal bearing system (Allweiler)

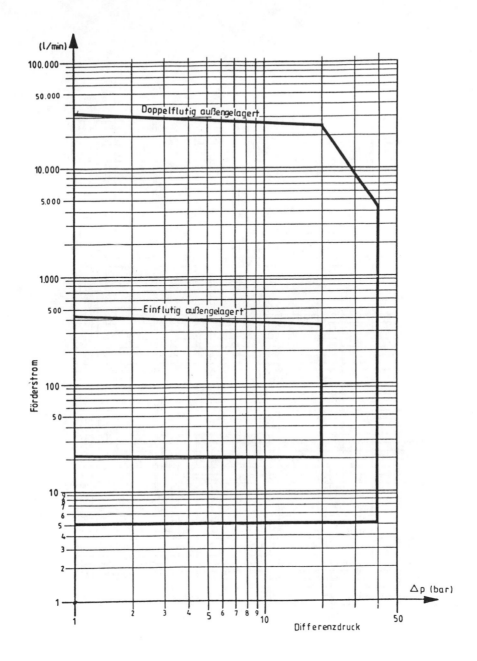

Doppelflutig außengelagert = Double suction, external bearing system
Einflutig auengelagert   = Single suction, external bearing system
Förderstrom   = Delivery flow
Differenzdruck   = Differential pressure

Fig. 4.116 Family of characteristics of screw pumps with external bearing systems (Allweiler)

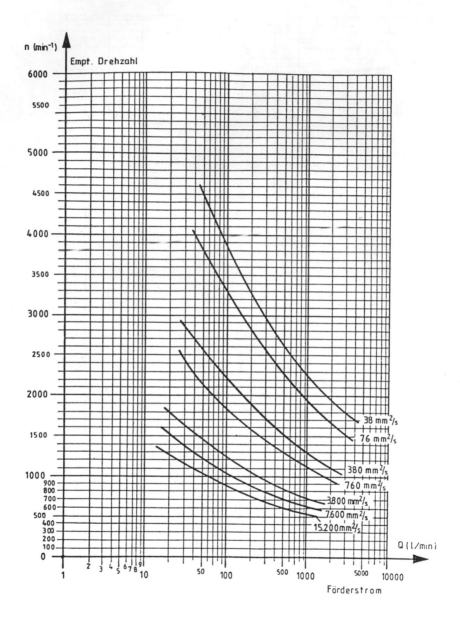

n (min⁻¹)

Empf. Drehzahl

Q (l/min)

Förderstrom

min⁻¹	=	r.p.m.
Empf. Drehzahl	=	Recommended speed
Förderstrom	=	Delivery flow

Fig. 4.117 Recommended speeds relative to delivery flow and viscosity of the fluid for screw pumps (Allweiler)

510

4.6 Progressive cavity pumps (single-screw pumps)

The progressive cavity pump is a very versatile single screw displacement machine (Fig. 4.118, 4.119 and 4.120). It can be used on both low and high viscosity fluids, even where these contain fibrous materials or abrasive additives or gases. Their application range extends to pastes provided that they can still be forced through a pipe system. The construction of the progressive cavity pump is very simple. A fixed stator, usually made of elastomers such as natural rubber of various Shore hardness ratings, synthetic rubbers such as neoprene, perbunane, ethylene propylene, hypalone, viton, polyurethane, butyl, silicone, thiocol or various plastics such as PTFE, polyamide, polypropylene, polyethylene, polyacetate or laminated fabric, incorporates a double-threaded, helical hollow chamber with a cross section resembling an elongated hole. An eccentric rotor of stainless steel, titanium, special alloy, or with a hard chrome plating or enamel finish, revolves within the stator. This has a single flight screw with the same depth of thread as the stator. The rotor is driven by a cardan shaft or flexible coupling.

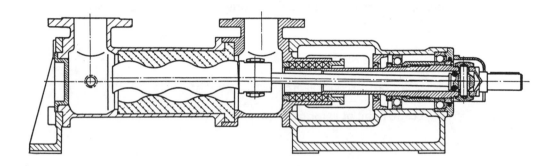

Fig. 4.118 Progressive cavity pump

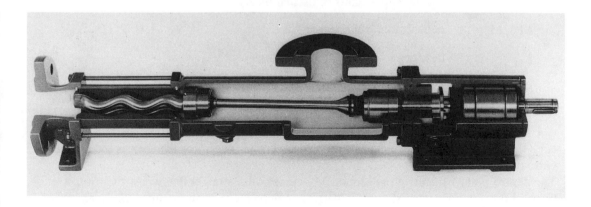

Fig. 4.119 Sectioned model of a progressive cavity pump (Netzsch)

511

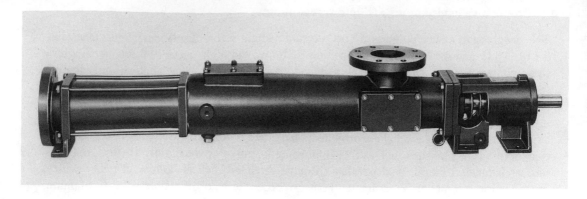

Fig. 4.120 View of a progressive cavity pump (Netzsch)

4.6.1 Functional description of pump [4-15]

Fig. 4.121 shows a cross section through a random point on the stator axis in which the motion of the related rotor cross section and the origin of the stator cross section can be seen:

F_R	=	rotor cross sectional area
F_s	=	stator cross sectional area
M_R	=	centre point of rotor cross section
M_{RH}	=	rotor main axis
K_{RH}	=	rotor maximum path
M_s	=	stator axis
e	=	eccentricity - radius of oscillation K_R
$M_{RO}\ M_{RX}$	$\triangleq$	degenerated hypocloid
K_G	=	base circle
K_R	=	rotating path with radius e; projection of the axis of the centre points of all rotor cross sections

(For the purpose of this example the ratio of eccentricity to rotor diameter is intentionally extremely large.)
The centre point of rotor cross section M_R moves on a straight line whilst the rotor itself turns about its main axis M_{RH}. This particular movement in an assumed circle K_G (base circle) occurs when rolling a circle K_R (rolling circle) on the circumference of which M_R is the centre point of the rotor cross section.

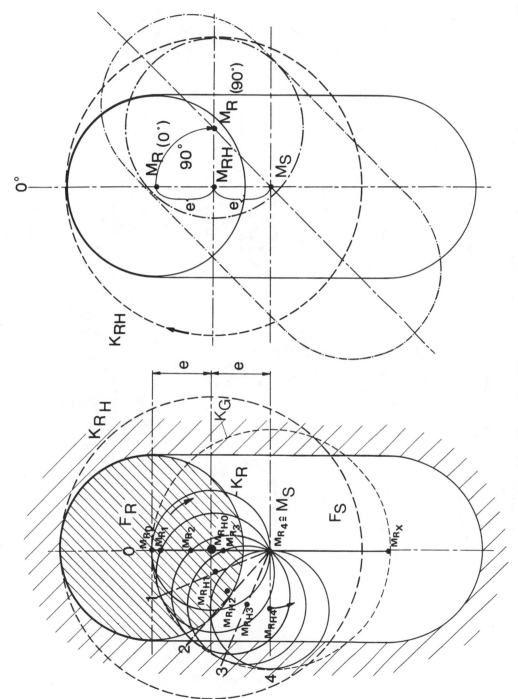

Fig. 4.121 Kinematics of the progressive cavity pump

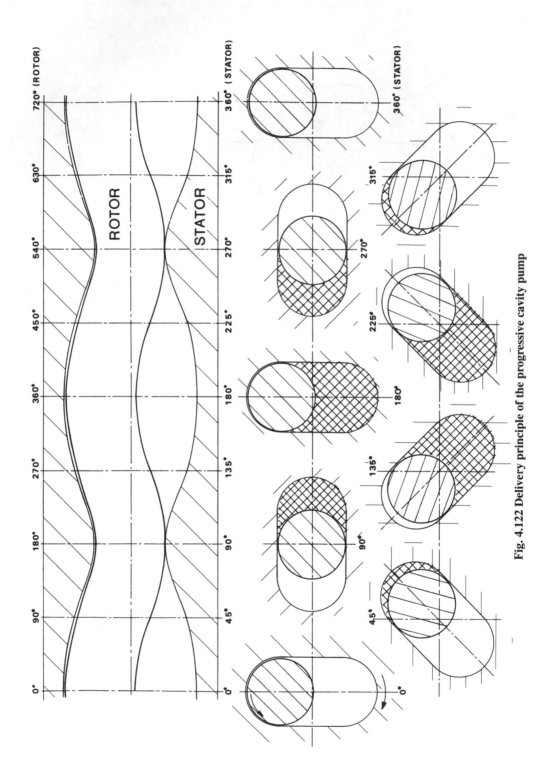

Fig. 4.122 Delivery principle of the progressive cavity pump

The sum of points M_{RO}, M_{R1}, M_{R2}, M_{R3}, ... to M_{RX} represents a hypocycloid in this kinematical process.

As the radius of the base circle K_G is identical to the diameter of the rolling circle K_R, the points M_R on the circumference of the rolling circle form a so-called degenerated or special hypocycloid, i.e. the line M_{RO} to M_{RX}.

The stator cross section is formed from both semicircles around the two end points M_{RO} and M_{RX} of the hypocycloid and the two connecting tangents with the length of two diameters of the rolling circle K_R i.e. four times the eccentricity e, whereby this may be seen as the distance between the centre points of the rotor cross section and the rotor maximum path. As rotation is the basis of the movement of the rotor cross section and the formation of the stator cross section shape, this part of the principle can be used to produce a pump system with a rotational i.e. continuous delivery and can produce both suction and pressure by forming a continuously moving seal line between the low pressure and high pressure side.

For this seal to be maintained at all stages there must always be a closed cavity. This closed cavity is made possible by the two rotor cross sections on a screw-like axis, the projection of which is rolling circle K_R, being arranged at a specific distance.

If between these two positions you introduce an infinite number of rotor cross sections then a screw (rotor) with extreme pitch results. Each of these rotor cross sections is within its own stator cross section. So that any rotor cross section can produce linear movement when the rotor turns, the longitudinal axis of the associated stator cross section must be related to a reference cross section (e.g. 0°) according to the position of the rotor cross section (Fig. 4.121).

This angle (e.g. 45°) is half the angle of the rotor cross section to the reference cross section (e.g. 0°) because the base circle K_G - which is the critical size of the stator cross section - is twice the diameter of rolling circle K_R on the circumference of which point M_R is the mid point of the rotor cross section so that each angle position of rotor and stator cross section must maintain an appropriate relationship to each other.

The sealed cavity between the rotor and stator is obtained if this space between two stator cross sections, which are offset 360° in the range of one pitch, is sealed by the corresponding rotor cross section. Accordingly, the rotor must have two pitches (720°) in order to meet these requirements (Fig. 4.122).

From the sum of the points of contact between 0° and 360° of the stator cross section a line results which constitutes the continuously advancing seal line. It divides the inside of the stator into two equal halves which during rotation change in shape and position but not in volume so that an even delivery results. This line also divides the inlet from the outlet side. So that this separation can be produced without interruption, there must be an effective seal line at all times i.e. as soon as this opens on the pressure side a fresh start must already have been made on the inlet side.

In Fig. 4.121 it can be seen that the rotor has a compound movement: firstly, rotating about the main axis M_{RH} and the eccentric rotation of this axis about stator axis M_S at the distance of eccentricity e. Both rotating movements result from the linear paths of the centre point resulting from the mathematical interrelationship of the hypocycloid. This particular movement is rendered possible in the pump by not connecting the rotor rigidly to the drive shaft but via linkages. Resulting from the explanations to Fig. 4.122 we have a formula to calculate the theoretical capacity V_{uth} for one rotation of the rotor.

$$V_{uth} = 4 \cdot e \cdot D_R \cdot H_s \qquad\qquad (4 - 63)$$

where

$$
\begin{array}{ll}
e & = \text{eccentricity of rotor} \\
D_R & = \text{diameter of rotor cross section} \\
H_s & = \text{pitch of stator/twice pitch of rotor}
\end{array}
$$

The theoretical delivery flow in the time unit is then obtained by

$$Q_{th} = V_{uth} \cdot n \qquad\qquad (4 - 64)$$

The actual rate of flow (at discharge) Q is less by the amount of the inner losses due to the gap geometry and pressure difference $p_2 - p_1$.
Therefore we get:

$$Q = Q_{th} - Q_{sp} \qquad\qquad (4 - 65)$$

The flow loss Q_{sp} cannot, however, be mathematically determined because the gap size cannot be established, as the casing (stator) wall consists mainly of an elastomer.

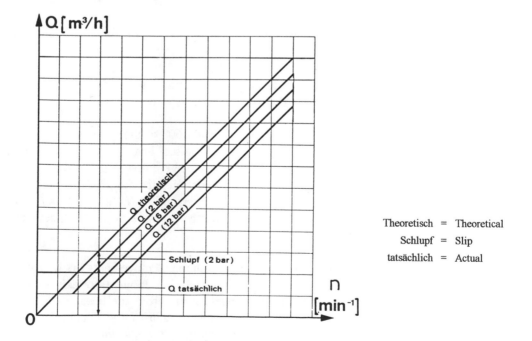

Theoretisch = Theoretical
Schlupf = Slip
tatsächlich = Actual

Fig. 4.123 Delivery curves of progressive cavity pumps Q relative to speed n

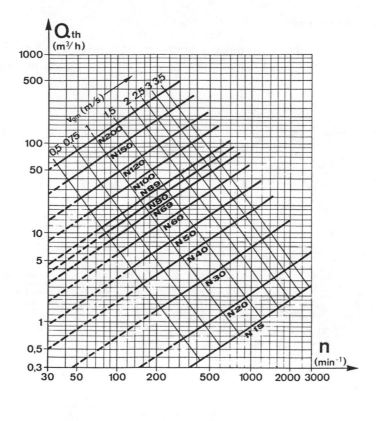

min-1 = r.p.m.

Fig. 4.124 Theoretical delivery flows Qth relative to speed and fluids in the 1 mPa·s area

The dimensions of the stator are such that the rotor rotates within it without a clearance. There is therefore no reverse flow at zero head. The rise Q_{sp} is approximately linear with increasing head so that the dimensional tolerances of the stator and rotor as well the flexibility of the stator are determinant (Fig. 4.123). The viscosity and temperature of the fluid are further influence variables.

The flexibility of the stator has the effect that no generally valid statements can be made regarding the flow rate Q at the given head for a particular viscosity of the fluid. The characteristic curves Q_{th} of the pump in a pressure-less state are therefore used for pump design and the actual delivery parameters for Q are calculated according to the given operating conditions (pressure, temperature, solid particles in fluid, viscosity). Fig. 4.124 is a diagram showing the theoretical delivery flow relative to rotor speed.

4.6.2. Power input of pump

The hydraulic power of the pump reduces by its overall efficiency, which has relatively high values in a large speed and delivery rate range. From this we get the required shaft power P_w.

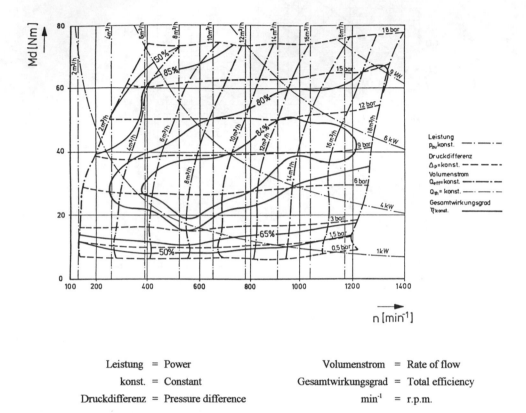

Leistung	= Power	Volumenstrom	= Rate of flow
konst.	= Constant	Gesamtwirkungsgrad	= Total efficiency
Druckdifferenz	= Pressure difference	min^{-1}	= r.p.m.

Fig. 4.125 Efficiencies of progressive cavity pumps (Netzsch)

$$P_W = \frac{\rho \cdot g \cdot H \cdot Q}{\eta_{ges}}$$

(4 - 66)

wheret

ρ = Density
g   = Acceleration due to gravity
Q   = Delivery flow
H   = Head
η_{ges} = Total efficiency formed from
η_h ; η_m = hydraulic and mechanical and volumetric efficiency η_v

Fig. 4.125 shows that the efficiency of progressive cavity pumps depends on torque M_d and speed n.

516

4.6.3. Suction capacity, NPSHR values

The hydrostatic pumping operation and the flexibility of the casing wall matched to the rotor give the progressive cavity pump a good self-priming capacity and good NPSHR values. Fig. 4.126 shows the curves for NPSHR values which were measured for water at 20 °C with a pressure reduction on the suction side and a free outlet of the fluid on the pressure side at the same time. The gap cavitation due to the pressure drop $p_2 - p_1$ was not allowed for. The $NPSHR_{SP}$ is, however, in this case also dependent on the Reynolds number i.e. the gap geometry, the speed and the viscosity of the fluid. Both NPSH values must be within the permissible range for proper operation of the pump.

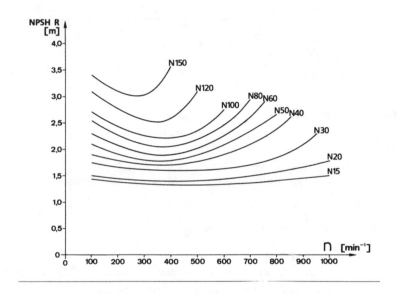

Fig. 4.126 NPSHRS values for progressive cavity pumps (Netzsch)

4.6.4 Dry-running protection

The susceptibility of the pump to run dry can be countered by various safety measures. Probes which react to electrically conductive fluids and those which build up an electrical high frequency field whose relative specific inductance constant ε_r exceeds 2,5 have proven their mesit.

These probes are best used as ring probes with the same internal diameter as the pump connection so that no additional pipe friction is caused. Fig. 4.127 shows the arrangement of such a probe.

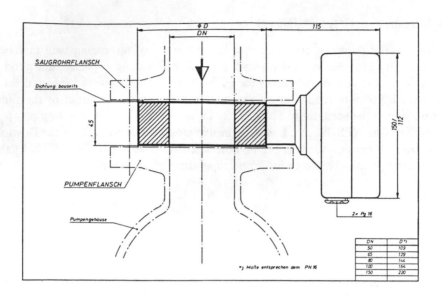

DN	D *)
50	109
65	129
80	144
100	164
150	220

*) Maße entsprechen dem PN 16

Pumpenflansch	= Pump flange	*) Maße entsprechen dem PN 16	= Dimensions correspond to PN 16
Saugrohrflansch	= Suction pipe flange	Dichtung bauseits	= Sealing provided by customer
Pumpengehäuse	= Pump casing		

Fig. 4.127 Electronic ring probe for dry running protection.

4.6.5 Progressive cavity pumps with inducing screw

Progressive cavity pumps are used mainly for fluids with a high viscosity. As viscosity increases the degree of admission of the pump reduces (danger of suction cavitation). To avoid it, or at least to reduce it to an acceptable level, use can made of ancillary devices which feed the product to the working screw. These consist mainly of an inducing screw

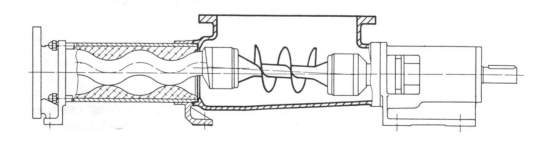

Fig. 4.128 Progressive cavity pump with inducing screw fitted at the pendulum rod

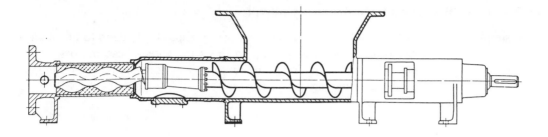

Fig. 4.129 Progressive cavity pump with an inducing screw on the centrically-rotating drive shaft

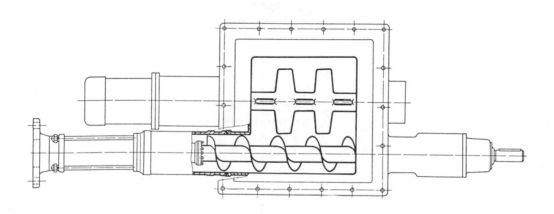

Fig. 4.130 Progressive cavity pump with a centrically-rotating inducing screw and additional paddle

which draws the product from a large capacity attachment on the casing (usually a rectangular funnel) and feeds it to the actual pump under a certain pressure. The inducing screw can be mounted on the pendulum connecting shaft between the rotor of the pump (Fig. 4.128) or on a centrically-rotating drive shaft. In this case the inducing screw terminates at the end of the pump in a circular stuffing box which covers at least one pitch of the screw (Fig. 4.129). This ensures a good infeed of the product. Where there are extremely high viscosities a paddle can be additionally fitted in the funnel attached to the casing (Fig. 4.130) to prevent "bridges" forming over the inducing screw and interrupting the feed. The paddle has its own drive and can be designed in various shapes corresponding to the particular product.

4.6.6 Submersible type progressive cavity pumps

The simple and slender construction of progressive cavity pumps enable them to be mounted vertically in vessels which require emptying. In this case, three-phase motors, gear motors (including controllable types), hydraulic motors or air motors can be used. Pumps of this kind are also suitable as sampling pumps (Fig. 4.131).

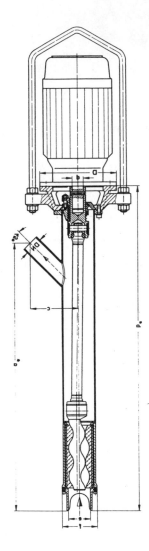

Fig. 4.131 Submersible progressive cavity pump

4.7 Sliding-vane pump

Sliding-vane pumps are mainly used for hydraulic oil, and to an increasing extent also in technical chemical processes and for foodstuffs. Their operating principle belongs to one of the oldest types of pump design. As early as 1588 Ramelli (Fig. 4.1) described this system of pumping. Their advantage is that they operate with low pulsation and good efficiency at low specific speeds, with the rotating vane design producing the suction and displacement of the fluid. A regular cylindrical rotor with radial slots carrying moving vanes (sliding segments) is mounted either eccentrically or concentrically inside a cylindrical casing within which it rotates . The vanes are held against the stroke curve of the casing wall by centrifugal force, assisted by the hydraulic pressure of the system and sometimes additionally by spring pressure. The stroke curves can be circular (with an eccentric bearing system) as shown in Fig. 4.132 or take the form shown in Fig. 4.133 for a concentric bearing system.

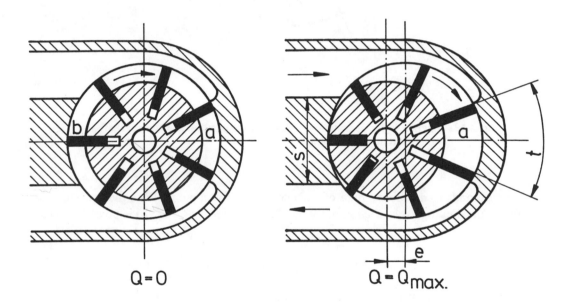

Fig. 4.132 Diagram of the working principle of a sliding-vane pump with an eccentric stator track

The flow is formed by chambers whose volume increases during aspiration and reduces during displacement as the vanes extend from the rotor or retract into it corresponding to their positive guidance. This produces an almost constant delivery. The unavoidable gap losses at the end of the vanes and the side gaps lead to a pulsation, which although small, reduces with the number of vanes. The pressure drop between the suction and pressure chamber is sealed by at least one cell of pitch t at the stroke curve of the casing. If the rotor is mounted eccentric to the casing, the vanes are pushed out of the rotor during aspiration by double the eccentricity e and are again retracted into the rotor by this amount during compression [4-16]

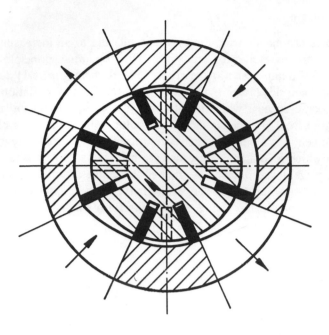

Fig. 4.133 Diagram of the operation of a sliding-vane pump with a central shaft arrangement and curved track

This corresponds to a double stroke of a single-action piston pump. With the vanes extended to the maximum a cross section of the working chamber of $(2 \cdot e \cdot b)$ is present where

$$e = \text{Eccentricity}$$
$$b = \text{Width of vanes and length of rotor}$$

From this we get a mean length during one rotation of

$$L = 2 \cdot \pi \cdot [R + e] - z \cdot x \qquad (4 - 67)$$

where

$$R = \text{Radius of rotor}$$
$$Z = \text{Number of vanes}$$
$$X = \text{Thickness of vanes}$$

The theoretical flow V_{th}/ rotation is:

$$V_{th} = [2 \cdot \pi[R + e] - z \cdot x] \cdot 2 \cdot e \cdot b \qquad (4 - 68)$$

and therefore the theoretical capacity:

$$Q_{th} = V_{th} \cdot n \qquad (4\text{ - }69)$$

and consequently the effective flow rate:

$$Q = Q_{th} \cdot \eta_v \qquad (4\text{ - }70)$$

η_v = volumetric efficiency
n = rotational speed

Where the eccentricity e changes, the rate of flow changes in accordance with (4-68) with the speed remaining the same so that with suitable design a stepless control of the flow rate of

$Q \leq Q_{max}$
can be achieved by changing
the eccentricity of $0 \leq e \leq e_{max}$

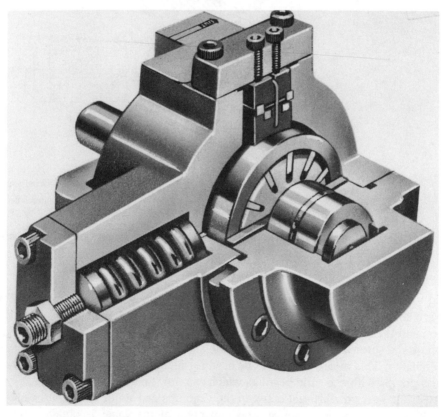

Fig. 4.134 Sliding-vane pump with automatic control of geometrical volume
(Racine USA)

523

Where e = 0 the rotor revolves without fluid being pumped (Fig. 4.132 left side). If the rotor axis is in line with the axis of the stator (Fig. 4.133) and the rotor dimensions are the same with the same track of the vanes as for the previously described eccentric bearing arrangement, then if the stroke of the stator casing follows a curve, double the rate of flow (in accordance with 4 - 68) is achieved on one rotation of the rotor, because the vanes perform two double strokes.

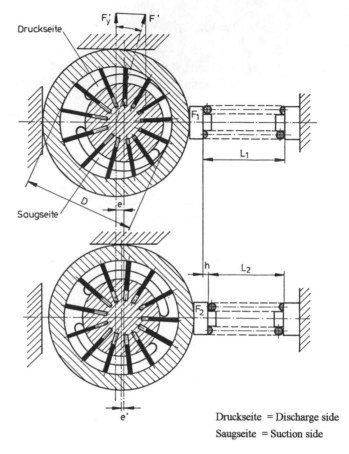

Fig. 4.135 Diagram of the stroke adjustment of a sliding-vane pump with a circular stator track

Druckseite = Discharge side
Saugseite = Suction side

Sliding-vane pumps with a regular circular stator and an eccentric bearing system for the rotor are subjected to radial stress by the delivery pressure building up over the angle of rotation and therefore the rotor bearings must be able to withstand these loads. For machines where the rotor and stator have the same axis as shown in Fig. 4.133 the radial forces are eliminated. This means that it can be used for higher pressures (up to 16 MPa) compared with pumps with eccentric bearings which can only be used up to 10 MPa.

To obtain the best possible seal between the vane tips and stator casing, pressure equalisation between the working chamber and underside of the vanes is obtained by means of bores in the side casing cover and a lateral annular chamber, so that the vanes are pressed against the track by full delivery pressure. The vanes are usually rounded off to improve hydrodynamic lubrication of the tips. The delivery pressure exerts a pulsating bending moment on the vanes which is at its highest when the vanes are fully extended from the stator. It is therefore important to have a good vane guide which is sufficiently long with the minimum side play in the rotor. The service life of the vanes therefore depends on the pressure stress and the acceleration of the oscillating movement of the vanes. Depending on the task, these are either radial or are slightly inclined in the direction of delivery (approximately 10° to 15°), with the latter type being mainly preferred where there are high bending stresses (Fig. 4.134)

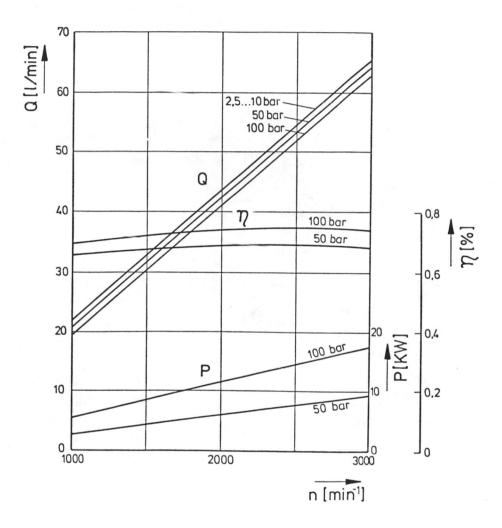

Fig. 4.136 Family of characteristics of a sliding-vane pump (Tewes)

With a spring-assisted pressure on the stator the eccentricity e can be controlled in line with the required flow (Fig. 4.135). This also provides protection against overload in that the control system acts as a safety valve. Fig. 4.136 shows the family of characteristics of a sliding-vane pump with the rotor concentrically mounted in the casing for delivery of class SAE 20 oil at 40 °C.

4.7.1 Reduction of the bending load on the vanes

As the delivery pressure increases so does the bending load on the vanes. The bending stress on the vanes is exerted on the rectangular surface formed by the vane width and the maximum extension from the rotor multiplied by the pressure difference $\dfrac{p_2 - p_1}{Z}$.

In addition to the flexing of the vanes, wear also occurs as they tilt within the rotor. This can be prevented by side guides of the vanes consisting of guide slots in guide discs arranged sideways on the rotor as shown in Fig. 4.137.

This arrangement enables wear to be reduced at higher delivery pressures.

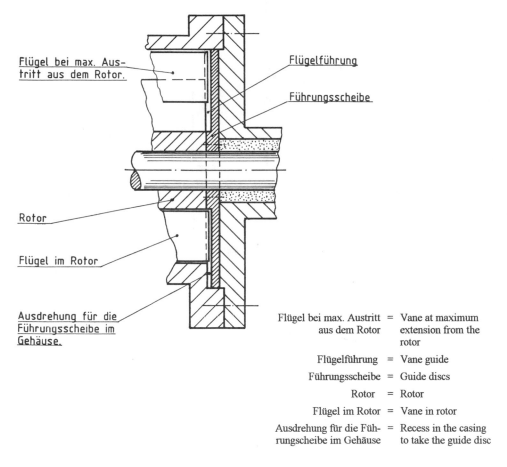

Flügel bei max. Austritt aus dem Rotor	=	Vane at maximum extension from the rotor
Flügelführung	=	Vane guide
Führungsscheibe	=	Guide discs
Rotor	=	Rotor
Flügel im Rotor	=	Vane in rotor
Ausdrehung für die Führungscheibe im Gehäuse	=	Recess in the casing to take the guide disc

Fig. 4.137 Vane guidance in side guide discs

4.7.2 Material of sliding-vane pumps

The choice of material depends upon the application. Most sliding-vane pumps are used in hydraulic oil systems so that corrosion problems have little effect on the choice of material. The most important thing is the abrasion resistance of the vanes and stator. For this reason most pumps of this kind are made from the following materials:

Rotor:	nodular cast iron
Side casing walls:	pearlite cast iron
Stator:	hard cast iron
Vanes:	hardened tool steel

For pumps used in chemical processes, the stators are made of stainless steel and various alloys, alloys with a high nickel content or plastic, depending on the exposure to corrosion, whilst the vanes are made of graphite, carbon fibre reinforced plastic or glass fibre reinforced PTFE.

4.7.3 Sliding-vane pumps with positively-controlled vanes

The hydrostatic working machines previously dealt with are capable of delivery of a wide spectrum of materials with a wide range of properties. A sliding-vane pump of the type shown in Fig. 4.138 has proved particularly suitable for handling food pulps. It was initially developed for pumping grape pulp both with and without seeds. Drawing and pumping stone fruit pulp requires great care. For seed fruit and berries on the other hand the pumping conditions are substantially simpler and they can also be handled by other hydrostatic pumps such as rotary piston and progressive cavity pumps.

4.7.3.1 Construction and operation of sliding-vane pumps with positive control

Within a cylindrical casing (1) as shown in Fig. 4.138, a rotor (3) is located which is also cylindrical and mounted on the same axis. Three barrier vanes (4), each offset 120° to each other, are mounted in the rotor so that they are free to move. The movement of the barrier vanes is positively controlled by a gate guide in the side of the casing cover. A strip (6) in the upper part of the pump casing forms a connection between the casing and rotor, so that together with the barrier vanes a closure between the suction and working chambers is always guaranteed.

The operation of the pump is as follows:

If one assumes that the rotor in the pump casing is stationary in the position shown in Fig 4.138 and the rotor is then set in motion, the barrier vane (a) moves corresponding to the angle of rotation of the rotor and of the gate track away from the centre of the axis and continues until the rotor has rotated approximately 90° counterclockwise. The barrier vane (a) is then at point I and the distance from the outer diameter of the barrier vane to the centre of the shaft at that point is only slightly less than that of the casing radius. The barrier vane remains at this point whilst the rotor continues up to point II. From then the barrier vane moves backwards towards the centre of the rotor axis until at point III the distance from its outer diameter to the centre of the axis is equal to the radius of the rotor (original position).

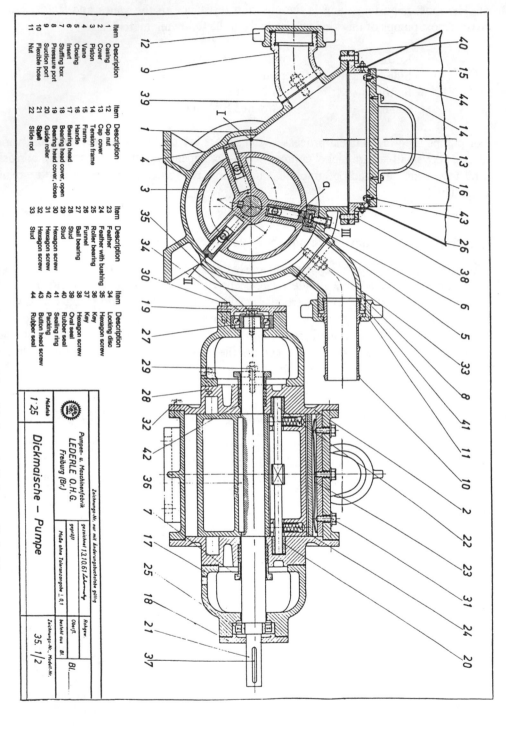

Fig. 4.138 Sliding-vane pump with positively controlled vanes (Lederle)

Item	Description		Item	Description		Item	Description
1	Casing		12	Cap nut		23	Feather
2	Cover		13	Cap cover		24	Feather with bushing
3	Piston		14	Tension frame		25	Roller bearing
4	Vane		15	Frame		26	Funnel
5	Closing		16	Handle		27	Ball bearing
6	Insert		17	Bearing head		28	Stud
7	Stuffing box		18	Bearing head cover, open		29	Stud
8	Pressure port		19	Bearing head cover, close		30	Hexagon screw
9	Suction port		20	Guide roller		31	Hexagon screw
10	Flexible hose		21	**Shaft**		32	Hexagon screw
11	Nut		22	Slide rod		33	Stud

Item	Description
34	Locking disc
35	Hexagon screw
36	Key
37	Key
38	Hexagon screw
39	Oval seal
40	Rubber seal
41	Sealing ring
42	Packing
43	Button head screw
44	Rubber seal

Pumpen- u. Maschinenfabrik
LEDERLE O.H.G.
Freiburg (Br.)

Dickmaische — Pumpe

Maßstab 1:25

gezeichnet 13.10.61 *Schwartz*

geprüft

Hate ohne Toleranzangabe ± 0,1

gezeichnet 13.10.61 ... gültig

Rohgew.

Oberfl.

bearbeit aus

Bl. ——

Zeichnungs-Nr., Modell-Nr.

35. 1/2

The product is drawn in the first third of the operation, conveyed in the second part and discharged in the third. The all-round surface seal between the chambers with various pressures gives this pump a particularly good volumetric efficiency and a high suction capacity. This is further enhanced by its low rotational speed and associated low flow velocity. The individual barrier vanes are a sliding fit in the rotor and are moved by the gate guide. They are also spring-loaded (20 and 24) this means that the barrier vanes can be moved outwards or inward along the gate track in the rotor or held in a specific position; but if a force is acting in the direction of the axis of the shaft they can still be moved. This internal movement is of great practical significance. If, for example, stone fruit is being handled and a fruit stone is caught between the casing walls and barrier vane it could lead to blockage of the pump or damage to the pump casing and barrier vane. The spring-loaded mounting of the vanes in the gate guide precludes blockage because the barrier vane can be deflected, avoiding damage to the pump.

A further advantage which has a decisive effect in increasing the service life of these pumps is the pressure-relieving movement of the barrier vane in the rotor. The centrifugal movement of the rotor takes place in the suction chamber with suction pressure being present on both sides of the sliding faces, whilst the centripetal movement takes place in the working chamber. In this case both sliding faces are subject to the same pressure i.e. exposed to the delivery pressure of the pump so that the pressure on the vane is balanced. In the area of the pressure increase where the barrier vane is subjected to differential pressure $p_2 - p_1$ the barrier vanes are held by the gate guide in their top end position and therefore perform no relative movement, and consequently suffer no wear due to side friction.

The suction part of the pump has to be designed to cope with a different pumping task (pulp of stone fruit, seed fruit and grapes). Stone fruit, seed fruit and berry pulp in de-stemmed form can be aspirated via a normal suction line similar to other liquids with suspensions. This type of aspiration fails when grape pulp with seeds is being handled as this type of product contains relatively large hollow spaces with air enclosures which reduce the suction effect - or cause it to fail completely. To be able to perform both pumping tasks, the suction part of the pump is designed such that a normal suction line with a quick-knuckle thread can be connected or it can be provided with a funnel through which the grape pulp is fed. Because only one or other type of feed is possible at one time, either the funnel section is closed with a lid (13) or the suction nozzle closed by a blanking cap (12).

The rotational speed of these pumps is mostly n = 50 r.p.m. Slow drive speed enables a gentle transport of the product but requires step-down gearing between the pump and drive motor. Gear motors are mainly used for this purpose. If it is necessary to match the delivery volume to specific operating conditions, a stepless controllable geared motor is provided instead of a normal gear motor. In this case care must be taken to ensure that the delivery amount and the speed are directly proportional to each other, pumps for thick pulp are used mainly in distilleries and wine growers' co-operatives. In many cases these pumps are mounted on a mobile chassis and can solve various pumping tasks. Fig 4.139 shows a standard pump for pumping thick pulp mounted on a trolley.

Fig 4.139 Sliding-vane pump as shown in Fig 4.138, mounted on a trolley (Lederle)

4.7.4. Vane pump with fixed vanes

The design as described in Patent DE 3514197 C2 opened up a whole new avenue of development for vane pumps. In contrast to all other pumps of this kind , the vanes on this type do not move but are instead fixed to a rotor connected to a pump drive. The vanes are fixed to the rotor both centrally and radially relative to the drive axis. They drive a rotor which is eccentrically offset by dimension e in the pump casing and has relatively wide lengthwise slots (Fig 4.140). This forms the sealing element between the suction and working chambers and also forms vane cells together with the vanes, whereby it has no force transfer function. As the drive rotor rotates, the vanes entering the lengthwise slots carry the inner driven rotor around in the direction of rotation whereby on the six fixed vane in turn one vane forms the seal between the suction and working chambers (block-sealing point). The delivery flow of this pump is theoretically continuous. The sealing gaps which are unavoidable with all rotating displacement pumps do, however, cause reverse flows relative to viscosity, pressure difference p_2 - p_1 and speed of the pumps.

The advantage of the construction of this type of pump is that no jamming of the vanes in the guide slots can take place from temperature increase due to the different thermal coefficients of expansion of the rotor and vanes. The slots in the driven rotor in this case only come into contact with one drive surface of the six vanes, whilst the side which is not pressurised has a large clearance from the adjacent wall. The pumps have been developed as inline types so that with a satisfactory arrangement of the suction and pressure lines they are easy to install.

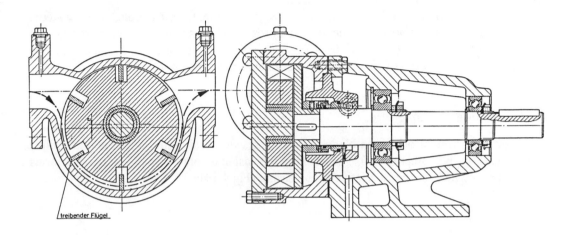

treibender Flügel = driving vane

Fig 4.140 Vane pump with fixed vanes

4.8 Liquid ring vacuum pumps/compressors

Large numbers of liquid ring vacuum pumps are used in the low and high vacuum range not only in the technical process plants of the chemical industry but also in other branches of industry and in public utilities. They have the advantage of being able to deliver large volumes with a relatively low structural size, there is no contact between casing and the rotary pump rotor at the circumference and they are largely tolerant of the entrained impurities in the pumped gas. As DIN 28400 classifies these under the rotating displacement pump group they are dealt with in this Section.

Liquid ring vacuum pumps have a wide working range so that, for example, where water is used as the liquid compressant, suction pressures of 33 mbar (up to 5 mbar for gas jet pumps) with delivery rates from 100 - 25 000 m³/h at a pressure of 1013 mbar can be achieved. As compressors they can achieve discharge pressures up to 2,5 bar.

4.8.1. Working principle

A rotor with straight or curved blades rotates in a cylindrical casing which is partly filled with liquid (liquid compressant). The liquid in the casing is rotated by the blades and forms a liquid ring which is coaxial relative to the casing (Fig 4.141).

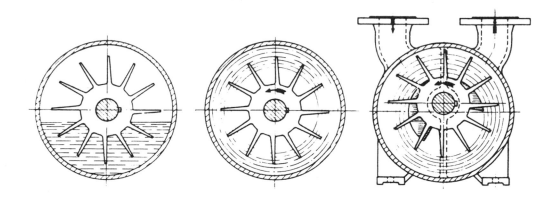

Fig 4.141 Development of a liquid and gas ring

Because the casing is only partly filled with liquid, an air/gas ring which is also concentric with respect to the casing or rotor hubs is established between the liquid ring and rotor hub due to the lower density. If the rotational axis is now shifted by dimension e (Fig 4.142) parallel to the casing, two sickle-shaped chambers filled with air/gas are formed., whereby the liquid ring touches the rotor hub at point B and rests against the casing at vertex A [4.17]. The eccentricity e and the length of the blades of the rotor must be matched so that on one hand the blades rotate past the housing with a small gap A and the blades at C are

532

immersed in the liquid ring. This causes chambers to form which draw air/gas from the suction nozzle through side ports in the casing coverplates (port plates) due to the volume of the chambers increasing in the direction of rotation. The gas flowing into the pumps is accelerated during the aspiration process.

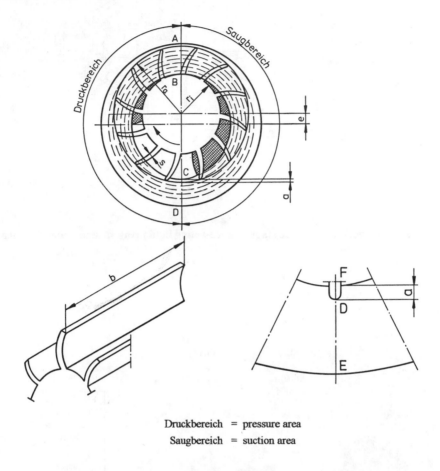

Druckbereich = pressure area
Saugbereich = suction area

Fig 4.142 Principle of operation of a liquid ring-vacuum pump/compressor

Finally, part of the flow energy is converted into pressure energy by the deceleration as the liquid ring dips into the impeller chambers. The internal contact between the gas and the ring liquid leads to an almost isothermal compression. The moist gases condense during compression. The sealing of the impeller chambers at the blade tips by the liquid ring have the advantage of enabling impurities also to be pumped in the aspirated gas. The axial inlet flow of aspirated fluid (air/gas) limits the maximum possible axial expansion of the rotor, to obtain a good degree of admission. Where large flow rates are desired these are therefore designed as double suction pumps (Fig 1.143 and Fig 1.144). Fluid can flow into the blade chambers from both sides.

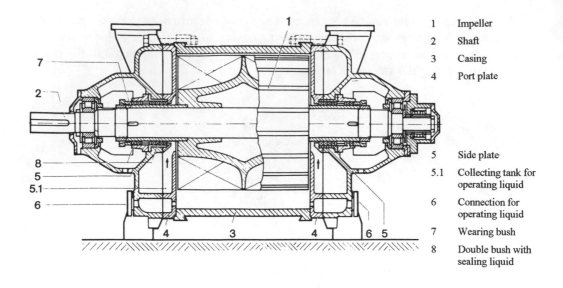

1	Impeller
2	Shaft
3	Casing
4	Port plate
5	Side plate
5.1	Collecting tank for operating liquid
6	Connection for operating liquid
7	Wearing bush
8	Double bush with sealing liquid

Fig 4.143 Construction of a single stage, double-acting liquid ring vacuum pump (Siemens)

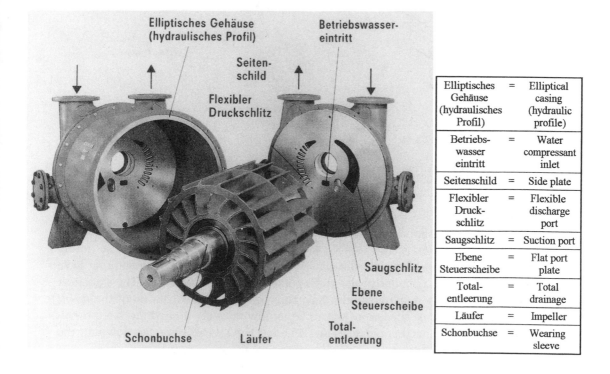

Elliptisches Gehäuse (hydraulisches Profil)	=	Elliptical casing (hydraulic profile)
Betriebs-wasser eintritt	=	Water compressant inlet
Seitenschild	=	Side plate
Flexibler Druck-schlitz	=	Flexible discharge port
Saugschlitz	=	Suction port
Ebene Steuerscheibe	=	Flat port plate
Total-entleerung	=	Total drainage
Läufer	=	Impeller
Schonbuchse	=	Wearing sleeve

Fig 4.144 Single stage, double-acting liquid ring vacuum pump, showing compression chamber with impeller and port plates (Siemens)

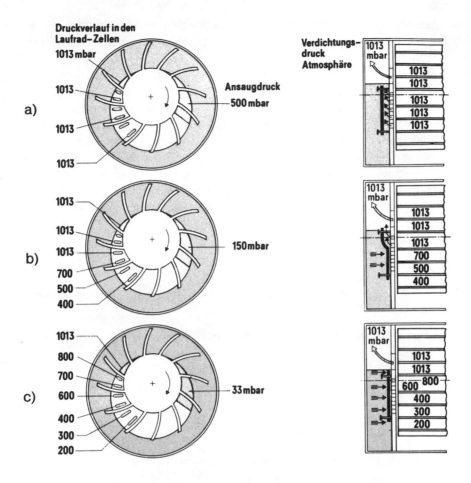

Fig 4.145 Functioning of flexible discharge port (Siemens)

For the double suction type, the impeller hub should be designed with a rising path towards the centre, and for single suction pumps it should rise towards the impeller shaft. Providing an impeller exit angle of more than 90° has the advantage of obtaining an absolute outlet velocity c_2 greater than at 90° and therefore obtaining a greater velocity of the liquid ring which increases the discharge energy. Furthermore, a favourable design of the casing by changing from the circular shape (curved projections) can produce a further increase in the ring velocity. The vapour pressure behaviour of the liquid compressant is particularly important, the lowest achievable suction pressure is a function of the vapour pressure of the liquid compressant. In the case of moist gases the latter has the effect of condensing them and thus improving the suction capacity. This is why the term liquid-contact condensers is used for these pumps. When dry air is being aspirated, the suction capacity therefore changes, and it also changes with increase in temperature of the ring liquid.

The entrainment of liquid presents no problems for liquid ring vacuum pumps. The inlet and outlet ports are designed so that they are capable to adapting to the particular pumping conditions with regard to position and size. This means that pumps of optimum design do not have rigid outlet ports, because if the specified pressure ratio, e.g. 1:7, were exceeded this would lead to overcompression and therefore to reverse flow.

An interesting solution to this problem is found by providing variable discharge ports whereby the actual discharge port is shortened and individual bores are positioned before it. To obtain stepless matching to the particular pressure conditions, these bores are provided with plate valves [4-18]. Where water is used as a ring liquid a discharge pressure of 1:30,7 (at a pressure ratio of 33 mbar to 1013 mbar) can be achieved. Fig 4.145 shows the operation of a flexible discharge port.

4.8.2. Determining the volume flow rate of a single chamber liquid ring vacuum pump [4-19]

The theoretical gas volume flow for a single chamber liquid ring pump is shown in Fig 4.142

$$Q_{th} = \underbrace{\pi \cdot \left[[r_a - a]^2 - r_i^2 \right] \cdot b \cdot \phi}_{\text{Volume flow per rotation}} \cdot n \qquad (4 - 71)$$

where

a = Minimum immersion depths of blade
ϕ = Overlap index due to blade thickness
n = Rotational speed
r_a = Outer radius
r_i = Inner radius

For a number of blades z we get the mean blade pitch

$$t = \frac{\pi \cdot [r_i + r_a - a]}{z} \qquad (4 - 72)$$

and therefore with a blade thickness s the overlap index

$$\phi = 1 - \frac{s}{t} \qquad (4 - 73)$$

The following are used in (4-71):

Hub ratio	$\nu = r_i / r_a$
Relative rotor width	$K_b = b/r_a$
Circumferential speed	$u_a = 2 \cdot \pi \cdot r_a \cdot n$

With this a minimum blade insertion depth a = 0 can be used:

$$Q_{th} = \frac{\varphi}{2} \cdot [1 - v^2] \cdot k_b \cdot r_a^2 \cdot u_a \qquad (4 - 71a)$$

and resolved according to the impeller outer radius we get

$$r_a = \sqrt{\frac{2 \cdot Q_{th}}{\varphi \cdot k_b \cdot [1 - v^2] \cdot u_a}} \qquad (4 - 71b)$$

It should be noted that due to the volumetric losses the theoretical delivery flow rate Q_{th} to be used in (4-71b) is greater than the required volume rate of flow Q. With a volumetric efficiency η_v we get:

$$Q_{th} = \frac{Q}{\eta_v} \qquad (4 - 74)$$

The following dimensional relationships are normal for single chamber liquid ring pumps:

Hub ratio	$v = \frac{r_i}{r_a}$	=	0,5 bis 0,55
Relative rotor width	$k_b = \frac{b}{r_a}$	=	2 bis 2,8
Relative eccentricity	$\frac{e}{r_a}$	=	0,15 bis 0,2
Overlap index	$\varphi = 1 - \frac{s}{t}$	$\approx$	0,85
Volumetric efficiency	$\eta_v = \frac{Q}{Q_{th}}$	=	0,7 bis 0,8
Number of blades	z	=	16 bis 24

4.8.3 Shaft power P_{is}

The required shaft power P_{is} for isothermal compression on vacuum pumps is obtained from the term [4-20].

$$p_{is} = p_s \cdot Q \cdot \ln \frac{p_D}{p_s} \qquad\qquad (4 - 75)$$

P_{is}	=	Isothermic work of compression
Q	=	Suction capacity at the suction pressure p_s in m³/h
p_s	=	Suction pressure p_{abs}
p_D	=	Discharge pressure

For compressors

$$P_{is} = p_s \cdot Q \cdot \ln \frac{p_D}{p_s} \qquad\qquad (4 - 76)$$

Q	=	Suction quantity at suction pressure p_s in m³/h
p_D	=	Discharge pressure

4.8.4 Liquid compressant

Water cannot be used as a liquid compressant in many technical chemical processes and also vacuums are frequently required which cannot be achieved using water. In such cases other liquid compressants such as oils, dimethyl sulphate, orthochlorbenzene, toluene, ethylene chloride or others which have a lower vapour pressure are used. Needless to say, because these fluids are often toxic and also very expensive they cannot usually be supplied in a through flow like water. It should be noted that in addition to providing the compression, the liquid compressant is also needed to draw off any heat of compression which occurs. When the pump is operating a portion of the ring liquid is drawn into the downstream liquid separator and the gas is separated from the liquid compressant. The gas is drawn off upwards and the liquid is returned to the compressor.

The various operating conditions under which liquid ring vacuum pumps/compressors can be used mean that different circuit arrangements, described in the following text, are necessary [4-21].

4.8.4.1 Fresh liquid operation

With this type of operation the liquid compressant is not reused. The total amount of liquid necessary for operating the pump is taken from a liquid network e.g. return cooling water from the supply system.

Frischflüssigkeit	=	Fresh liquid
Gas	=	Gas
Gas-Flüssigkeits-Gemisch	=	Gas-liquid mixture
Ablauf	=	Drain

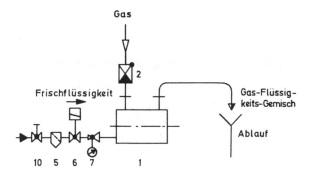

Fig 4.146 Fresh liquid operation

If it is not necessary for the pumped gas and the liquid compressant expelled from the discharge nozzle to be separately drawn off there is no need for a liquid separator. If the pressure of the liquid (e.g. the water pressure in the pipeline) fluctuates strongly, the liquid should not be fed directly into the pump for this mode. The fresh liquid should then be supplied to the pump through a pressure reducer with a solenoid valve fitted up stream (Fig 4-146).

Umlaufflüssigkeit = Recirculating liquid
Frischflüssigkeit = Fresh liquid
Ablaufflüssigkeit = Drain liquid

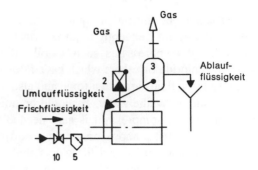

Fig 1.147 Combined liquid operation (partial circuit)

Frischflüssigkeit = Fresh liquid
Ablaufflüssigkeit = Drain liquid
Umlaufflüssigkeit = Recirculating liquid

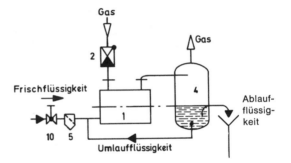

Fig 4.148 Combined liquid operation (large pump)

Frischflüssigkeit = Fresh liquid
Ablaufflüssigkeit = Drain liquid

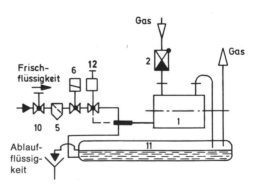

Fig 4.149 Combined liquid operation (compacted system with temperature control)

4.8.4.2 Combined liquid operation

Combined operation (preferably with water as a liquid compressant) is the system most frequently used (Fig 4.147). Just sufficient fresh water as is necessary for return cooling is added to the recirculating liquid which is automatically drawn from the liquid separator of the machine. An amount of liquid approximately equal to the amount of fresh liquid added leaves the separator via the liquid drain. This mode enables the amount of fresh liquid required be limited to the amount absolutely necessary for removal of the heat of compression.

4.8.4.3 Recirculating fluid operation using a closed circuit

This mode of operation is frequently used in the chemical industry. All of the liquid compressant separated from the gas (or vapour) in the liquid separator is reused. The liquid compressant heated in the pump must, however, be cooled in the heat exchanger. If condensing vapour forms during evacuation, liquid is drawn off at the overflow of the tank (Fig 4.148 and Fig 4.150).

Flüssigkeitsergänzung = Liquid replenishment
Ablaufflüssigkeit = Drain liquid
Kühlflüssigkeit = Cooling liquid
Umlaufflüssigkeit = Recirculating liquid

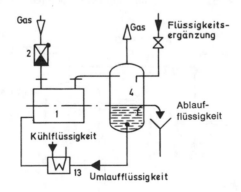

Fig 4.150 Recirculating liquid operation (closed circuit)

A resistance to flow, which must be minimised, is felt by the circulating liquid compressant. If it exceeds 20% of the difference in pressure between the suction and pressure nozzle of the vacuum pump, the circulation of the liquid compressant can be maintained using a circulating pump (Fig 4.152). In systems which are provided with hermetic vacuum pumps/compressors (with a canned motor drive or permanent-magnet coupling), the circulating pump must also be hermetically sealed to ensure that the complete system is leak-free.

541

Flüssigkeitergänzung	=	Liquid replenishment
Umlaufflüssigkeit	=	Circulating liquid
Ablaufflüssigkeit	=	Drain liquid
Kühlflüssigkeit	=	Cooling liquid

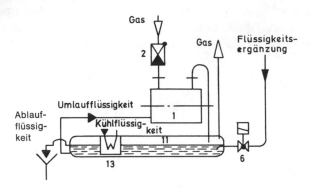

1	Liquid ring vacuum pump	7	Pressure reducer and gauge
2	Ball type check valve	10	Shut-off valve
3	Liquid build-up separator	11	Liquid separator in compact vacuum system
4	Liquid separator	12	Thermostatic control valve
5	Dirt strainer	13	Heat exchanger
6	Solenoid valve		

Fig 4.151 Recirculating liquid operation (closed circuit)

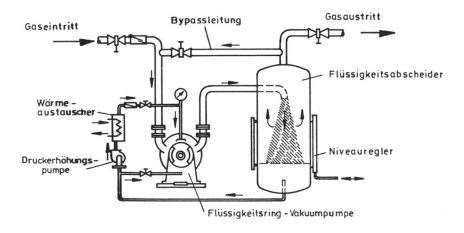

Gaseintritt	=	Gas inlet	Niveauregler	=	Level control
Bypassleitung	=	Bypass line	Flüssigkeit-Vakuumpumpe	=	Liquid ring vacuum pump
Gasaustritt	=	Gas outlet	Wärmeaustauscher	=	Heat exchanger
Flüssigkeitsabscheider	=	Liquid separator	Druckerhöhungs pumpe	=	Boost pump

Fig 4.152 Vacuum system with circulating pump for liquid compressant [4-22]

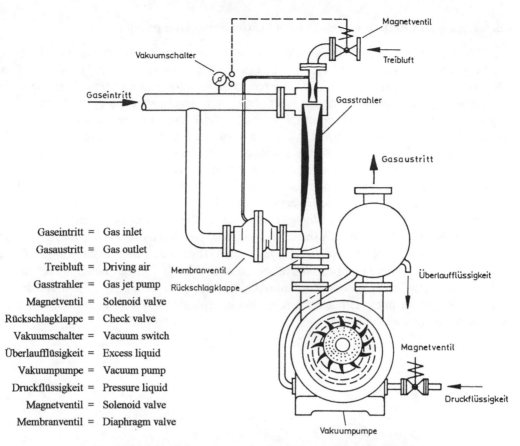

Gaseintritt	=	Gas inlet
Gasaustritt	=	Gas outlet
Treibluft	=	Driving air
Gasstrahler	=	Gas jet pump
Magnetventil	=	Solenoid valve
Rückschlagklappe	=	Check valve
Vakuumschalter	=	Vacuum switch
Überlaufflüsigkeit	=	Excess liquid
Vakuumpumpe	=	Vacuum pump
Druckflüssigkeit	=	Pressure liquid
Magnetventil	=	Solenoid valve
Membranventil	=	Diaphragm valve

Fig 4.153 Liquid ring vacuum pump with a gas jet pump [4-22]

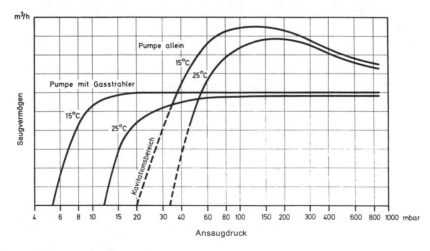

Saugvermögen	=	Suction capacity			
Pumpe mit Gasstrahler	=	Pump with gas jet pump	Kavitationsbereich	=	Cavitation range
Pumpe allein	=	Pump alone	Ansaugdruck	=	Suction pressure

Fig 4.154 Suction capacity of a liquid ring vacuum pump at different operating temperatures, with and without a gas jet pump [4-21]

4.8.4.4 Gas jet pump as an inlet pump for the liquid ring pump

To achieve a high vacuum, the suction conditions can be substantially improved by using an inlet gas jet pump which uses air or gas from the liquid separator as a driving gas (Fig 4.153). This does, however, mean a reduction in volume rate of flow from the receiver because the aspirated driving gas has to be compressed to atmospheric pressure.

Fig 4.154 shows the difference in the characteristic curve of a gas jet pump-liquid ring pump combination relative to the temperature of the liquid compressant (water) of the pump. The shift in the curves of the combination of both pumps which occurs when the pump curve changes is characteristic. The operation of gas jet pumps is independent of the temperature of the driving gas because fluctuations in the temperature of the gas or of the gas pumped at atmospheric pressure have practically no effect because of the magnitude of the adiabatic thermal gradient of the gas jet pump. In the low vacuum range (1013 to 100 mbar) the gas jet pump acts as a throttling device so that if a bypass line is fitted it is possible to operate initially without the gas jet pump. When a suction pressure of 40 mbar is reached (e.g. at a liquid compressant temperature of 55 °C), "gas jet pump mode" can be selected manually or by means of an automatic control system (Fig 4.153). If the driving gas for the gas jet pump is taken from the liquid separator this makes an enclosed, more environmentally friendly circuit.

References for Chapter 4

[4-1]   Neumaier R.:   Rotierende Verdràngerpumpen (Rotating displacement pumps)
Lederle GmbH, Company Literature

[4-2]   Stieβ. W.:   Verdrängerpumpen- und Motoren, Pumpen- Atlas Teil I (Displacement pumps and motors, pump atlas Part I),
AGT Verlag, Ludwigsburg 1966

[4-3]   Vetter, G.:   NPSHR of Rotary Piston Pumps FED-Vol 154 Pumping Machinery
Zimmermann, G.:   ASME Washington 1993 p. 347-353

[4-4]   Vetter, G.:   Zur Ermittlung von NPSHR rotierender Verdrängerpumpen an Beispiel der
Zimmermann, G.:   Kreiskolbenpumpe 3R (Determination of the NPSHR of rotating displacement pumps using rotary piston pumps 3R)
International 32 (193) vol. 7, p. 376-383

[4-5]   Surek, D.:   Spaltströme in Rotierenden Verdrängerpumpen (Flow through narrow passages in rotating displacement pumps)
Konstruktion 45 (Design 45) (1993) p. 10-15 Springer-Verlag 1993

[4-6]   Alber, H.:   Untersuchung einer Kreiskolbenpumpe (Tests on a rotary piston pump)
Prof. Dr.-Ing. W. Stieβ Laboratory,
Staatliche Ingenieur-Schule (State Engineering School) Constance, 1954

[4-7]   Krämer, H.:   Untersuchung einer Kreiskolbenpumpe (Tests on a rotary piston pump)
Prof. Dr.-Ing. W. Stieβ Laboratory,
Staatliche Ingenieur-Schule (State Engineering School) Constance, 1956

[4-8]   Molly, H.:   Die Zahnradpumpe mit evolventischen Zähnen (Spur gear pumps with involute gears),
Ölhydraulik und Pneumatik (Hydraulics and pneumatics) 1958 vol. 1,
p. 24-26

[4-9]   Stieβ, W.:   Vorlesungsmanuskript (Lecture manuscript), FH Constance

[4-10]   Leiber, W.:   Die Funktion verschiedener rotierender Verdrängerpumpen als Basis für den anlagen- und pumpengerechten Einstaz (The function of various rotating displacement pumps has a basis for selecting the correct systems and pumps)
Pumpen Vakuumpumpen Kompressoren '90 (Pumps, Vacuum Pumps, Compressors '90)
Dr. Harnisch Verlagsgesellschaft mbH, Nürnberg.

[4-11]   Wegener, L.:   Rotierende Verdrängerpumpen, Stand und Entwicklung (Rotary displacement pumps, state and development), Pump Conference, Karlsruhe '78, Sect. V.

[4-12]   Pfund, M.:   Analyse der den hydraulischen Wirkungsgrad von Schraubenspindelpumpen bzw. Schraubenspindelmotoren bestimmenden Einflußgrößen (Analysis of the influence variables which determine the hydraulic efficiency of screw spindle pumps or screw spindle motors)
Pump Conference '73, Karlsruhe, Section V.

[4-13]   Chaimowitsch, E.M.:   Ölhydraulik, Grundlagen und Anwendung (Hydraulics, basic principles and application)
Verlag Technik Berlin 1958

[4-14]   Hamelberg, F.W.:   Läuferkräfte bei Schraubenpumpen (Rotor forces in screw pumps)
Dissertation, Technical University Hannover 1966

[4-15]   Bauernfeind, H.:   Exzenterschnecken-Pumpen, Kollektiv: Pumpen und Pumpenanlagen (Helical rotor pumps, Collective: Pumps and pump systems)
Band 27 Kontaktstudium Maschinenbautechnik (Contact Study, Mechanical Engineering Technology, Volume 27)
Expertverlage Grafenau

[4-16]   Blaha, I; Brade K.:   Hydraulickñ Stroje SNTL-Nakladatelstvi Technickñ Literatnvy,
Praha 1992

[4-17]   Schulz, H.:   Die Pumpen (Pumps)
Springer Verlag Berlin, Heidelberg, New York,
13th edition 1977

[4-18]   Siemens Company   Works Literature

[4-19]   Troskolanski , A.T.; Tazarkiewicz, St.:   Kreiselpumpen Berechnung und Konstruktion (Centrifugal pumps, design and construction),
Birkenhäuser Verlag, Basel and Stuttgart 19776

[4-20]   Siemens Company   ELMO-Gaspumpen für Vakuum und Überdruck (ELMO gas pumps for vacuum and positive pressures) Siemens, Company Paper 1970

[4-21]   Faragallah, W.:   Liquid ring vacuum pumps and compressors
Gulf Publishing Company 1988

[4-22]   Bannwarth, H.:   Flüssigkeitsring- Vakuumpumpen, -Kompressoren und- Anlagen (Liquid ring vacuum pumps, compressors and systems),
VCH-Verlag Weinheim, New York, Basel, Cambridge 1991

Part IV

Hermetic rotary displacement pumps

5. Hermetic rotary displacement pumps [5-1]

When planning or modernising technical process plants in which harmful substances within meaning of MAC (Maximum Workplace Concentration) or BWT (Biological Workplace Threshold) lists have to be stored or moved, particular attention has to be paid to official regulations. This also includes locations where highly viscous fluids have to be pumped.

When changing over from the "open" to the hermetic type, priority was given to the centrifugal pumps normally used in the low-viscosity range, as they are used in large numbers in plants in which toxic, carcinogenic, malodorous substances or substances which form explosive mixtures, e.g. liquid gases, are being pumped. The performance data of hermetic centrifugal pumps in the normal speed range (n = 750 to 3000 r.p.m.) is usually sufficient to meet the delivery parameters required of standard chemical pumps to DIN 24256/ISO 2858 or of the usual multistage types. It was therefore understandable that work on the development of hermetic pumps for such tasks was begun early, at the start of the fifties. Today, these are state of the art and are included in various companies' standards.

The first designs of leak-free rotary displacement pumps took place after a delay of 15 years, if one ignores pumps operating on the well-known peristaltic principle. The reason for this was that there was less of a problem of tolerable leak rates because these pumps are used mainly for high viscose fluids with a low toxic effect and on the other hand their high friction losses and heat dissipation in the rotor chamber were factors against economic use. After design solutions to the problems where found in both systems and using special motor windings or modern magnetic materials with high coercivity field strengths, hermetic rotary displacement pumps have today become standard machines for the many tasks in the chemical, petrochemical, pharmaceutical and other industries of this kind, as well as for food processing.

The canned motor as a drive unit for rotary displacement pumps and the permanent-magnet coupling for hermetic power transfer guarantee leak-free operation for both, whereby the magnet coupling is the preferred option. As viscosity reduces, the rotational speed of the pump must increase in order to still obtain an acceptable admittance (danger of working chamber cavitation) of the working chambers, so that normal motor speeds are no longer suitable for the drive. A geared unit must be fitted between the pump and the motor. For hermetic torque transmission a permanent magnet coupling is then necessary, which compared with a canned motor has the advantage of a lower heat development and therefore presents fewer problems with regard to heat removal.

5.1 The permanent magnet coupling for leak-free power transmission

Where a magnetic coupling (Fig. 5.1) is used for leak-free power transmission from a drive motor to a rotary displacement pump, the starting characteristics are of great importance. In particular, this determines the break-off torque depending on the ratio between the driven and driving masses, especially for high viscosity fluids. If magnetic couplings are integrated directly into the product flow their use in highly viscous products is limited by the friction loss of the rotor. The extent to which the tolerance limit can be shifted in the direction of any increase in viscosity requires the economic efficiency to be calculated. This should include particularly factors such as the initial cost of the pump set (pump with permanent magnet coupling, flexible coupling between the pump and motor, coupling pro-

549

tection, drive motor and base plate), electricity costs and operating time. It is therefore not possible clearly to stipulate a maximum viscosity limit. Fig. 5.2 provides an overview of the friction losses of a magnetic coupling occurring on the rotor.

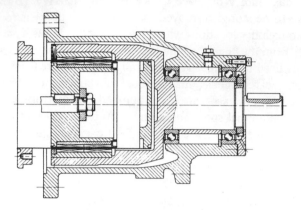

Fig. 5.1 Permanent magnet coupling for rotary displacement pumps

If the anticipated friction losses are unacceptable, if the fluid contains solids, fibres or gas or has a tendency to solidify, such products can be pumped without difficulty using the same design measures as described for the canned motor, i.e. injecting a sealing fluid into the rotor chamber. A precondition in this case is a throttling gap or shaft seal (Fig. 5.11) to separate the working chamber of the pump from the magnetic clutch. The sealing liquid should preferably be injected by a piston diaphragm metering pump, so that a hermetic seal of the complete pump set is guaranteed and the amount of liquid which is dosed can also be quantitatively determined and controlled. In addition to performing a sealing function, the dosed fluid removes the heat loss from the pump flow. It is also possible to separate the pump working chamber from the rotor chamber of the magnetic coupling by a single-action mechanical seal. The coupling rotor chamber must then be pressurised to 1 to 2 bar above the delivery pressure of the pump. The coupling heat loss can be removed either by a thermosiphon effect through a cooling device in the pressure tank or by forced circulation through a pipe bundle cooler with the aid of an impeller fitted in the rotor chamber. The sealing-cooling system to be chosen depends on the operating parameters, the properties of the fluid, its temperature and the compatibility of the sealing fluid with the product. Permanent magnetic couplings provide an ideal overload protection, which is of particular importance with hydrostatic pumps. The new magnetic materials based on samarium cobalt prevent separation of the magnetic coupling leading to permanent damage. After the drive is shut down in the no-load state, the north and south poles of the magnets of the drive and driven halves of the coupling shaft are opposing so that the coupling can again be positively engaged. Today, magnetic couplings are used for power transfer for pumps at temperatures up to 300 °C.

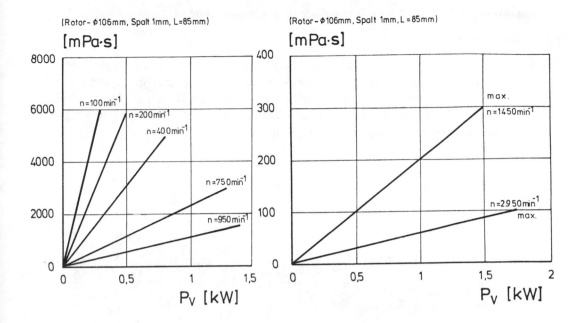

Fig. 5.2 Friction losses in the can of a magnetic coupling, relative to viscosity and speed

In principle, all rotating displacement pumps with the exception of the peristaltic pump can be fitted with either a canned motor or central magnetic coupling. A overview of the main leak-free rotary displacement pumps in use should serve to show the state of development and provide an impetus to further development, as this branch of pump design is still comparatively new and suitable for further development.

5.2 Examples of designs of leak-free rotary displacement pumps

5.2.1 Gear pumps

Leak-free gear pumps are used for a variety of tasks - mainly in chemical plants. Depending on the task, the delivery gears are provided with either external (Fig. 5.3 and 5.4) or internal teeth (Fig. 5.7 and 5.8). Their construction is particularly simple. The hydraulic data corresponds to gear pumps of conventional design. Fig. 5.9 shows the performance characteristics of a series of pumps as shown in Fig. 5.3.

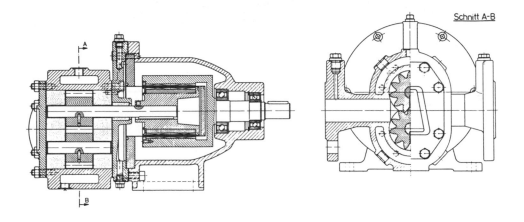

Schnitt A-B = Section A-B

Fig. 5.3 Gear pump with magnetic coupling (Lederle)

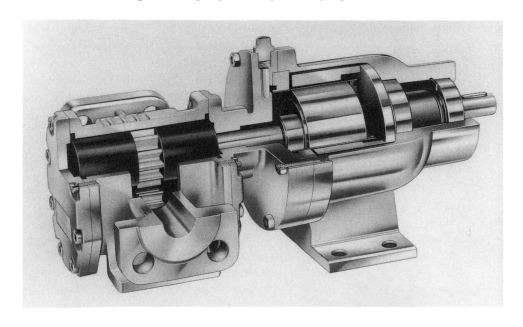

552

Fig. 5.5 Gear pump set with a magnetic clutch and three-phase motor (Lederle)

Fig. 5.6 Gear pumps with magnetic clutch and compressed air motor drive (Lederle)

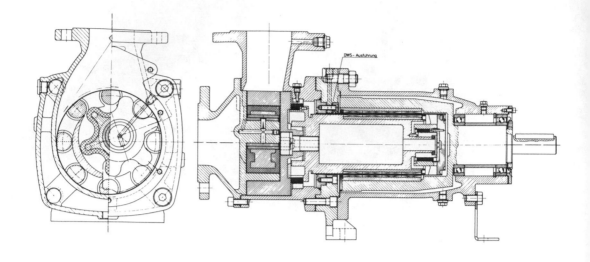

Fig. 5.7 Internal gear pump with magnetic clutch (Lederle)

Fig. 5.8 Sectioned model of an internal gear pump as shown in Fig. 5.7 (Lederle)

554

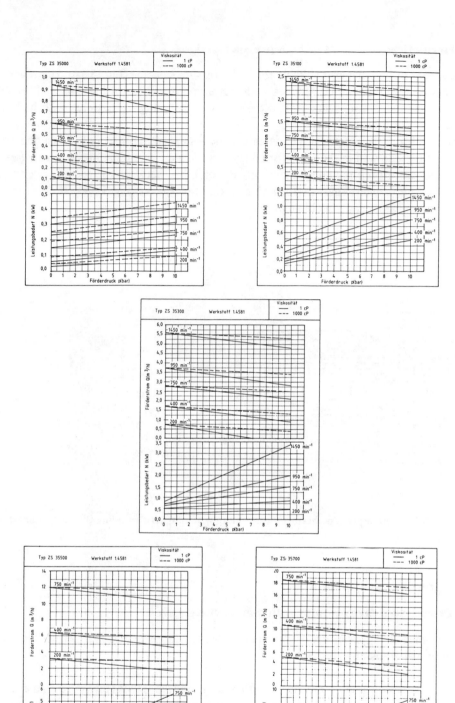

Fig. 5.9 Performance parameters of a hermetic gear pump series with external gearing, as shown in Fig. 5.3 (Lederle)

5.2.1.1 Hermetic gear pumps of high pressure design

Leak-free gear pumps are also suitable for high pressure systems (Fig. 5.10). However, the torque that can be transmitted reduces with increasing thickness of the can wall. For this reason, the canned motor is superior to the magnetic coupling for systems with high delivery heads and/or pressures.

For canned motors the electrical data is independent of system pressure because the 1 mm thickness of the can wall is the same for all pressure stages. The high internal pressures are taken by the stator pack acting as a support. Beyond this, between the stator pack and the casing enclosure, support rings perform this task and they have only little relevance for the electrical values, assuming suitable materials.

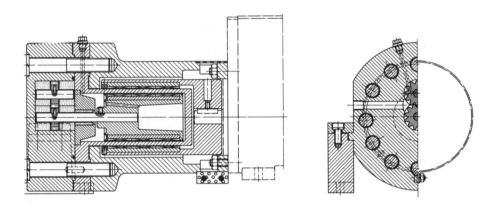

Fig. 5.10 High pressure type hermetic gear pump with magnetic coupling (PN 260) (Lederle)

5.2.2 Progressive cavity pumps

5.2.2.1 Progressive cavity pumps with a magnetic clutch

The wide range of applications of progressive cavity pumps demands a hermetic drive or leak-free power transmission.

If progressive cavity pumps are fitted with a magnetic coupling (Fig. 5.11 and 5.12), a wide range of applications with homogenous and heterogenous fluids can be covered with regard to viscosity and solids, gas or fibre content. In this case it must be decided, considering the composition of the product, whether the pump can be operated with a normal magnetic coupling without a special additional device or whether special sealing measures have to be taken, such as dosing a sealing fluid as with canned motor pumps (refer to Section 3.5.7 "Pumping of suspensions").

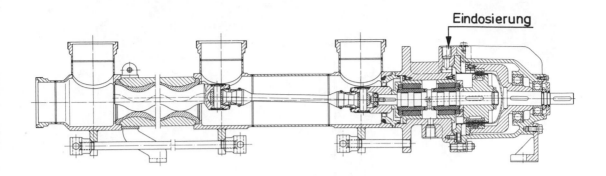

Eindosierung = Dosing inlet

Fig. 5.11 Progressive cavity pump with a magnetic coupling (Netsch)

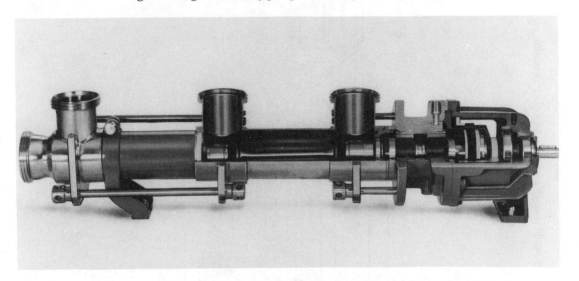

Fig. 5.12 Sectioned model of the progressive cavity pump shown in Fig. 5.11 (Netsch)

5.2.2.2 Progressive cavity pumps with a canned motor drive

This practical example should serve to illustrate the possible applications of such pumps for particularly difficult tasks [5-2].

A highly toxic fluid which is corrosive and contains solid particles has to be continuously pumped under severely fluctuating viscosities (400 - 2500 mPa·s) and delivery rates (750 - 3000 l/h) at temperatures of 50 - 300 °C and a system pressure of 150 bar. This problem is to be solved using a metal progressive cavity pump with an integrated, externally cooled, speed-controlled canned motor (Fig. 5.13).

Fig. 5.14 is a flow diagram of the task under consideration.

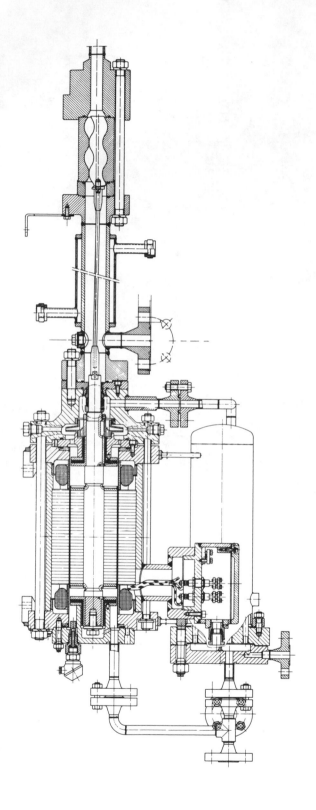

Fig. 5.13 Progressive cavity pump with a canned motor drive (Lederle)

558

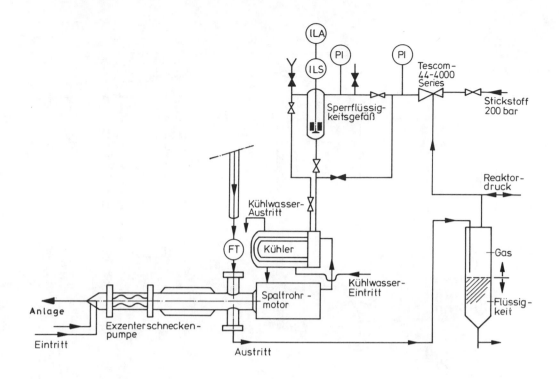

Sperrflüssigkeitsgefäß	=	Sealing liquid tank
Stickstoff	=	Nitrogen
Reaktordruck	=	Reactor pressure
Kühlwasser-Austritt	=	Cooling water outlet
Kühler	=	Cooler
Spaltrohrmotor	=	Canned motor

Kühlwasser-Eintritt	=	Cooling water inlet
Flüssigkeit	=	Liquid
Austritt	=	Outlet
Exzenterschneckenpumpe	=	Progressive cavity pump
Anlage	=	System
Eintritt	=	Inlet

Fig. 5.14 Flow diagram of an application of a progressive cavity pump with a canned motor drive as shown in Fig. 5.13

5.2.3 Sliding-vane screw pumps

This type of pump is particularly suitable for power transmission using a permanent magnet because of its single-shaft design and coaxial rotary motion (Fig. 5.15 and 5.16) and it is also suitable for direct drive from a canned motor. Where a special displacement principle is used (refer to Section 4.4. "Sliding-vane pumps") the sliding-vane pumps are particularly suitable due to their short length, especially when a canned motor is used as a drive. The pumped material must however be free of solids but may contain gas. The particular measures for removal of heat and sealing the rotor chamber, dealt with in Section 5.5.7 ("Pumping of suspensions"), can equally well be used for these pumps.

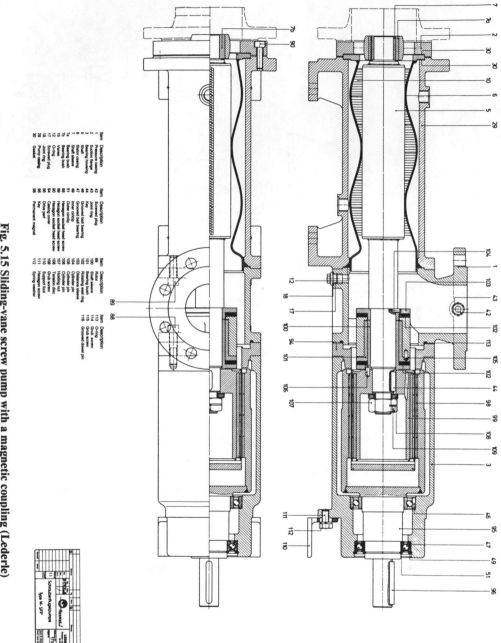

Fig. 5.15 Sliding-vane screw pump with a magnetic coupling (Lederle)

Item	Description
1	Pressure casing
2	Suction flange
3	Bearing housing
5	Shaft
6	Stator casing
7	Shaft sleeve
7a	Bearing bush
7b	Bearing bush
10	Vane
11	O-ring
12	Screwed plug
17	Joint ring
18	Pump casing
29	Pump casing
30	Gasket

Item	Description
42	Screwed plug
43	Joint ring
44	Key
46	Grooved ball bearing
47	Grooved ball bearing
49	Shaft sleeve
51	Inner circlip
88	Outer circlip
89	Hexagon socket head screw
90	Hexagon socket head screw
94	Casing screw
95	Hexagon socket head screw
96	Key
98	Permanent magnet

Item	Description
99	Rotor
100	Shaft sleeve
101	Bearing bush
102	Rotating seal ring
103	Distance sleeve
104	Cylinder pin
105	Cylinder pin
106	Cylinder pin
107	Holding nut
108	Torsion disc
109	Grub screw
110	Support foot
111	Hexagon screw
112	Spring washer

Item	Description
113	O-ring
114	Grub screw
115	Grub screw
116	Grooved dowel pin

560

Fig. 5.16 View of the sliding-vane screw pump shown in Fig 5.15 (Lederle)

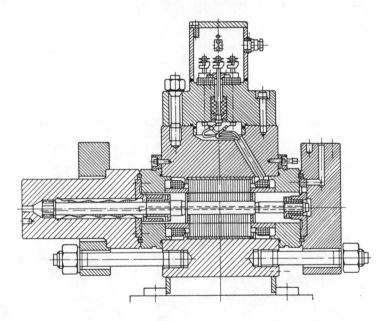

Fig 5.17 High pressure sliding-vane screw pump with a canned motor drive

5.2.4 Screw pumps

5.2.4.1 Screw pumps with a magnetic coupling

Pumps of this kind are particularly suitable for lubricating fluid without abrasive constituents. They are suitable for leak-free operation in either the single or multiscrew designs, but for screws, with an internal bearing system which have multiple screws only those which are mutually driven can be used. The product must however be free of solids. Pumps with standard, unmodified magnetic couplings are preferred (Fig. 5.18 and 5.19) because the fluid being pumped is in most cases still within an acceptable viscosity range.

561

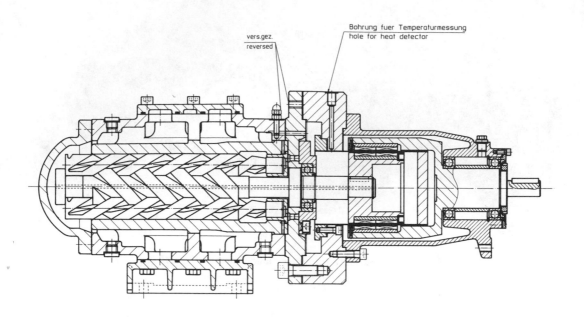

Fig. 5.18 Three-screw pump with magnetic coupling (Allweiler)

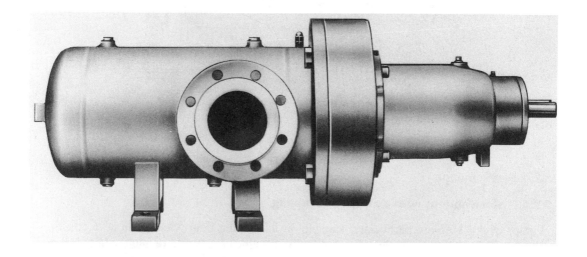

Fig. 5.19 View of the screw pump shown in Fig. 5.18 (Allweiler)

5.2.4.2 Screw pumps with a canned motor drive

Screw pumps are preferred for use in oil hydraulic systems because of their quiet running (Section 4.5 "Screw pumps"). Even quieter running and a reduction in sound emission can be achieved with the aid of a canned motor drive (Fig. 5.20), so that this kind of pump can be regarded as particularly suitable for such tasks.

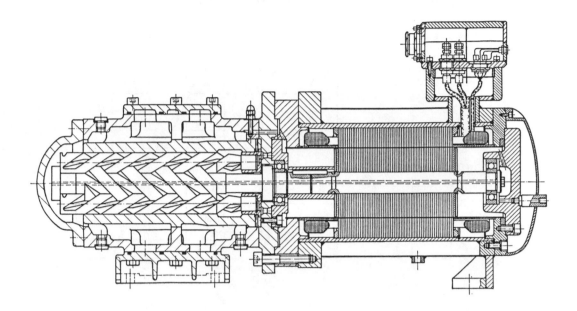

Fig. 5.20 Three-screw pump with canned motor drive

5.2.5 Sliding-vane pumps with magnetic coupling and canned motor drive

Sliding-vane pumps are used particularly where low volume rates of flow against high pressure are required without sacrificing efficiency (Fig. 5.21). Their low power/weight ratio, their short length and also the possibility of having block construction (Fig. 5.22 and 5.23) give these pumps a special price advantage.

The new type of displacement pump described in the "Vane pump with fixed vanes" Section can also be fitted with a magnetic coupling (Fig. 5.24) or a canned motor (Fig. 5.25). This enables the advantages of the vane pump with fixed vanes to be combined with hermetic construction to produce an interesting working machine.

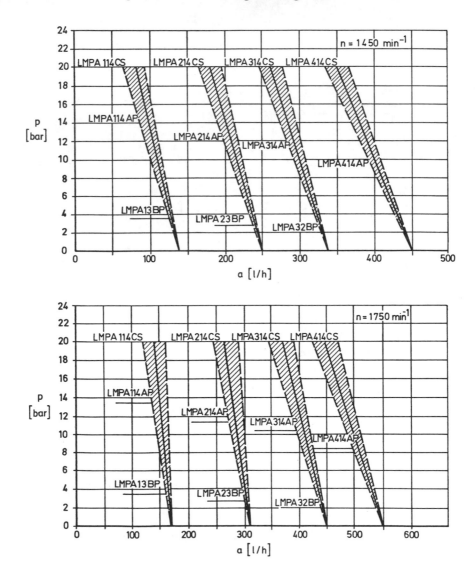

Fig. 5.21 Family of characteristics of a sliding-vane pump series with magnetic couplings

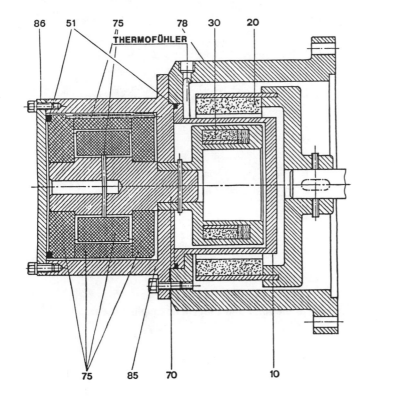

10	Can
20	Drive section
30	Rotor
51	O-ring
70	Rotor shaft
75	Graphite parts + pins
78	Pedestal
85	Casing
86	Cover

Thermofühler = Temperature sensor

Fig. 5.22 Theoretical diagram of a vane pump with magnetic coupling in monobloc construction

Fig. 5.23 Monobloc sliding-vane pump with a magnetic coupling (Lederle)

565

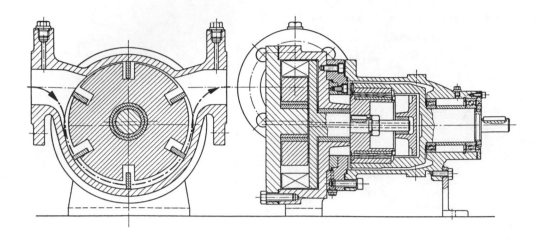

Fig. 5.24 Vane pump with vanes fixed in the rotor and a magnetic coupling (Lederle)

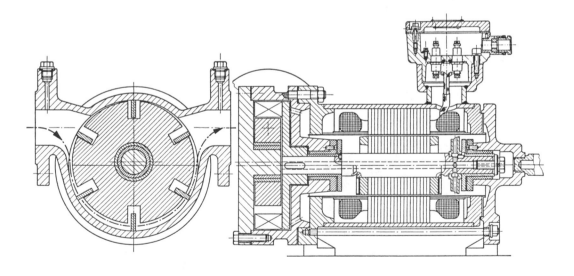

Fig. 5.25 Vane pump with vanes fixed in the rotor and a canned motor drive (Lederle)

5.2.6 Rotary piston pumps with a magnetic coupling

Single-shaft types are preferred where a rotary displacement pump is fitted with a permanent magnet coupling or canned motor. An exception is the screw pump where the main screw drives the auxiliary screw. The situation is the other way round for rotary piston pumps where no force is imparted by a positive contact to the rotary piston. In this case a separate synchronous geared unit provides the counterrotation of the piston. Only one drive shaft is brought out of the displacer chamber. The torque for hermetic power transmission is produced from a magnetic coupling and is transmitted via the synchronous gearing to a parallel, driven shaft. It is usually not possible to use a canned motor as a drive unit because the maximum rotational speed of the drive is n = 700 r.p.m.

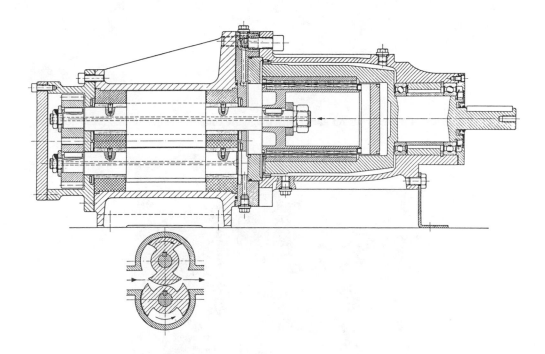

Fig. 5.26 Rotary piston pump with a magnetic coupling (Lederle)

The synchronous gears are fitted in a casing in contact with the product and must therefore be made of a corrosion-resistant material which does not cause problems when the gears mesh (Fig. 5.26). New solutions have had to be found to this problem.

Chemically-resistant special materials with good sliding properties and high thermal load capacity are used for shaft bearings located in the product chamber [5-3].

Rotary piston pumps are used in many cases for drawing-off products in vacuum-vaporisation plants. And they are also able to handle highly viscose fluids without difficulty. Their hermetic construction is provided by the permanent magnet drive so that both during operation and shutdown at high vacuums no air can be drawn into the product chamber. This is necessary in more than a few cases in order to avoid oxidation of the product. There are also instances where the entry of air causes the product to adhere, and in this case a hermetic drive is also necessary.

Fig. 5.27 View of a rotary piston pump with magnetic coupling shown in Fig. 5.26 (Lederle)

Fig. 5.28 Hermetic rotary piston pump in a process plant [5 - 3]

568

5.2.7 Liquid ring vacuum pumps/compressors

Hermetic-type liquid ring-vacuum pumps/compressors have proved useful particularly where toxic gases have to be evacuated or compressed. In this case, vacuum pumps/compressors both with magnetic couplings and canned motor drives are used.

5.2.7.1 Liquid ring vacuum pumps/compressors with magnetic coupling

These pumps can be of either of single-suction (Fig. 5.29 and 5.30) or double-suction (Fig. 5.32 and 5.33) design.

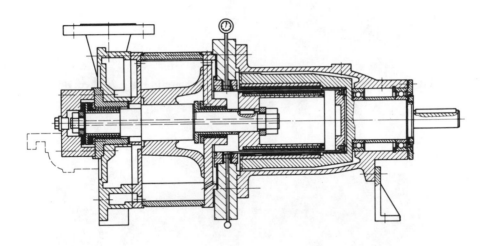

Fig. 5.29 Single suction liquid ring vacuum pump with a magnetic coupling (Lederle)

Fig. 5.30 View of the liquid ring vacuum pump shown in Fig. 5.29 (Lederle)

Fig. 5.31 Vacuum pumps with a magnetic coupling as shown in Fig. 5.30, in a unit package system (Lederle)

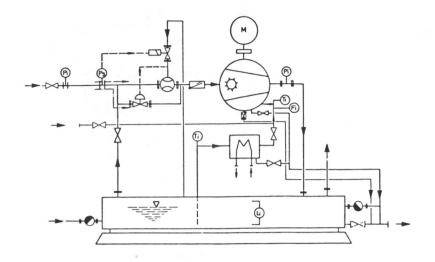

Fig. 5.31.1 Functional and installation diagram of unit package system shown in Fig. 5.31
        (Lederle)

Fig. 5.32 Double-suction liquid ring vacuum pumps with magnetic couplings, in a unit
        package system (Lederle)

571

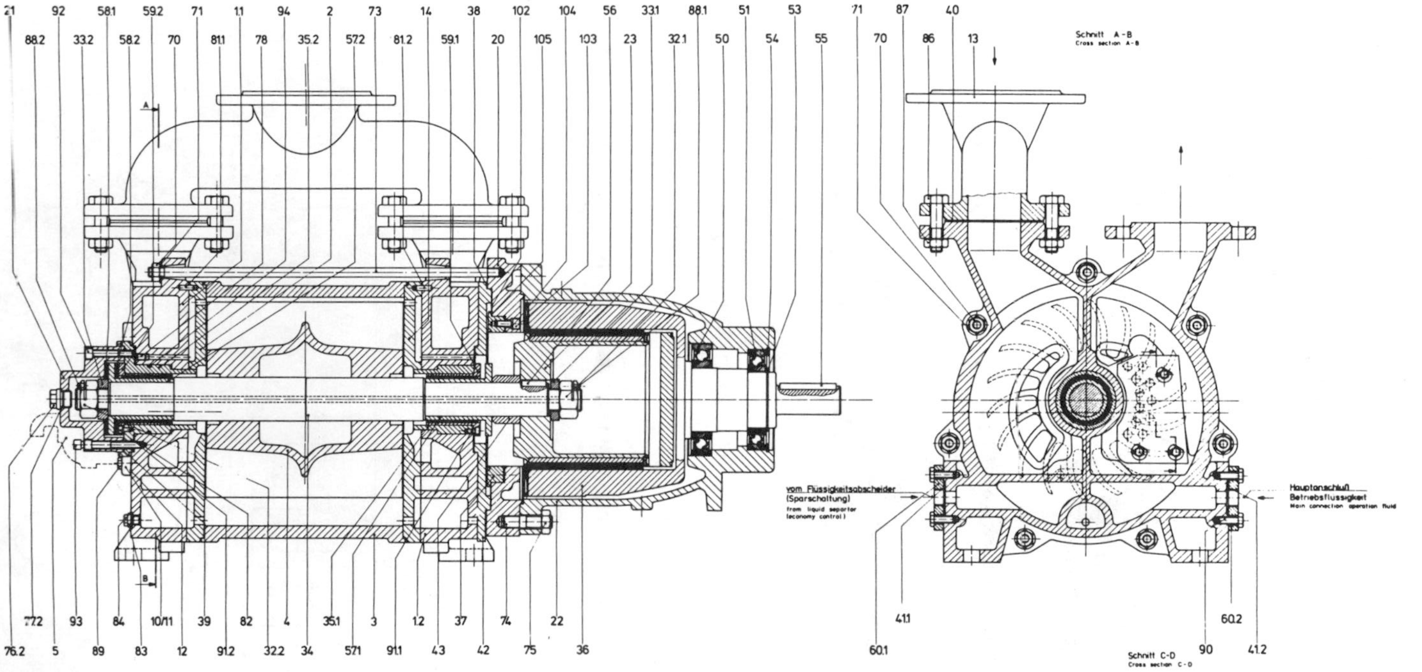

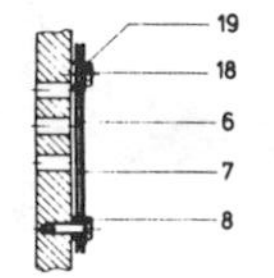

Fig. 5.33 Double-suction liquid ring vacuum pump with a magnetic coupling (Lederle)

Item	Description
1.1	Lateral plate
1.2	Lateral plate
2	Port plate (left)
3	Casing
4	End plate
5	Impeller
6	Automatic drain valve
7	Catch plate
8	Valve plate
9	Distance plate
10	Hand hole cover
11	Hand hole cover with bore
12	Gasket
13	Manifold
14	Port plate (right)

Item	Description
18	Hexagon screw
19	Top washer
38	Gasket
20	Casing cover
21	End plate
22	Pump bracket
23	Rotor
32.1	Hexagon nut
32.2	Hexagon nut
33.1	Tension disc
33.2	Tension disc
34	Impeller shaft
35.1	Bush
35.2	Reciprocating bush
36	Drive shaft

Item	Description
37	Shroud
38	Gasket
38	Gasket
40	Gasket
41.1	Joint ring
41.2	Joint ring
42	Centering disc
43	Distance sleeve
50	Radial ball bearing
51	Radial ball bearing
53	Outer circlip
54	Inner circlip
55	Key
56	Key

Item	Description
57.1	Bearing sleeve
57.2	Bearing sleeve
58.1	Rotating seal ring
58.2	Rotating seal ring
59.1	Bearing bush
59.2	Bearing bush
60.1	Flange
60.2	Flange
70	Hexagon nut
71	Washer
73	Stud
74	Tie bolt
75	Hexagon nut
76.1	Screwed plug

Item	Description
76.2	Screwed plug
81.1	Joint ring
81.2	Joint ring
78	Grub screw
79	Grub screw
81.1	Cylinder pin
81.2	Cylinder pin
82	Screwed plug
83	Joint ring
84	Hexagon nut
86	Grub screw
88.1	Grub screw
89	Hexagon screw
90	Hexagon screw
91.1	Hexagon socket head screw

Item	Description
91.2	Hexagon socket head screw
92	Hexagon socket head screw
93	Hexagon socket head screw
94	Hexagon socket head screw
102	O-ring
103	Cover sheet
104	Hexagon socket head screw
105	Gasket
601	Spring washer

Schnitt A–B
Cross section A–B

Schnitt C–D
Cross section C–D

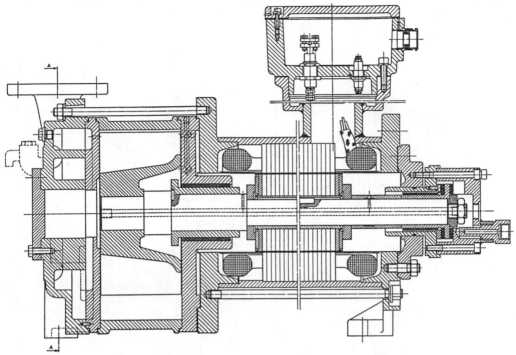

Fig. 5.34 Single-suction vacuum pump with a canned motor

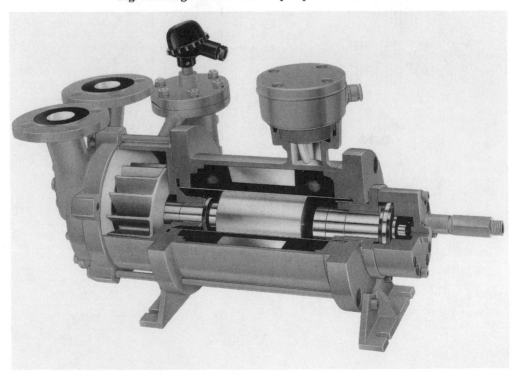

Fig. 5.35 Sectioned model of a single-suction vacuum pump with a canned motor as shown in Fig. 5.34

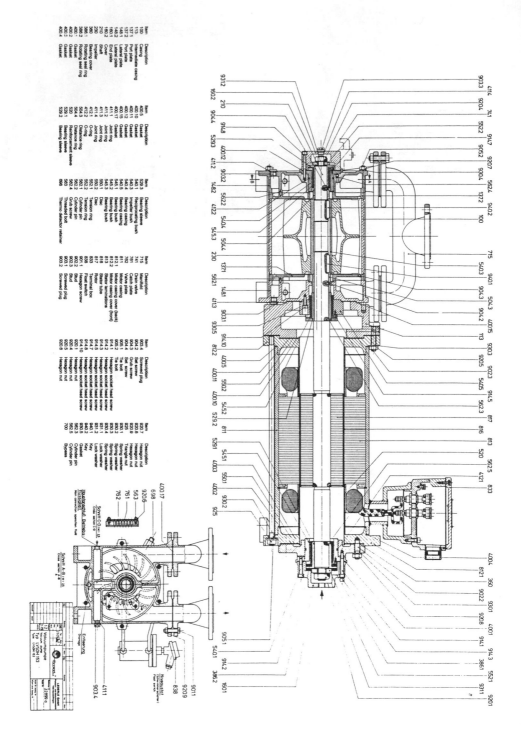

Fig. 5.36 Double-suction liquid ring vacuum pump with a canned motor drive (Lederle)

574

5.2.7.2 Liquid ring vacuum pumps/compressors with a canned motor drive

Particularly good starting characteristics, compact construction and low noise make these pumps particularly suitable for systems where safety is important. In this case also the pumps can be single-suction (Fig. 5.34 and 5.35) or double-suction (Fig. 5.36). A tandem design as shown in Fig. 5.37 and 5.38 is also common for this kind of pump. Fig. 5.39 shows the construction of a unit package system (CAD illustration).

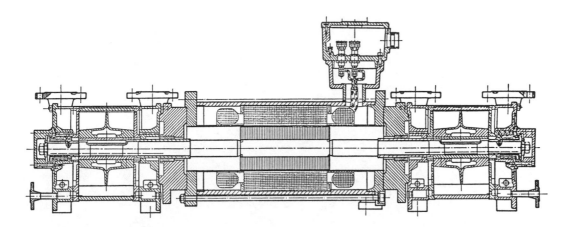

Fig. 5.37 Liquid ring vacuum pump of tandem design (Lederle)

Fig. 5.38 View of a liquid ring vacuum pump of tandem design (Lederle)

575

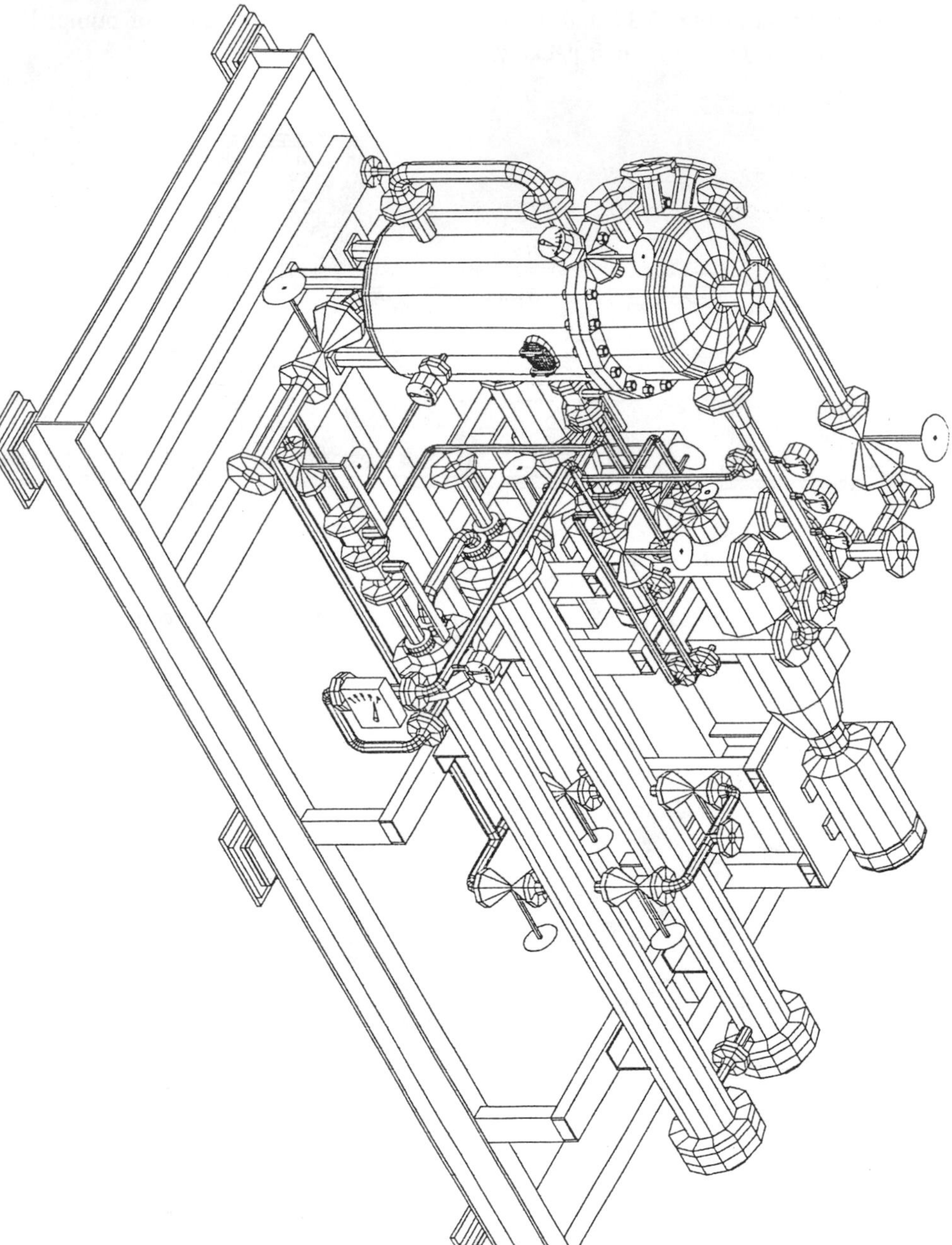

Fig. 5.39 Perspective view of a unit package vacuum pump system (Lederle)

576

5.2.8 Peristaltic pumps

The peristaltic pump is a special kind of leak-free rotary displacement pump. Its construction and operating principle are completely different from the other pumps dealt with.

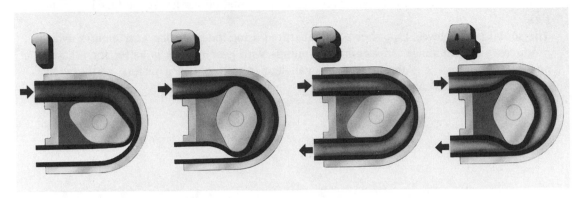

Fig. 5.40 Operating principle of a peristaltic pump (KWW Gesellschaft für Verfahrenstechnik)

Peristaltic pumps have a wide range of applications from medicine to clarification tasks, with delivery flows from the millilitre range per day up to 85 m³/h. The maximum achievable delivery pressure is 15 bar. They are able to deliver liquids with a wide variety of consistencies including those containing constituents which are abrasive or consist of long fibres. The liquid is transported in a tube so that no other part of the pump comes in contact with the product and there is complete separation of the working chamber from the drive part. The tube is periodically squeezed by sliding shoes or rollers so that the squeeze points move in the direction of delivery (Fig. 5.40). Control overlap is solved by the fact that the succeeding squeeze point in each case has closed before the preceding one opens. This produces a pulsing, but leakfree, delivery with the rate of flow remaining proportional to the rotational speed.

Fig. 5.41 Peristaltic pump set for the mobile use of dangerous substances (KWW Gesellschaft für Verfahrenstechnik)

6. Noise emissions from rotary displacement pumps

The noise emission limits for working machines specified by law must also be complied with for rotary displacement pumps in accordance with the DIN guidelines "TA-Lärm" (TA Noise). Fig. 6.42 provides an overview of the sound power levels produced by pump sets.

The sound power levels L_{WA} shown are measured using the particular gear motors, which produces a scatter range. Measuring the pumps with gear motors is better for practical purposes because the planning engineer can immediately obtain a picture of the sound pressure level of the complete pump set.

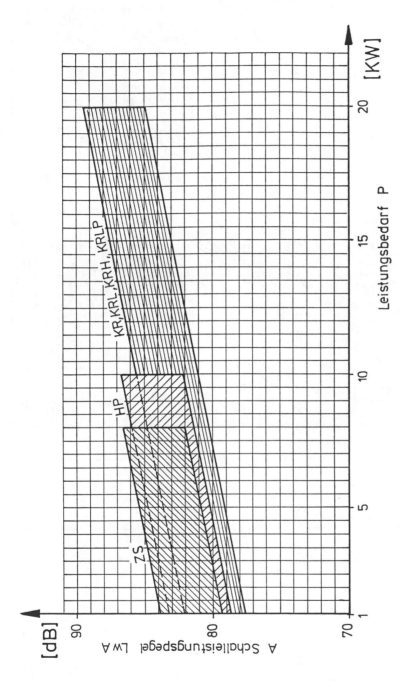

Leistungsbedarf = Power requirement
Schalleistungspegel = Sound power level
ZS Zahnradpumpen = Gear pumps
HP Innenzahnradpumpen = Internal gear pumps
KR, KRL, KRH, KRLP = Rotary piston pumps

Fig. 5.42 Sound power levels L $_{w}$ of rotary displacement pump sets (pump with gear motor) (Lederle)

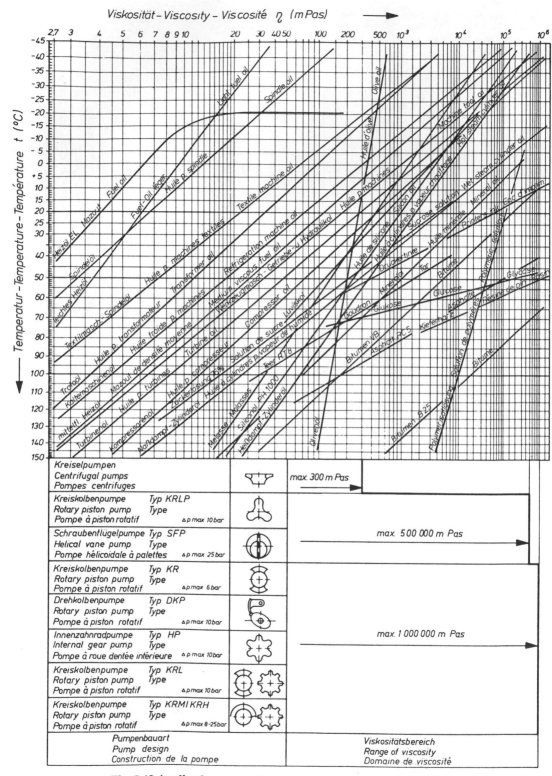

Fig. 5.43 Application range of various pump systems relative to viscosity

7. Concluding remarks on the subject of hermetic rotary displacement pumps

Leak-free rotary displacement pump technology - still a growth area - increases in importance together with growth in environmental awareness. New applications are always being found to make processes safer and emission-free. This presents a specific challenge to the pump manufacturer, as each application must be thoroughly analysed to find the optimum type of drive. It is obvious that the standard production models and complete pump sets available offer only limited solutions. Special designs, or at least modifications of existing pump designs, are necessary. We should not, however, hesitate to take this option in the interests of increased safety and a higher quality of life in today's industrial society.

References for Chapter 5

[5-1]   Neumaier, R.:   Leckfreie rotierende Verdrängerpumpen (Leak-free rotating displacement pumps), Pumpen, Vakuumpumpen, Kompressoren '91 (Pumps, Vacuum Pumps, Compressors '91)
Publisher: Fachgemeinschft Pumpen, Kompressoren und Vakuumpumpen im VDMA Frankfurt (Technical Association for Pumps, Compressors and Vacuum Pumps in the VDMA Frankfurt),
Dr. Harnisch Verlagsgesellschaft mbH, Nürnberg

[5-2]   Kamminga, J.:   Leckfreie, hydrostatische Pumpen für ungewöhnliche Förderparameter (Leak-free
                Neumaier, R.:   hydrostatic pumps for unusual delivery parameters),
Pump Conference, Karlsruhe '88
0.4 - 0.6 October 1988

[5-3]   Bannwarth, H.:   Eine zukunftsorientierte Neuentwicklung; Leckfreie Kreiskolbenpumpen mit Magnetkupplung (A futuristic new development; leak-free rotary piston pumps with a magnetic coupling)
Chemie-Technik (Chemical Technology) vol. 10.93

Index

Index

Index

Index

Index

Index

Index

Index

Index

Index

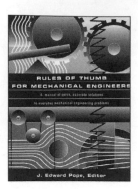

HERMETIC PUMPS

I wish to extend my thanks to all who have cooperated in the production of this book.

My special thanks go to Lederle GmbH and Hermetic-Pumpen GmbH for their unstinting technical support and the provision of illustrations.

I also wish to thank the staff of both companies who have devoted much for their free time to the preparation of numerous drawings and graphics and the production of camera-ready copy.

My thanks also go to all of the other companies whose helpful provision of illustrations and documents has enabled a wide range of pumps to be presented and discussed.

I also wish to extend well-deserved thanks to Professor Dr.-Ing. F. Engleberg, Dr. Rer. Nat. R. Krämer and Professor Dr.-Ing. Habil. D. Surek for reading the manuscript and offering valuable suggestions and remarks.

And finally, I wish to extend my gratitude to Verlag und Bildarchiv W. H. Faragallah for their splendid cooperation which has enabled this first edition to be published in what is a very short time for such a work.

Robert Neumaier
Glottertal, June 1994

HERMETIC PUMPS

The latest innovations and industrial applications of sealless pumps

Robert Neumaier

Gulf Publishing Company
Houston, Texas

© W. H. Faragallah, D-65843 Sulzbach Verlag und Bildarchiv 1994
Layout and formulae: Dipl.-Ing. Th.Andreas Schmiß, 79211 Denzlingen
Translation: Addison-Emert & Partner, Sulzbacher Str. 95, 65835 Liederbach
Assistance on translation: Dr. R. Krämer and S. G. Boalch
Cover design/illustration: Daniell Christian McCleney

Library of Congress Cataloging-in-Publication Data
Neumaier, Robert.
    [Hermetische Pumpen. English]
    Hermetic pumps : the ecological solution, centrifugal pumps and rotary displacement pumps / Robert Neumaier.
      p.   cm.
    Includes index.
    ISBN 0-88415-801-2 (alk. paper)
    1. Centrifugal pumps.   2. Rotary pumps.   3. Sealing (Technology)
I. Title.
TJ919.N3613   1997
621.6'7—dc21   97-11477
                                                             CIP

Printed on acid-free paper (∞).

ML